Smart Innovation, Systems and Technologies

Volume 102

Series editors

Robert James Howlett, Bournemouth University and KES International,
Shoreham-by-sea, UK
e-mail: rjhowlett@kesinternational.org

Lakhmi C. Jain, University of Technology Sydney, Broadway, Australia;
University of Canberra, Canberra, Australia; KES International, UK
e-mail: jainlakhmi@gmail.com; jainlc2002@yahoo.co.uk

The Smart Innovation, Systems and Technologies book series encompasses the topics of knowledge, intelligence, innovation and sustainability. The aim of the series is to make available a platform for the publication of books on all aspects of single and multi-disciplinary research on these themes in order to make the latest results available in a readily-accessible form. Volumes on interdisciplinary research combining two or more of these areas is particularly sought.

The series covers systems and paradigms that employ knowledge and intelligence in a broad sense. Its scope is systems having embedded knowledge and intelligence, which may be applied to the solution of world problems in industry, the environment and the community. It also focusses on the knowledge-transfer methodologies and innovation strategies employed to make this happen effectively. The combination of intelligent systems tools and a broad range of applications introduces a need for a synergy of disciplines from science, technology, business and the humanities. The series will include conference proceedings, edited collections, monographs, handbooks, reference books, and other relevant types of book in areas of science and technology where smart systems and technologies can offer innovative solutions.

High quality content is an essential feature for all book proposals accepted for the series. It is expected that editors of all accepted volumes will ensure that contributions are subjected to an appropriate level of reviewing process and adhere to KES quality principles.

More information about this series at http://www.springer.com/series/8767

Anna Esposito · Marcos Faundez-Zanuy
Francesco Carlo Morabito · Eros Pasero
Editors

Neural Advances in Processing Nonlinear Dynamic Signals

 Springer

Editors
Anna Esposito
Dipartimento di Psicologia
Università della Campana Luigi Vanvitelli
Caserta, Italy

and

International Institute for Advanced
  Scientific Studies (IIASS)
Vietri sul Mare, Italy

Marcos Faundez-Zanuy
Fundació Tecnocampus
Pompeu Fabra University
Mataro, Barcelona, Spain

Francesco Carlo Morabito
Department of Civil, Environmental,
  Energy, and Material Engineering
University Mediterranea of Reggio Calabria
Reggio Calabria, Italy

Eros Pasero
Laboratorio di Neuronica, Dipartimento
  Elettronica e Telecomunicazioni
Politecnico di Torino
Torino, Italy

ISSN 2190-3018   ISSN 2190-3026 (electronic)
Smart Innovation, Systems and Technologies
ISBN 978-3-030-06977-3   ISBN 978-3-319-95098-3 (eBook)
https://doi.org/10.1007/978-3-319-95098-3

Preface

This book proposes neural network algorithms and advanced machine learning techniques for processing nonlinear dynamic signals such as audio, speech, financial signals, feedback loops, waveform generations, filtering, equalization, signals from arrays of sensors, and perturbation in the automatic control of industrial production processes.

Computational intelligence (CI) and information communication technologies (ICT) are research fields providing sophisticated processing methodologies able to develop appropriate transmission codes, detection, and recognition algorithms to manage promptly, efficiently, and effectively huge amount of data. To this aim, neural networks, deep learning networks, genetic algorithms, fuzzy logic, and complex artificial intelligence designs are favored because their easy handling of nonlinearities while discovering new data structure and new original patterns to enhance the efficiency of industrial and economic applications.

The growing interest in these research areas favors the presentation of a book devoted to collect the current progress in *Neural Advances in Processing Nonlinear Dynamic Signals* and shows the latest evolvements in these communities.

Key aspects considered are the integration of neural adaptive algorithms for the recognition, analysis, and detection of dynamic complex structures and the implementation of systems for discovering patterns in industrial and economic data. The primary goal is to exploit the commonalities between computational intelligence (CI) and information communication technologies (ICT) to promote transversal skills and sophisticated processing techniques in industrial and economic application.

The proposed contributions introduce computational intelligence thematic in financial, industrial, and ICT engineering research. These fields are closely connected to the problems they afford and provide fundamental insights for cross-exchanges among these disciplines.

The chapters composing this book were first discussed at the international workshop on neural networks (WIRN 2017) held in Vietri Sul Mare from June 14 to June16, 2017, in the regular and special sessions. In particular, it is worth to mention the special session on: "**Computational Intelligence and related**

techniques in industrial and ICT engineering" organized by Giovanni Angiulli, Mario Versaci, and the special session on "**Intelligent tools for decision making in economics and finance**" organized by Marco Corazza.

The scientists contributing to this book are specialists in their respective disciplines. We are indebted to them for making (through their chapters) the book a meaningful effort. The coordination and production of this book have been brilliantly conducted by the Springer project coordinator for books production Mr. Ayyasamy Gowrishankar, the Springer executive editor Dr. Thomas Ditzinger, and the editor assistant Mr. Holger Schaepe. They are the recipient of our deepest appreciation. This initiative has been skillfully supported by the editors in chief of the Springer series Smart Innovation, Systems and Technologies, Profs. Jain Lakhmi C. and Howlett Robert James, to whom goes out deepest gratitude.

Caserta, Itlay/Vietri sul Mare, Italy   Anna Esposito
Mataro, Spain   Marcos Faundez-Zanuy
Reggio Calabria, Italy   Francesco Carlo Morabito
Torino, Italy   Eros Pasero

The chapters submitted to this book have been carefully reviewed by the following technical committee to which the editors are extremely grateful.

Technical Reviewer Committee

Altilio Rosa, Università di Roma "La Sapienza"
Alonso-Martinez Carlos, Universitat Pompeu Fabra
Angiulli Giovanni, Università Mediterranea di Reggio Calabria
Bevilacqua Vitoantonio, Politecnico di Bari
Bramanti Alessia, ISASI-CNR "Eduardo Caianiello" Messina
Brandenburger Jens, VDEh-Betriebsforschungsinstitut GmbH, BFI, Dusseldorf
Buonanno Amedeo, Università degli Studi della Campania "Luigi Vanvitelli"
Camastra Francesco, Università Napoli Parthenope
Carcangiu Sara, University of Cagliari
Campolo Maurizio, Università degli Studi Mediterranea Reggio Calabria
Capuano Vincenzo, Seconda Università di Napoli
Cauteruccio Francesco, Università degli Studi della Calabria
Celotto Emilio, Ca' Foscari University of Venice
Ciaramella Angelo, Università Napoli Parthenope
Ciccarelli Valentina, Università di Roma "La Sapienza"
Cirrincione Giansalvo, UPJV
Colla Valentina, Scuola Superiore S. Anna
Comajuncosas Andreu, Universitat Pompeu Fabra
Commimiello Danilo, Università di Roma "La Sapienza"
Committeri Giorgia, Università di Chieti

Cordasco Gennaro, Seconda Università di Napoli
De Carlo Domenico, Università Mediterranea di Reggio Calabria
De Felice Domenico, Università Napoli Parthenope
Dell'Orco Silvia, Università degli Studi della Basilicata
Diaz Moises, Universidad del Atlántico Medio
Droghini Diego, Università Politecnica delle Marche
Ellero Andrea, Ca' Foscari University of Venice
Esposito Anna, Università degli Studi della Campania "Luigi Vanvitelli" and IIASS
Esposito Antonietta Maria, sezione di Napoli Osservatorio Vesuviano
Esposito Francesco, Università di Napoli Parthenope
Esposito Marilena, International Institute for Advanced Scientific Studies (IIASS)
Faundez-Zanuy Marcos, Universitat Pompeu Fabra
Ferretti Paola, Ca' Foscari University of Venice
Gallicchio Claudio, University of Pisa
Giove Silvio, University of Venice
Giribone Pier Giuseppe, Banca Carige, Financial Engineering and Pricing
Kumar Rahul, University of South Pacific
Ieracitano Cosimo, Università degli Studi Mediterranea Reggio Calabria
Inuso Giuseppina, University Mediterranea of Reggio Calabria
Invitto Sara, Università del Salento
La Foresta Fabio, Università degli Studi Mediterranea Reggio Calabria
Lenori Stefano, University of Rome "La Sapienza"
Lo Giudice Paolo, University "Mediterranea" of Reggio Calabria
Lupelli Ivan, Culham Centre for Fusion Energy
Maldonato Mauro, Università di Napoli "Federico II"
Manghisi Vito, Politecnico di Bari
Mammone Nadia, IRCCS Centro Neurolesi Bonino-Pulejo, Messina
Maratea Antonio, Università Napoli Parthenope
Marcolin Federica, Politecnico di Torino
Martinez Olalla Rafael, Universidad Politécnica de Madrid
Matarazzo Olimpia, Seconda Università di Napoli
Mekyska Jiri, Brno University
Micheli Alessio, University of Pisa
Militello Carmelo, Consiglio Nazionale delle Ricerche (IBFM-CNR), Cefalù (PA)
Militello Fulvio, Culham Centre for Fusion Energy
Monda Vincenzo, Università degli Studi della Campania "Luigi Vanvitelli"
Morabito Francesco Carlo, Università Mediterranea di Reggio Calabria
Nardone Davide, Università di Napoli "Parthenope"
Narejo Sanam, Politecnico di Torino
Neffelli Marco, University of Genova
Parisi Raffaele, Università di Roma "La Sapienza"
Paschero Maurizio, University of Rome "La Sapienza"
Pedrelli Luca, University of Pisa
Portero-Tresserra Marta, Universitat Pompeu Fabra
Principi Emanuele, Università Politecnica delle Marche

Josep Roure, Universitat Pompeu Fabra
Rovetta Stefano, Università di Genova (IT)
Rundo Leonardo, Università degli Studi di Milano-Bicocca
Salvi Giampiero, KTH, Sweden
Sappey-Marinier Dominique, Université de Lyon
Scardapane Simone, Università di Roma "La Sapienza"
Scarpiniti Michele, Università di Roma "La Sapienza"
Senese Vincenzo Paolo, Seconda Università di Napoli
Sesa-Nogueras Enric, Universitat Pompeu Fabra
Sgrò Annalisa, Università Mediterranea di Reggio Calabria
Staiano Antonino, Università Napoli Parthenope
Stamile Claudio, Université de Lyon
Statue-Villar Antonio, Universitat Pompeu Fabra
Suchacka Grażyna, Opole University
Taisch Marco, Politecnico di Milano
Terracina Giorgio, Università della Calabria
Theoharatos Christos, Computer Vision Systems, IRIDA Labs S.A.
Troncone Alda, Seconda Università di Napoli
Vitabile Salvatore, Università degli Studi di Palermo
Xavier Font-Aragones, Universitat Pompeu Fabra
Uncini Aurelio, Università di Roma "La Sapienza"
Ursino Domenico, Università Mediterranea di Reggio Calabria
Vasquez Juan Camilo, University of Antioquia
Vesperini Fabio, Università Politecnica delle Marche
Vitabile Salvatore, Università degli Studi di Palermo
Wesenberg Kjaer Troels, Zealand University Hospital
Walkden Nick, Culham Centre for Fusion Energy
Zucco Gesualdo, Università di Padova

Sponsoring Institutions

International Institute for Advanced Scientific Studies (IIASS) of Vietri S/M (Italy)
Department of Psychology, Università degli Studi della Campania "Luigi Vanvitelli" (Italy)
Provincia di Salerno (Italy)
Comune di Vietri sul Mare, Salerno (Italy)
International Neural Network Society (INNS)
Università Mediterranea di Reggio Calabria (Italy)

Contents

Part I
Introduction

Chapter 1
Processing Nonlinearities

Anna Esposito, Marcos Faundez-Zanuy, Francesco Carlo Morabito and Eros Pasero

Abstract The problem of non-linear data is one of the oldest in experimental science. The solution to this problem is very complex, since the exact mechanisms that describe a phenomenon and its nonlinearities, are often unknown. At the same time, environmental factors such as the finite precision of the processing machine, noise, and sensor limitations—among others—produce further inaccuracies making even more unfitting the description of the phenomenon described by the collected data. In this context, while developing complex systems, with optimal performance, capable of interacting with the environment in an autonomous way, and showing some form of intelligence, the ultimate solution is to process, identify and recognize such non-linear dynamics. Problems and challenges in Computational Intelligence (CI) and Information Communication Technologies (ICT) are devoted to implement sophisticated detection, recognition, and signal processing methodologies, to promptly, efficiently and effectively manage such problems. To this aim, neural networks, deep learning networks, genetic algorithms, fuzzy logic, and complex artificial intelligence designs, are favored because of their easy handling of nonlinearities while discovering new data structure, and new original patterns to enhance the efficiency

A. Esposito (✉)
Dipartimento di Psicologia, Università della Campania "Luigi Vanvitelli", Viale Ellittico 34,
81100 Caserta, Italy
e-mail: iiass.annaesp@tin.it

M. Faundez-Zanuy
Pompeu Fabra University, Barcelona, Spain
e-mail: faundez@tecnocampus.cat

F. C. Morabito
Università degli Studi "Mediterranea" di Reggio Calabria, Reggio Calabria, Italy
e-mail: morabito@unirc.it

E. Pasero
Dip. Elettronica e Telecomunicazioni, Politecnico di Torino, Turin, Italy
e-mail: eros.pasero@polito.it

A. Esposito
International Institute for Advanced Scientific Studies (IIASS), Salerno, Italy

© Springer International Publishing AG, part of Springer Nature 2019
A. Esposito et al. (eds.), *Neural Advances in Processing Nonlinear Dynamic Signals*, Smart Innovation, Systems and Technologies 102,
https://doi.org/10.1007/978-3-319-95098-3_1

of industrial and economic applications. The collection of chapters presented in this book offer a scenery of the current progresses in such scientific domain.

Keywords Financial and industrial process · Speech enhancement
Machine leaning methods · Neural networks · Artificial intelligence
Nonlinear signal processing · Decision making in economic and finance
Evolutionary algorithms

1.1 Introduction

Research on Computational Intelligence (CI) and Information Communication Technologies (ICT) aims to improve the quality of life of the end users by implementing user friendly complex autonomous systems that simplify the solution of problems related to their everyday needs. The difficulties associated with the dealing of nonlinear data is very challenging in these fields for practical and intellectual reasons.

From the practical point of view, solving such hindrances will improve productivity and facilitate the user interaction with engineering systems.

Intellectually, to accurately process nonlinear data hold considerable promises as well as challenges, in the years to come, for scientists and developers alike. Considering the immense amount of research over the last three decades, one may wonder why the processing of nonlinear data is still an unsolved problem.

This is because, when dealing with different nonlinear data, from an engineering design point of view, there is the need to define a model for each data category. When the number of data increase, the computational effort to process and handle relations among them, increase exponentially. Basic research may continue to provide mathematical models for distinct nonlinear categories. However, the exponential complexity required by large amount of data with sophisticated relations among them cannot be easily solved both in the short and long time period. The challenges are to handle complex relations among the data and ensure optimality in the system design.

Since automatize the nonlinear processing of large amount of data cannot be solved only specifying data categories and relationships among them, the new approach is to observe how living creatures successfully interact among them and within an unpredictable environment. The principles governing such interactions, on the basis of the daily experience, seem to be that of building iteratively changing models of the data at the hand, and stores them in terms of iteratively extracted changing features, and iteratively derived features representations. Then, when required from the circumstances, choose the model and associated representation that best fit the task at the hand and assess its likelihood of success on the basis of the limited resource and available information.

Neural, adaptive systems seem to be the solution more appropriate under such circumstances. These systems are able to solve the challenge to process nonlinear data since they show characteristics of adaptation to the environment, distributed processing abilities due to their large number of computational elements, and resources, in term of algorithms and mathematical models, to provide optimal solutions for the constantly changing data. They allow continuously data learning, and when correctly applied, provide solutions that outperform sophisticated conventional linear systems. The revolutionary distinctiveness of such systems with respect to the traditional engineering design approach is in discarding a priori specifications of their parameters and exploit external data to fix them automatically. To do so they apply learning algorithms with feedback loops and cost functions that will converge to the desired output through a continuously training that modify the system's parameters exploiting information on unremittingly changing external data. What remain to the designer is to specify the architecture of the system (essentially the neural network topology), define the most effective cost function, and decide the extent to which admit an error from the system. Even these decisions are not left to the designer alone, since an extensive research has been conducted on evaluating learning algorithms, training procedures, and allowed errors for several neural network models and several application contexts, such as signal recognition, separations, and enhancement, system predictions, noise cancellation, patterns' classification, and control [1–4, 6].

1.2   Content of This Book

The research reported in this book bring up the latest advances on neural processing of nonlinear dynamic signals. The content of the book is organized in sections, each dedicated to a specific topic, including peer-reviewed chapters, not published elsewhere. The inspiring content of each chapters was discussed for the first time at the International Workshop on Neural Networks (WIRN 2017) held in Vietri sul Mare, Italy, from the 14th to 16th of June 2017. The workshop, is nowadays a historical and traditional scientific event gathering together researchers from Europe and overseas.

Part I describes the main motivations for a book on neural advances in processing nonlinear dynamic signals through an introductory chapter proposed by Anna Esposito, Marcos Faundez-Zanuy, Francesco Carlo Morabito and Eros Pasero, which are the editors of the book.

Part II is dedicated to nonlinear data and techniques for data mining. This is because the need for mining huge and complex data has become essential in education, web mining, social network analysis, security, medicine, and health management, as well as in many other engineering fields. Neural adaptive systems seems to fail to handle these large amount of data, because they are too big to a certain extent and hardly scalable. Solutions are however proposed in this section, always exploiting neural systems and machine learning algorithms which succeed to scale and mine such data.

Part III is dedicated to computational intelligence and related techniques in industrial and ICT engineering. This section proposes neural adaptive systems able to manage in a quick, effective and efficient manner the increasing amount of information to be shared and processed by organizations and industry, since previously offered solutions were not able to handle the transformation of large amounts of data (involving uncertainties) into knowledge. In this context, computational intelligence techniques are proposed, mainly including artificial neural network, learning systems, evolutionary computation, and fuzzy logic, to exactly extract (hidden) information from data and make it useful for application such as fault diagnosis and forecasting, product quality and process forecasting, smart maintenance systems, big data and analytics transforms, knowledge management, production, planning and scheduling, evolutionary big data interaction for predictive product design and marketing.

Part IV is intended to report on neural intelligent tools for decision making in economics and finance, and the current exploitation of intelligent systems for modeling economic behaviors which involve either evolutionary optimizers showing collective intelligence—such as artificial bee colonies, genetic algorithms, particle swarm optimizations—or machine learning techniques based on supervised and reinforcement learning procedures, or unconventional fuzzy logic and rough sets. The contributions of this section embrace these research fields and highlight the benefits of using intelligent tools in the investigation of economic/financial phenomena.

1.3 Conclusion

When Mc Carthy and colleagues [5] suggested the basic idea of providing a machine of "intelligence" he may have not been thinking to the processing of nonlinear data and the detection, recognition, and analysis of complex data structures. He may have even not been thinking to neural adaptive systems for processing nonlinear dynamics even though his statement was *that every aspect of learning or any other feature of intelligence can in principle be so precisely described that a machine can be made to simulate it*. Currently, neural adaptive processing techniques, have capacities and skills, such as—reacting flexibly to situations, taking advantage of fortuitous circumstances, recognizing the relative importance of different elements in a situation, finding similarities in different situations, synthesizing new concepts, taking old concepts and linking them in a different way—which although are not sufficient to define "intelligence" wisely, certainly have revolutionized the design and modeling of complex engineering problems, offering better and more "intelligent" solutions to them.

Acknowledgements The research leading to the results presented in this paper has been conducted in the project EMPATHIC (Grant N: 769872) that received funding from the European Union's Horizon 2020 research and innovation programme.

References

1. Bassis, S., Esposito, A., Morabito, F.C., Pasero, E. (eds.): Advances in Neural Networks: Computational Intelligence for ICT. Smart Innovation, Systems and Technologies (SIST), vol. 54, pp. 1–539. Springer International Publishing, Switzerland (2016)
2. Duda, R.O., Hart, P.E., Stork, D.G.: Pattern Classification, pp. 1– 680. Wiley (2012)
3. Esposito, A., Faundez-Zanuy, M., Morabito, F.C., Pasero, E. (eds.): Multidisciplinary Approaches to Neural Computing. Smart Innovation, Systems and Technologies (SIST), vol. 69, pp. 1–388. Springer International Publishing, Switzerland (2017)
4. Haykin, S. (ed.): Kalman Filtering and Neural Networks, pp. 1– 284. Wiley (2004)
5. McCarthy, J., Minsky, M.L., Rochester, N., Shannon, C.E.: A proposal for the Dartmouth summer research project on artificial intelligence. AI Mag. **27**(4), 12–14 (2006) (© AAAI)
6. Ripley, B.D.: Pattern Recognition and Neural Networks, pp. 1– 403. Cambridge University Press (2007)

Part II
Processing Nonlinearities

Chapter 2
Temporal Artifacts from Edge Accumulation in Social Interaction Networks

Matt Revelle, Carlotta Domeniconi and Aditya Johri

Abstract There has been extensive research on social networks and methods for specific tasks such as: community detection, link prediction, and tracing information cascades; and a recent emphasis on using temporal dynamics of social networks to improve method performance. The underlying models are based on structural properties of the network, some of which we believe to be artifacts introduced from common misrepresentations of social networks. Specifically, representing a social network or series of social networks as an accumulation of network snapshots is problematic. In this paper, we use datasets with timestamped interactions to demonstrate how cumulative graphs differ from activity-based graphs and may introduce temporal artifacts.

2.1 Introduction

The modeling of social networks is an expansive and active area of research. While models may incorporate other network features such as node attributes [4, 16, 25], nearly all rely on network structure. Many methods are now also incorporating temporal dynamics [10, 12, 20, 22], but how the temporal information is integrated varies. There are various approaches [20, 21] to representing a dynamic social network as a series of networks, but until recently [15] all have lacked theoretical foundation.

Dynamic network representations which capture edge deactivation [20] have shown to improve task-specific performance. However, many state-of-the-art methods [16, 24] are based on *cumulative graphs* and ignore edge deactivation. The findings presented in this paper suggest that some existing models may be designed

M. Revelle · C. Domeniconi (✉) · A. Johri
George Mason University, Fairfax, VA 22030, USA
e-mail: carlotta@cs.gmu.edu

M. Revelle
e-mail: revelle@cs.gmu.edu

A. Johri
e-mail: ajohri3@gmu.edu

© Springer International Publishing AG, part of Springer Nature 2019
A. Esposito et al. (eds.), *Neural Advances in Processing Nonlinear Dynamic Signals*, Smart Innovation, Systems and Technologies 102,
https://doi.org/10.1007/978-3-319-95098-3_2

to accommodate temporal artifacts introduced by not including edge deactivation in the processing of network data.

There are two social network phenomena which motivate our analysis: *social capacity* [3] and *bursty events* [1]. Social capacity can be viewed as a per-node limit on the number of incident edges active at any given time and thus conflicts with the claim of densification and shrinking diameters in social networks [8, 11] unless additional conditions are met. For example, a network where every new node has a larger social capacity would lead to densification and shrinking diameters. While variation in social capacity based on demographics has been observed [15] there has been no evidence presented that would indicate social capacity is a function of when a node joins the network.

In order to measure the existence of densification and shrinking diameters, we first must construct a series of network snapshots which more accurately captures network structure than simply accumulating all edges over time. We do this by using communication activity between nodes as evidence that an edge is active. The bursty dynamics of social communication are accounted for by measuring the inter-event times and selecting an observation window large enough to minimize incorrectly deactivating an active edge. Thus we are able to construct a series of *activity graphs* which provide a more accurate approximation of the network state at a given point in time. This method of graph construction has been used previously on a mobile phone network [15] to improve understanding of communication strategies. We can then measure and compare evidence of densification and shrinking diameters in both a cumulative graph series and an activity graph series.

Densification and diameter shrinking are accepted as basic characteristics of dynamic social networks. However, this paper presents results which contradict those findings. When edge deactivation is incorporated, we do not find evidence of densification and diameter shrinking appears to be dependent on the rate of new nodes entering the network. We suggest this may be an effect of social capacity.

2.2   Related Work

Existing methods for social network tasks have either ignored temporal dynamics [16, 25] or proposed methods to filter edges with a decay function [20] or sliding window [21]. While these attempts to account for temporal dynamics may be effective, they are ad-hoc and lack a theoretical justification. The work by Miritello et al. [14] proposes the selection of an observation window size based on inter-event statistics and a simple method for identifying edge activation and deactivation. While similar to existing sliding window approaches, this method is motivated by social interaction patterns (bursty events). This approach is used to construct the activity graph series for our experiments.

Models of dynamic social networks based on node interaction activity [9, 17] have been introduced. These models are capable of generating a final network which resembles a real world network but they are unable to construct a network series which corresponds to a real-world network series. However, their ability to generate networks with realistic structure indicates they are an alternative to previous models which heavily rely on preferential attachment [2] or community affiliation [24] and ignore social interaction patterns. There are many types of temporal networks [6] and this paper presents observations on dynamic social networks, specifically person-to-person communication networks.

2.3 Background

The concepts of social capacity [3] and bursty communications [1, 19] have been considered separately and more recent literature [13–15] has attempted to measure and use these to determine the state of edges in a large social network.

Social capacity captures the maximum number of relationships one prefers to maintain at any given time and there is evidence that social capacity is conserved over time [5, 7, 15]. The term bursty is used to describe the temporal patterns of social interactions between pairs of nodes. That is, humans tend to interact in bursts and these patterns must be considered in order to correctly identify the activation/deactivation of edges.

The observation of social capacity and burstiness of human interaction in some networks suggests careful consideration is required to construct accurate static views of these networks. In fact, accepted claims of graph evolution [8, 11] appear to fail when graph series are constructed based on timestamped interactions rather than accumulated without regard for edge deactivation.

Previous literature [24] introduced densification and diameter shrinking as common network characteristics and we briefly describe them here. Densification is the super-linear growth of edges relative to nodes and results in a network becoming denser over time. Diameter shrinking is the reported tendency for network diameters to decrease over time as more edges are accumulated. We can see both how densification contradicts the notion of social capacity and might account for diameter shrinking.

2.4 Evidence of Temporal Artifacts

2.4.1 Dataset Descriptions

A dataset with timestamped interactions is required to construct an accurate temporal series of networks. We use data from Scratch [18], an online community where users may write and share programming projects, and Facebook [23].

In Scratch, there are several ways by which users may interact: project comments, project remixes, gallery curation, and user following. More information about Scratch

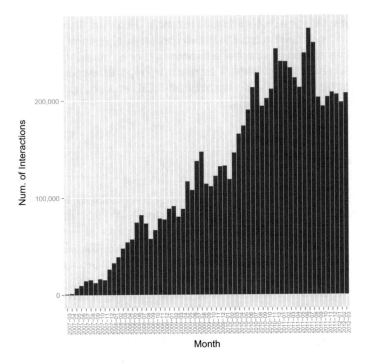

Fig. 2.1 The number of interaction events occurring by month in the Scratch dataset

and these interactions may be found in [18]. We selected a single type of interaction to simplify analysis. Project comments are a natural choice as they are the most-frequent interaction between Scratch users and thus a better approximation of edge status (active/inactive). These project comments serve as a means for users to communicate within the context of a project. The comments in the Scratch dataset are timestamped and thus we can create timestamped edges from comment authors to project authors.

The Scratch dataset spans over March 2007 to December 2011 and includes a large period of rapid growth in Scratch users, shown in Fig. 2.1, which does not slow until towards the end of the dataset. There are a total of 7,788,000 project comment interactions between 164,205 users. There are many short-term interactions and we filter out directed interactions between pairs which only occur once or twice when measuring communication behavior. Such interactions have undefined or trivial inter-event statistics as there are zero or one inter-event observations when only one or two interactions are observed. There are a total of 1,799,050 of such interactions which were removed, leaving 5,988,950 interactions. The Scratch dataset used to construct the networks may be obtained from the MIT Media Lab website.[1]

---

[1]https://llk.media.mit.edu/scratch-data.

Facebook allows users to interact by posting on each other's wall and these posts are typically comments, photos, and web links. Each of these posts is recorded as an interaction with a source user (the post author), a destination user (the owner of the wall), and a timestamp.

The Facebook data includes wall post interactions between a subset of Facebook users over October 2004 to January 2008. There are a total number of 876,993 wall posts between 46,952 users. The Facebook networks were prepared similarly to the Scratch networks, edges are only formed between node pairs that have at least a total of three interactions over the entire dataset. The Facebook dataset used to construct the networks may be obtained from the KONECT website.[2]

2.4.2 Methodology

As the relationships in the interaction networks are based on communication events between nodes, we check for evidence of bursty patterns. Bursty communication can be identified by the dispersion of inter-event times between node pairs. If communication is bursty then the standard deviation of inter-event time will be larger than the mean. The ratio of the mean and standard deviation of inter-event times is the coefficient of variation (cv) and used to measure dispersion. When $cv > 1$, there is evidence of bursty communication. The use of dispersion to identify burstiness is further discussed by Miritello et al. [14].

We hypothesize the observation of densification and diameter shrinking [8, 11] may be attributed to the inclusion of deactivated edges in a network. To test this we construct two graph series from each dataset. Each network in all the series captures network activity over consecutive and non-overlapping periods. The size of the observation windows are based on the inter-event times of the datasets. A three-month observation window was selected for the Scratch series because it is large enough to account for the majority of inter-event times (97% of inter-event times are <62 days) and conveniently maps to annual quarters. The first series is a *cumulative graph series* where new nodes and edges are added at each consecutive snapshot to the previous network in the series. The second series is based on node interaction activity and we refer to it as the *activity graph series*. The two Facebook series are based on a six-month observation window that was calculated similarly.

Edge activity is determined by tracking the activation and deactivation of edges between consecutive observation windows. A similar approach has been used in previous literature [14]. An edge is considered to activate if it is not present in the preceding observation window but an interaction event occurs in the current observation window. Similarly, an edge is deactivated if an event occurs in the current observation window but not in succeeding period. Only edges active in each observation window are used to construct the activity graph series of both datasets.

---

[2]http://konect.uni-koblenz.de/networks/facebook-wosn-wall.

The edge-node ratio ($\frac{num.\ of\ edges}{num.\ of\ nodes}$) is calculated for each graph in all series and used to measure densification. If densification is present, we expect the number of edges to grow super-linearly in the number of nodes [11]. We also measure the effective diameter of every graph in both series to determine whether diameter shrinking is observed. The effective diameter is the smoothed count of the smallest number of hops at which at least 90% of all connected pairs of nodes can be reached [11]. We prefer the effective diameter over the standard diameter because it was used in [11] to report diameter shrinking and is more robust to degenerative graph structures.

2.4.3 Results

As shown in Fig. 2.2, bursty communication patterns are observed in the Scratch dataset as the cv values are frequently greater than 1 ($log(cv) > 0$). Bursty communication patterns are also observed in the Facebook dataset, though somewhat less frequently with just under half of interacting node pairs having cv values >1. The reduced frequency of bursty communication patterns in Facebook may be due to node pairs having either a single burst of interactions or interactions around a regular event (e.g., posting on a friend's Facebook wall for their birthday).

For Scratch, we see evidence of densification in the cumulative series but not in the activity series—shown in Fig. 2.3. The accumulation of edges, without removal of deactivated edges, appears to introduce densification as a temporal artifact in the

Fig. 2.2 The $log(cv)$ for node pairs in the Scratch interaction network with at least three events. A small number of node pairs (1,038) were removed for this plot as they had a cv of zero and thus were undefined

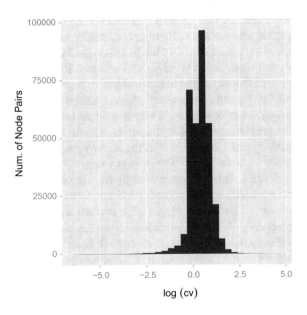

Fig. 2.3 The edge-node ratio over time in the Scratch cumulative and activity graph series

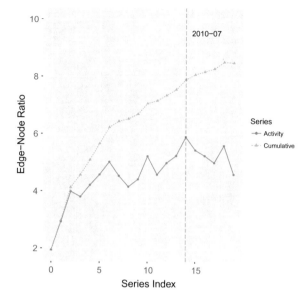

Fig. 2.4 The effective diameter over time in the Scratch cumulative and activity graph series

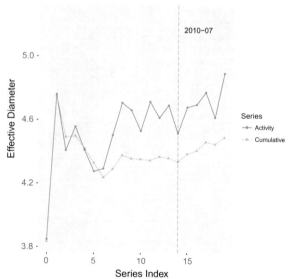

Scratch interaction network. This is especially clear when the number of interactions stops growing around July 2010, denoted by dashed vertical line in both Figs. 2.3 and 2.4.

An overall trend of diameter shrinking is not clearly observed in either Scratch network series. After an initial decrease, Fig. 2.4 shows a generally increasing diameter for both series and a larger variance in diameter for the activity series. The lack

Fig. 2.5 The edge-node
ratio over time in the
Facebook cumulative and
activity graph series

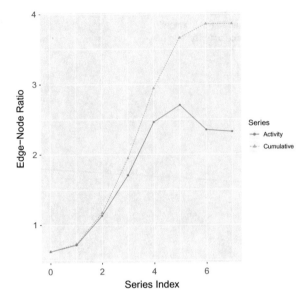

Fig. 2.6 The effective
diameter over time in the
Facebook cumulative and
activity graph series

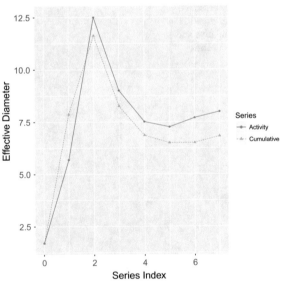

of diameter shrinking may be due to the growth of the Scratch website during the
period of time covered in the dataset. However, the effective diameter of the cumulative series is generally smaller than that of the activity series. This is unsurprising
given that the cumulative snapshots contain additional edges which would reduce
the distances between pairs of nodes.

Fig. 2.7 The node counts over time in the Scratch cumulative and activity graph series

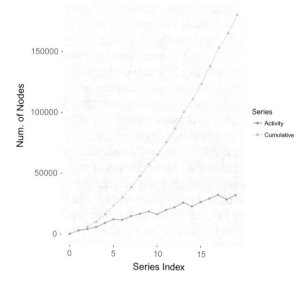

Fig. 2.8 The node counts over time in the Facebook cumulative and activity graph series

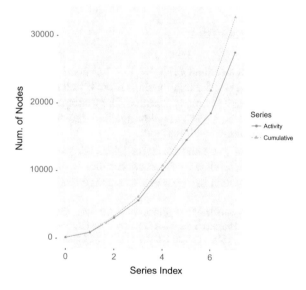

In the Facebook network series we see trends similar to that of the Scratch network series. As in Scratch, Fig. 2.5 shows that the edge density continues to increase in the cumulative series but not in the activity series. While an initial decrease in the diameter is observed in both Facebook network series, Fig. 2.6 depicts the diameter of both series slowly increasing in the later network snapshots.

These findings are not unexpected but they are contrary to previous literature [8, 11] which has served as the basis for state-of-the-art network models. The edge-node ratio in the cumulative graphs is monotonically increasing over time and social capacity is ignored. In contrast, the edge-node ratio in activity graphs may decrease or stabilize as inactive edges are detected and removed.

Figures 2.7 and 2.8 reveal another interesting artifact caused by accumulating inactive edges. In Fig. 2.7, we see that the number of nodes in the cumulative series is increasing at a much larger rate than the activity series. The total number of nodes, along with edge density, are exaggerated by not removing inactive edges. While this is the case for the Scratch dataset, the snapshots in both of the Facebook network series have a similar number of nodes.

This contrast between the Scratch and Facebook datasets indicates many Scratch users do not remain active while most users in the Facebook dataset do stay active. The short membership of some nodes in the Scratch network is likely due to Scratch being used as a teaching aid in classrooms. Students will create an account to participate in class and then become inactive at the end of a school semester or year. Ignoring edge deactivation may mask the actual node interaction patterns and contribute to mistaken findings which appear reasonable, as demonstrated by the difference in node lifetimes of the Scratch and Facebook datasets.

2.5 Conclusion

This paper presents evidence that temporal artifacts may be introduced in social networks when the relationships represented by edges require allocation of inelastic resource such as time or attention. Our findings suggest more accurate social networks may be derived from ongoing dyadic interactions rather than one-time events such as "following" or "friending".

We plan to extend this work to include other datasets, explore how community affiliation correlates to interaction patterns, and ultimately provide a model of social networks which incorporates knowledge from these findings.

Acknowledgements We appreciate the Lifelong Kindergarten group at MIT for publicly sharing the Scratch datasets. This work is partly based upon research supported by U.S. National Science Foundation (NSF) Awards DUE-1444277 and EEC-1408674. Any opinions, recommendations, findings, or conclusions expressed in this material are those of the authors and do not necessarily reflect the views of NSF.

References

1. Barabasi, A.-L.: The origin of bursts and heavy tails in human dynamics. Nature **435**(7039), 207–211 (2005)
2. Barabâsi, A.-L., Jeong, H., Néda, Z., Ravasz, E., Schubert, A., Vicsek, T.: Evolution of the social network of scientific collaborations. Phys. A Stat. Mech. Appl. **311**(3), 590–614 (2002)
3. Gonçalves, B., Perra, N., Vespignani, A.: Modeling users activity on twitter networks: validation of dunbars number. PloS One **6**(8), e22656 (2011)

4. Günnemann, S., Boden, B., Färber, I., Seidl,T.: Efficient mining of combined subspace and subgraph clusters in graphs with feature vectors. In: Advances in Knowledge Discovery and Data Mining, pp. 261–275. Springer (2013)
5. Hidalgo, C.A., Rodriguez-Sickert, C.: The dynamics of a mobile phone network. Phys. A Stat. Mech. Appl. **387**(12), 3017–3024 (2008)
6. Holme, P., Saramäki, J.: Temporal networks. Phys. Rep. **519**(3), 97–125 (2012)
7. Kossinets, G., Watts, D.J.: Empirical analysis of an evolving social network. Science **311**(5757), 88–90 (2006)
8. Kumar, R., Novak, J., Tomkins, A.: Structure and evolution of online social networks. In: Link Mining: Models, Algorithms, and Applications, pp. 337–357. Springer (2010)
9. Laurent, G., Saramäki, J., Karsai, M.: From calls to communities: a model for time varying social networks (2015). arXiv preprint arXiv:1506.00393
10. Leskovec, J.: Social media analytics: tracking, modeling and predicting the flow of information through networks. In: Proceedings of the 20th International Conference Companion on World Wide Web, pp. 277–278. ACM (2011)
11. Leskovec, J., Kleinberg, J., Faloutsos, C.: Graph evolution: densification and shrinking diameters. ACM Trans. Knowl. Discov. Data (TKDD) **1**(1), 2 (2007)
12. Matsubara, Y., Sakurai, Y., Prakash, B.A., Li, L., Faloutsos, C.: Rise and fall patterns of information diffusion: model and implications. In: Proceedings of the 18th ACM SIGKDD International Conference on Knowledge Discovery and Data Mining, pp. 6–14. ACM (2012)
13. Miritello, G., Lara, R., Cebrian, M., Moro, E.: Limited communication capacity unveils strategies for human interaction. Sci. Rep. **3** (2013)
14. Miritello, G., Lara, R., Moro, E.: Time allocation in social networks: correlation between social structure and human communication dynamics. In: Temporal Networks, pp. 175–190. Springer (2013)
15. Miritello, G., Moro, E., Lara, R., Martínez-López, R., Belchamber, J., Roberts, S.G., Dunbar, R.I.: Time as a limited resource: communication strategy in mobile phone networks. Soc. Netw. **35**(1), 89–95 (2013)
16. Moser, F., Colak, R., Rafiey, A., Ester, M.: Mining cohesive patterns from graphs with feature vectors. Proceedings of the SIAM International Conference on Data Mining (SIAM) **9**, 593–604 (2009)
17. Perra, N., Gonçalves, B., Pastor-Satorras, R., Vespignani, A.: Activity driven modeling of time varying networks. Sci. Rep. **2** (2012)
18. Resnick, M., Maloney, J., Monroy-Hernández, A., Rusk, N., Eastmond, E., Brennan, K., Millner, A., Rosenbaum, E., Silver, J., Silverman, B., et al.: Scratch: programming for all. Commun. ACM **52**(11), 60–67 (2009)
19. Rivera, M.T., Soderstrom, S.B., Uzzi, B.: Dynamics of dyads in social networks: assortative, relational, and proximity mechanisms. Ann. Rev. Sociol. **36**, 91–115 (2010)
20. Rossi, R., Neville, J.: Modeling the evolution of discussion topics and communication to improve relational classification. In: Proceedings of the First Workshop on Social Media Analytics, pp. 89–97. ACM (2010)
21. Rossi, R.A., Gallagher, B., Neville, J., Henderson, K.: Modeling dynamic behavior in large evolving graphs. In: Proceedings of the Sixth ACM International Conference on Web Search and Data Mining, pp. 667–676. ACM (2013)
22. Sun, Y., Tang, J., Han, J., Gupta, M., Zhao, B.: Community evolution detection in dynamic heterogeneous information networks. In: Proceedings of the Eighth Workshop on Mining and Learning with Graphs, pp. 137–146. ACM (2010)
23. Viswanath, B., Mislove, A., Cha, M., Gummadi, K.P.: On the evolution of user interaction in facebook. In: Proceedings of the 2nd ACM Workshop on Online Social Networks, pp. 37–42. ACM (2009)
24. Yang, J., Leskovec, J.: Community-affiliation graph model for overlapping network community detection. In: 2012 IEEE 12th International Conference on Data Mining (ICDM), pp. 1170–1175. IEEE (2012)
25. Yang, J., McAuley, J., Leskovec, J.: Community detection in networks with node attributes. In: IEEE 13th International Conference on Data Mining, pp. 1151–1156. IEEE (2013)

Chapter 3
Data Mining by Evolving Agents for Clusters Discovery and Metric Learning

Alessio Martino, Mauro Giampieri, Massimiliano Luzi and Antonello Rizzi

Abstract In this paper we propose a novel evolutive agent-based clustering algorithm where agents act as individuals of an evolving population, each one performing a random walk on a different subset of patterns drawn from the entire dataset. Such agents are orchestrated by means of a customised genetic algorithm and are able to perform simultaneously clustering and feature selection. Conversely to standard clustering algorithms, each agent is in charge of discovering well-formed (compact and populated) clusters and, at the same time, a suitable subset of features corresponding to the subspace where such clusters lie, following a local metric learning approach, where each cluster is characterised by its own subset of relevant features. This not only might lead to a deeper knowledge of the dataset at hand, revealing clusters that are not evident when using the whole set of features, but can also be suitable for large datasets, as each agent processes a small subset of patterns. We show the effectiveness of our algorithm on synthetic datasets, remarking some interesting future work scenarios and extensions.

3.1 Introduction

In the era of information, data mining and knowledge discovery turned out to be two of the most critical disciplines as they aim at the extraction and creation of knowledge from observed data.

A. Martino (✉) · M. Giampieri · M. Luzi · A. Rizzi
Department of Information Engineering, Electronics and Telecommunications,
University of Rome "La Sapienza", Via Eudossiana 18, 00184 Rome, Italy
e-mail: alessio.martino@uniroma1.it

M. Giampieri
e-mail: mauro.giampieri@uniroma1.it

M. Luzi
e-mail: massimiliano.luzi@uniroma1.it

A. Rizzi
e-mail: antonello.rizzi@uniroma1.it

© Springer International Publishing AG, part of Springer Nature 2019
A. Esposito et al. (eds.), *Neural Advances in Processing Nonlinear Dynamic Signals*, Smart Innovation, Systems and Technologies 102,
https://doi.org/10.1007/978-3-319-95098-3_3

Amongst the several techniques which may regard data mining and knowledge discovery, *cluster analysis* certainly is one of the most acclaimed, since it is the very basic approach to search for regularities in a dataset. In a general sense, data clustering aims at discovering clusters (groups) of patterns such that, according to some (dis)similarity measure, similar pattern will lie in the same cluster whereas dissimilar patterns will fall into different clusters. Indeed, throughout the years, several families of clustering algorithms have been proposed in literature, such as partitional clustering (e.g. *k*-means [14]), density-based clustering (e.g. DBSCAN [8], OPTICS [2]) and hierarchical clustering (e.g. BIRCH [22], CURE [10]).

Further, in recent years the multi-agent clustering paradigm emerged (see Sect. 3.1.1): in such paradigm, a set of agents cooperates in either synchronous or asynchronous manner in order to cluster the dataset at hand. The agent per-se is an entity whose definition (i.e. tasks to be performed by the agent itself) strictly depends on the specific considered algorithm, as well as the way agents cooperate for a common objective. In this paper, such paradigm will be followed.

3.1.1 Contribution and State of the Art Review

We propose a novel agent-based clustering algorithm (*Evolutive Agent Based Clustering*, hereinafter E-ABC) in which each agent runs a very simple clustering procedure on a small subsample of the whole dataset. A genetic algorithm [9] orchestrates the evolution of such agents in order to return a set of well-formed clusters, thus possible regularities in the dataset at hand. Moreover, many clustering algorithms deal with a global metric; that is, the set of patterns to be processed are modelled as vectors in $\mathbb{R}^d$ and the same distance measure is adopted for discovering all possible clusters. Conversely, in E-ABC, each agent is in charge of discovering clusters and, at the same time, an instance of the parameters defining the considered parametric distance measure, i.e. the subspace where these clusters are well-formed.

As the agent's definition is vague and algorithm-dependent, there are several agent-based clustering algorithms proposed in literature. To the best of our knowledge, albeit some Authors proposed some evolutive algorithms (mainly Ant Colony Optimisation) in order to orchestrate agents, none of them implied such evolutive mechanisms to deal simultaneously with (a) searching for compact and populated clusters and (b) searching for the subspace in which such clusters exist.

Indeed, in [1] a multi-agent approach has been used for local graph clustering in which each agent performs a random walk on a graph with the main constraint that such agents are "tied" together by a rope, forcing them to be close to each other. In [5] a set of self-organising agents by means of Ant Colony Optimisation has been applied to anomaly detection and network control. In [6] each agent runs a different clustering algorithm in order to return the best one for the dataset at hand. In [7] agents negotiate one another rather than being governed by a master/wrapper process (e.g. evolutive algorithm). In [11] Ant Colony Optimisation has been used in order to organise agents, where each ant "walks" on the dataset, building connections amongst points. In [17] each agent consists in a set of data points and agents link to each other, thus leading to clustering. In [18] a genetic algorithm has been used where the agents' genetic code is connection-based: each agent is a clustering result whose genetic code builds a (sub)graph and, finally, such subgraphs can be interpreted as clusters. In [19] the multi-agent approach collapses into two agents: a first agent runs a cascade of Principal Component Analysis, Self Organizing Maps and k-means in order to cluster data and a second agent validates such results: the two agents interactively communicate with each other. Finally, in [3] a multi-agent algorithm has been proposed in which agents perform a Markovian random walk on a weighted graph representation of the input dataset. Each agent builds its own graph connection matrix amongst data points, weighting the edges according to the selected distance measure parameters, and performs a random walk on such graph in order to discover clusters. This algorithm has been employed in [4] to identify frequent behaviours of mobile network subscribers starting from a set of call data records. Conversely, the E-ABC approach consists in deploying many agents, each performing very simple tasks.

The remainder of this paper is structured as follows: Sect. 3.2 introduces the proposed algorithm, describing the agent behaviour (Sect. 3.2.1), the whole evolutive framework (Sect. 3.2.2) and the output collection (Sect. 3.2.3); Sect. 3.3 will describe some experimental results, by introducing the datasets (Sect. 3.3.1) and the evaluation metrics (Sect. 3.3.2) used in order to address E-ABC performances (Sect. 3.3.3). Finally, Sect. 3.4 will draw some conclusions, discussing some future works and some further applications.

3.2   Proposed Algorithm

3.2.1   Agent Definition

In E-ABC each agent is in charge of running a core clustering algorithm on a randomly chosen subset R of the total number of patterns N_P.

Specifically, it runs a variant of the Basic Sequential Algorithmic Scheme (BSAS) algorithm [21], namely RL-BSAS, which adds some Reinforcement Learning-based behaviour [20] to the standard BSAS by means of two additional parameters (reward factor α and forgetting factor β), together with the maximum allowed cluster radius θ. Basically, each new cluster discovered by RL-BSAS starts with a strength value $S = 1$. As in BSAS, the first pattern is used to initialise the first cluster by setting its centroid; successively, for each new pattern, its distance with respect to the existing centroids should be evaluated in order to check whether to assign such pattern to one of these clusters, or to initialise a new cluster with the pattern at hand. Then, the strength of the cluster to which the new pattern has been possibly assigned to is incremented by α, whereas the strengths of the other clusters is decremented by β. All clusters whose strength vanishes will be removed from the agent.

BSAS (and, by extension, RL-BSAS) does not need to know a-priori the number of clusters to be found (e.g. as in k-means) but, at the same time, BSAS is very sensitive to the parameter θ, as well as the order in which patterns in the dataset are presented to the clustering algorithm. Specifically, very low θ values usually lead to a huge number of clusters. In order to avoid such solutions, each agent is entitled to discover at most M clusters (with M a-priori defined). However, RL-BSAS is characterised by a low computational cost, well suited to be considered as the core clustering algorithm in this agent-based approach, where many agents are designed to perform low computational cost tasks.

E-ABC must be capable of discovering clusters in a given subspace; to this end, the dissimilarity measure between patterns will be evaluated according to a weighted Euclidean distance:

$$d(\mathbf{a}, \mathbf{b}, \mathbf{w}) = \sqrt{\sum_{i=1}^{N} \mathbf{w}_i (\mathbf{a}_i - \mathbf{b}_i)^2} \qquad (3.1)$$

where $\mathbf{a}, \mathbf{b} \in \mathbb{R}^N$ are two generic patterns drawn from the dataset and $\mathbf{w} \in \{0, 1\}^N$, where N is the number of features, which basically states whether a given feature is considered (1, true) or not (0, false). The latter has a strong impact on the dissimilarity measure as it acts as a feature selector.

Regarding the two Reinforcement Learning parameters (α and β), since β can be seen as clusters' death rate, one shall have $\alpha, \beta \in [0; 1]$ with $\alpha > \beta$. Also, it is possible to define a linear relationship between the two parameter (i.e. $\alpha = m \cdot \beta$) which might, however, differ from dataset to dataset and should be estimated a-priori. This is the reason why only the β parameter can be considered to be included in the agent genetic code, together with the metric's parameters $\mathbf{w}$ and the $\boldsymbol{\theta}$ vector:

$$\begin{bmatrix} \beta & \boldsymbol{\theta} & \mathbf{w} \end{bmatrix} \qquad (3.2)$$

At the end of the clustering procedure, each agent will return a set of clusters which will lie in the subspace given by $\mathbf{w}$ where, in turn, each cluster is described by its centroid, radius, strength and the list of patterns belonging to it.

The reason why $\boldsymbol{\theta}$ in (3.2) has been defined as a vector strictly depends on the agent's behaviour which must be discussed in more detail. Each new agent starts with a single value in $\boldsymbol{\theta}$ which actually indicates the maximum allowed radius, unique for all clusters discovered by the agent. At the end of the clustering procedure an agent can discover at most M clusters whose actual radii might be smaller; thus, their actual radii will be re-evaluated and stored in $\boldsymbol{\theta}$ which, at this stage, is a proper vector whose length is at most M. After the evolutive steps (Sect. 3.2.2), agents with more than one value in $\boldsymbol{\theta}$ can survive to the next generation, thus receiving a new random dataset shard to be processed. As $|\boldsymbol{\theta}| > 1$, such agents will not run a vanilla RL-BSAS; rather, they will bootstrap using such already known clusters and they will use the new random dataset shard to either:

(a) if $|\boldsymbol{\theta}| < M$, try to include its patterns in one of the already known clusters or create new clusters
(b) if $|\boldsymbol{\theta}| = M$, just check whether such new patterns can be included in one of the already known M clusters.

At the end of its clustering task, in order to shrink the output size, each agent will perform the *intra-agent fusion* procedure: if the distance between two centroids of two generic clusters A and B is below a user-defined threshold θ_{fus}:

1. incorporate patterns from cluster B to cluster A
2. re-evaluate centroid and radius for cluster A
3. since A is now a brand new cluster, its strength will be re-set to its original, starting value
4. remove cluster B from the agent's pool.

3.2.2 Evolutive Environment

The first generation will consist in N_A agents whose genetic codes of the form (3.2) will be randomly generated.

Each agent will operate according to Sect. 3.2.1 returning its own list of (possibly fused) clusters. After all agents have finished their run, a further fusion step is triggered: indeed, different agents can share the very same weights vector **w**, thus working in the very same subspace. Such additional *inter-agents fusion* procedure is defined as follows:

1. group agents by subspace
2. identify clusters across all agents that can be merged
3. try to merge clusters (i.e. using the θ_{fus} procedure as in Sect. 3.2.1), removing duplicate patterns, if any
4. assign the resulting clusters to the agents with higher fitness value, re-evaluate clusters' parameters and re-set their strength values
5. remove all empty agents.

The inter-agents fusion procedure is important as it reduces the number of agents per metric (freeing up some space—in terms of individuals—so that some more agents can be re-spawned, allowing a deeper subspaces exploration) while keeping all relevant (and removing all redundant) information collected so far.

At this point the output will be built and the stopping criterion will be verified (Sect. 3.2.3): if such check has a negative result, the whole procedure in order to produce the next generation is triggered.

As in every evolutionary optimisation algorithm, each individual is evaluated by a fitness function. To this end, each agent i will compute (for each cluster j in its pool) the following measures, namely normalised compactness (f_1) and normalised cardinality (f_2):

$$f_1(i, j) = 1 - \frac{1}{\sqrt{N}} \frac{\sum_{\mathbf{x} \in C_{i,j}} d(\mathbf{x}, \mathbf{c}_{i,j})}{|C_{i,j}|} \tag{3.3}$$

$$f_2(i, j) = |C_{i,j}|/N_P \tag{3.4}$$

The former (normalised compactness) is defined as the complement of dispersion, where the dispersion is in turn defined as the normalised average distance between each pattern $\mathbf{x}$ in the jth cluster of ith agent ($C_{i,j}$) and its centroid ($\mathbf{c}_{i,j}$). The latter (normalised cardinality) is defined as the ratio between the cardinality of the cluster and the total number of patterns. Normalisations in Eqs. (3.3) and (3.4) ensure that $f_1, f_2 \in [0; 1]$. Since each agent i will have at most M cluster(s), it will evaluate the average cardinality ($\bar{f}_1$) and the average compactness ($\bar{f}_2$) amongst its clusters, which in turn are used to compute a fitness function of the form:

$$F(i) = \lambda \cdot \bar{f}_1(i) + (1 - \lambda) \cdot \bar{f}_2(i) \tag{3.5}$$

where $\lambda \in [0; 1]$ is a trade-off parameter. After all individuals have been evaluated by the fitness (3.5), they will be sorted according to it and customised versions[1] of the standard genetic operators (elitism, selection, crossover, mutation) will be applied in order to form the next generation as follows:

1. copy the elite unaltered to the next generation
2. given the number of individuals to be selected,[2] perform the selection operator in order to gather the list of individuals to be mutated and the list of individuals to be crossovered (parents)
3. perform crossover
4. perform mutation, either big or small depending on the fitness value
5. merge elite and new offsprings to form the next generation.

3.2.3 Stopping Criterion and Output Collection

E-ABC iteratively builds the final output, properly collecting each generation's results (i.e. after the inter-agents fusion procedure, Sect. 3.2.2). Starting from the second generation, the output will be built by merging all clusters in the current generation in a unique list of clusters, regardless of the agent who identified them. In this manner, each entry in the output list contains the whole set of clusters (along with their parameters) for a given metric. Each cluster in the output list will be mapped with the very same fitness function[3] of the form (3.5) in order to verify the stopping criterion. E-ABC will run up to a maximum, a-priori defined, number of generations

---

[1]In the sense that they must be able to deal with an heterogeneous genetic code such as (3.2).

[2]Such value is user-configurable and it is defined as a percentage of the number of survived agents.

[3]Such value now acts more like a cluster quality measure rather than a fitness value, as we are evaluating a list of clusters rather than individuals from a genetic population.

N_G with a stopping criterion based on the average fitness values amongst the output items: if for an a-priori defined number of generations N_G^{STOP} the average fitness does not change significantly, the algorithm halts.

At the end of the whole evolution the entire final output list (after all fusion phases) will be collected, not just the best item. That is because:

- instead of using a genetic algorithm to solve a standard, numerical optimisation problem, it serves as a proper orchestrator with the final goal of evolving the population as a whole
- collecting just the best item will return clusters in just one of the possible subspaces, restricting the local metric learning capabilities.

3.3   Experimental Results

In order to validate our approach, a set of experiments is proposed in order to address E-ABC efficiency with respect to the dataset size (both in terms of number of patterns and number of features) and its effectiveness in terms of clusters identification in given subspaces and accurate estimation of their parameters (namely, centroid and radius).

3.3.1   Dataset Description

To this end, eight synthetic datasets built as follows will be considered:

1. let $G_1, G_2, ..., G_8$ be eight Gaussian distributions with mean values linearly spaced in $\mu \in [0.15; 0.85]$ and standard deviation $\sigma = 0.03$
2. four datasets by concatenating four clusters in $\mathbb{R}^4$ are built as follows:

$$
\begin{aligned}
&\text{cluster1}: G_1\ G_2\ \sim\ \sim \\
&\text{cluster2}: G_3\ G_4\ \sim\ \sim \\
&\text{cluster3}: \sim\ \sim\ G_5\ G_6 \\
&\text{cluster4}: \sim\ \sim\ G_7\ G_8
\end{aligned}
$$

where features marked as $\sim$ are "noisy" features, drawn from a uniform distribution in $[0; 1]$. Specifically, these four datasets contain 1000, 10000, 100000 and 1000000 patterns per cluster
3. four additional datasets are built by adding 6 features (i.e. columns) of pure random noise to the previous four datasets and then random shuffling by columns; after which the Gaussian features for cluster 1 were in columns 1 and 6, for cluster 2 were in columns 3 and 9, for cluster 3 were in columns 2 and 5 and, finally, for cluster 4 were in columns 7 and 10.

3.3.2 Evaluation Metrics

In order to address E-ABC ability in discovering clusters and, more importantly, in estimating their parameters, the following two evaluation measures are proposed:

- Estimation Quality (EQ): whose task is to judge radius and centroid estimation accuracy as

$$EQ = 1 - [(1 - \varepsilon) \cdot \delta_c + \varepsilon \cdot \delta_\theta] \qquad (3.6)$$

  where δ_c is the Euclidean distance between true and estimated centroids, δ_θ is the absolute difference between true and estimated radii and $\varepsilon \in [0; 1]$ is a trade-off parameter. Obviously $EQ \in [0; 1]$ and as $EQ \to 1$, the more accurate the results.

- Identified Patterns (IP): which basically counts how many patterns within the estimated cluster are indeed part of the ground-truth cluster. Such value (N_C) is then normalised by the size of the cluster itself ($|C|$), in order to have $IP \in [0; 1]$ and as $IP \to 1$, the more accurate the results:

$$IP = N_C / |C| \qquad (3.7)$$

3.3.3 Tests Results

Figures 3.1 and 3.2 show the average results (both in terms of efficiency and effectiveness) amongst five runs (changing the random number generator seed) obtained by E-ABC on the eight datasets from Sect. 3.3.1 under two different scenarios. As the dataset size increases:

1. both the number of agents and the number of patterns per agent have been kept constant
2. the number of agents has been kept constant whilst the number of patterns per agent increases

All of these tests have been performed on a Linux CentOS machine with an Intel(R) Xeon(R) E5520 CPU @ 2.27 GHz and 32 GB RAM. E-ABC has been developed in Python using the NumPy library for efficient numerical computation.

In Fig. 3.1 the average results regarding the first scenario are shown. Clearly, since the number of patterns per generation does not change even if the dataset size does, the running time is rather constant. Coherently, the percentage of identified patterns slightly decreases as the dataset size increases. Despite that, the estimation quality is in both cases ($\mathbb{R}^4$ and $\mathbb{R}^{10}$) always above 90%. Such results have been obtained with $N_A = 25$, $|R| = 100$ in $\mathbb{R}^4$ and $N_A = 100$, $|R| = 150$ in $\mathbb{R}^{10}$.

In Fig. 3.2 the average results regarding the second scenario are shown. Obviously, as the dataset shard size increases, the running time increases as well. For the same reason, the percentage of identified patterns is rather constant. As in the previous case, the estimation quality is in both cases ($\mathbb{R}^4$ and $\mathbb{R}^{10}$) always above 90%. Such

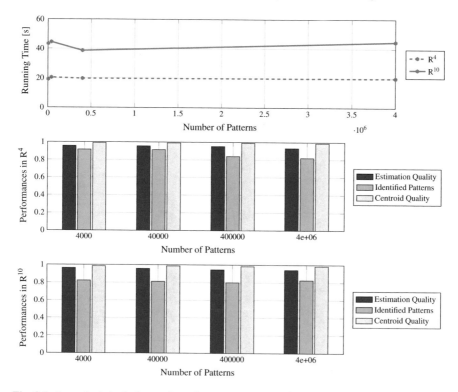

Fig. 3.1 Scenario 1: both the number of agents and the number of patterns per agent have been kept constant

results have been obtained with $N_A = 25$, $|R| = [50; 100; 200; 300; 400]$ in $\mathbb{R}^4$ and $N_A = 100$, $|R| = [100; 200; 300; 400; 500]$ in $\mathbb{R}^{10}$.

Leaving N_A and $|R|$ aside, all the other parameters have been kept constant amongst such different experiments: $\varepsilon = 0.5$, $\lambda = 0.2$ and $M = 3$.

From Figs. 3.1 and 3.2 it is possible to see that:

(a) EQ values are rather high, which means that both centroids and radii have been accurately estimated. Recall that in all cases $\varepsilon = 0.5$, thus correct estimations of centroids and radii have the same importance

(b) IP values are generally lower than EQ which should not surprise: indeed, E-ABC does not guarantee processing of the whole dataset as it will be randomly sub-sampled. In other words some patterns might be frequently selected by several, different agents and, similarly, some patterns might not be selected at all

In order to further explore the final E-ABC output (i.e. the final list of clusters) in Table 3.1 part of the genetic code is shown (namely, the weights vector and the

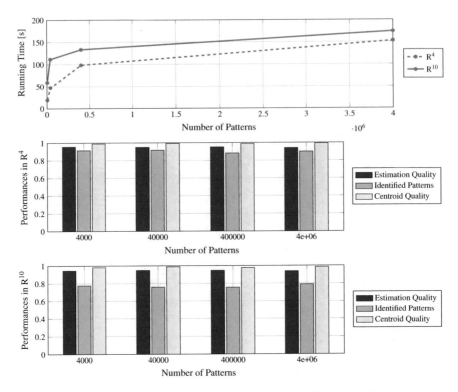

Fig. 3.2 Scenario 2: the number of agents has been kept constant whilst the number of patterns per agent increases

estimated cluster radius[4]) associated to each cluster, for four of the eight datasets for the first scenario only. Also, we will show the estimated centroids and, for the sake of ease, we will match the estimated values (marked with a hat) with the true values from the ground-truth datasets (Sect. 3.3.1).

Finally, from Table 3.1a, b, c and d it is possible to see that:

(a) true and estimated radii and centroids are very close, coherently to the high EQ values in Figs. 3.1 and 3.2
(b) centroids shrunk down from $\mathbb{R}^4$ ($\mathbb{R}^{10}$) to $\mathbb{R}^2$ as only two features have been selected (see their respective **w**)
(c) clusters in the output lists exactly corresponds to the four clusters from the ground-truth datasets: not only the clusters parameters have been accurately estimated, but also the subspaces in which they exist have been correctly detected; indeed, it is easy to match the 1's in **w** with the Gaussian components and 0's with noisy features (Sect. 3.3.1).

---

[4]We omit the penalty factor β (and, by extension, the reward factor α) as they mainly drive RL-BSAS rather then describe the final clusters and/or agents.

Table 3.1 E-ABC Final Output. For the sake of readability, the notation $\{0\}^x$ indicates a sequence of x zeros

	$\theta / \hat{\theta}$	$c / \hat{c}$	w
cluster 1	0.222 / 0.105	$\begin{bmatrix}0.162 & 0.250 \\ 0.156 & 0.249\end{bmatrix}$	$[1\ 1\ 0\ 0]$
cluster 2	0.198 / 0.233	$\begin{bmatrix}0.360 & 0.451 \\ 0.365 & 0.460\end{bmatrix}$	$[1\ 1\ 0\ 0]$
cluster 3	0.219 / 0.166	$\begin{bmatrix}0.551 & 0.629 \\ 0.548 & 0.629\end{bmatrix}$	$[0\ 0\ 1\ 1]$
cluster 4	0.212 / 0.168	$\begin{bmatrix}0.750 & 0.820 \\ 0.752 & 0.819\end{bmatrix}$	$[0\ 0\ 1\ 1]$

(a) 4000-patterns in $\mathbb{R}^4$

	$\theta / \hat{\theta}$	$c / \hat{c}$	w
cluster 1	0.229 / 0.117	$\begin{bmatrix}0.187 & 0.250 \\ 0.188 & 0.255\end{bmatrix}$	$[1\ \{0\}^4\ 1\ \{0\}^4]$
cluster 2	0.236 / 0.215	$\begin{bmatrix}0.350 & 0.450 \\ 0.354 & 0.451\end{bmatrix}$	$[\{0\}^2\ 1\ \{0\}^5\ 1\ 0]$
cluster 3	0.222 / 0.233	$\begin{bmatrix}0.550 & 0.651 \\ 0.549 & 0.650\end{bmatrix}$	$[0\ 1\ \{0\}^2\ 1\ \{0\}^5]$
cluster 4	0.230 / 0.231	$\begin{bmatrix}0.751 & 0.813 \\ 0.738 & 0.827\end{bmatrix}$	$[\{0\}^6\ 1\ \{0\}^2\ 1]$

(b) 40000-patterns in $\mathbb{R}^{10}$

	$\theta / \hat{\theta}$	$c / \hat{c}$	w
cluster 1	0.259 / 0.211	$\begin{bmatrix}0.215 & 0.250 \\ 0.212 & 0.241\end{bmatrix}$	$[1\ 1\ 0\ 0]$
cluster 2	0.250 / 0.213	$\begin{bmatrix}0.400 & 0.450 \\ 0.392 & 0.456\end{bmatrix}$	$[1\ 1\ 0\ 0]$
cluster 3	0.250 / 0.322	$\begin{bmatrix}0.550 & 0.591 \\ 0.548 & 0.590\end{bmatrix}$	$[0\ 0\ 1\ 1]$
cluster 4	0.266 / 0.234	$\begin{bmatrix}0.750 & 0.772 \\ 0.739 & 0.756\end{bmatrix}$	$[0\ 0\ 1\ 1]$

(c) 400000-patterns in $\mathbb{R}^4$

	$\theta / \hat{\theta}$	$c / \hat{c}$	w
cluster 1	0.288 / 0.225	$\begin{bmatrix}0.246 & 0.276 \\ 0.252 & 0.280\end{bmatrix}$	$[1\ \{0\}^4\ 1\ \{0\}^4]$
cluster 2	0.303 / 0.196	$\begin{bmatrix}0.350 & 0.450 \\ 0.366 & 0.455\end{bmatrix}$	$[\{0\}^2\ 1\ \{0\}^5\ 1\ 0]$
cluster 3	0.287 / 0.233	$\begin{bmatrix}0.550 & 0.650 \\ 0.555 & 0.633\end{bmatrix}$	$[0\ 1\ \{0\}^2\ 1\ \{0\}^5]$
cluster 4	0.278 / 0.197	$\begin{bmatrix}0.737 & 0.766 \\ 0.740 & 0.748\end{bmatrix}$	$[\{0\}^6\ 1\ \{0\}^2\ 1]$

(d) 4000000-patterns in $\mathbb{R}^{10}$

3.4 Conclusion

In this paper we presented a novel agent-based clustering technique in which each agent independently processes a subsample of patterns randomly drawn from the whole dataset and their results will be properly merged together. In order to orchestrate agents' evolution, a genetic algorithm has been used. Part of the agent's genetic code is a binary vector whose length is equal to the number of features: such vector acts as a feature selector and therefore each agent can discover clusters in a suitable (agent-dependent) subspace. Thus, the genetic algorithm orchestrates agents' evolution in (simultaneously) discovering clusters and selecting features. As a straightforward consequence, some more knowledge can be extracted from the dataset at hand since each agent will discover one or more clusters in a given subspace without using any exhaustive or bruteforce-like technique, which can be computationally unfeasible (e.g. in $\mathbb{R}^d$, $2^d - 1$ possible subspaces exist, leaving the all-zero combination aside). Experimental results on synthetic datasets demonstrated that E-ABC is very capable of discovering clusters in subspaces of $\mathbb{R}^d$ and estimating their main characteristics with high accuracy.

We will further investigate this approach, testing E-ABC in real-life scenarios. Moreover, several further improvements and E-ABC variants can be developed: first, it is rather straightforward to distribute agents' execution across several computational units in order to tackle massive datasets in a parallel and distributed fashion; second, we will try to expand the E-ABC philosophy (i.e. swarm of very simple agents performing very simple tasks on a small subset of data) in clustering non-metric spaces [16]: indeed, E-ABC can be considered as a somewhat "template algorithm" and BSAS-based algorithms do not strictly require an algebraic structure in order to compute clusters' representatives (e.g. as in k-means). Indeed, by changing the (dis)similarity measure (and its parameters, if any, in the genetic code) and the evaluation of clusters' representatives (e.g. medoid rather than centroid), as discussed in [15], it is possible to (ideally) use any ad-hoc (dis)similarity measures for the input space at hand. Seminal examples include the so-called edit distances for sequences (e.g. [12]) and graphs (e.g. [13]).

Acknowledgements The Authors would like to thank Daniele Sartori for his help in implementing and testing E-ABC.

References

1. Alamgir M., Von Luxburg, U.: Multi-agent random walks for local clustering on graphs. In: 2010 IEEE 10th International Conference on Data Mining (ICDM), pp. 18–27. IEEE (2010)
2. Ankerst, M., Breunig, M.M., Kriegel, H.P., Sander, J.: Optics: ordering points to identify the clustering structure. ACM Sigmod Rec. ACM **28**, 49–60 (1999)
3. Bianchi, F.M., Maiorino, E., Livi, L., Rizzi, A., Sadeghian, A.: An agent-based algorithm exploiting multiple local dissimilarities for clusters mining and knowledge discovery. Soft Comput. **5**(21), 1347–1369 (2015)
4. Bianchi, F.M., Rizzi, A., Sadeghian, A., Moiso, C.: Identifying user habits through data mining on call data records. Eng. Appl. Artif. Intel. **54**, 49–61 (2016)
5. Carvalho, L.F., Barbon, S., de Souza Mendes, L., Proença, M.L.: Unsupervised learning clustering and self-organized agents applied to help network management. Expert Syst. Appl. **54**, 29–47 (2016)
6. Chaimontree, S., Atkinson, K., Coenen, F.: Clustering in a multi-agent data mining environment. Agents Data Min. Interact., 103–114 (2010)
7. Chaimontree, S., Atkinson, K., Coenen, F.: A multi-agent based approach to clustering: harnessing the power of agents. In: ADMI, pp. 16–29. Springer (2011)
8. Ester, M., Kriegel, H.P., Sander, J., Xu, X., et al.: A density-based algorithm for discovering clusters in large spatial databases with noise. Kdd **96**, 226–231 (1996)
9. Goldberg, D.E.: Genetic Algorithms in Search, Optimization and Machine Learning. Addison-Wesley, Reading (1989)
10. Guha, S., Rastogi, R., Shim, K.: Cure: an efficient clustering algorithm for large databases. ACM Sigmod Rec. ACM **27**, 73–84 (1998)
11. Inkaya, T., Kayalıgil, S., Özdemirel, N.E.: Ant colony optimization based clustering methodology. Appl. Soft Comput. **28**, 301–311 (2015)
12. Levenshtein, V.I.: Binary codes capable of correcting deletions, insertions, and reversals. Sov. Phys. Dokl. **10**, 707–710 (1966)
13. Livi, L., Rizzi, A.: The graph matching problem. Pattern Anal. Appl. **16**(3), 253–283 (2013)
14. MacQueen, J.: Some methods for classification and analysis of multivariate observations. Proceedings of the Fifth Berkeley Symposium on Mathematical Statistics and Probability, Oakland, CA, USA **1**, 281–297 (1967)
15. Martino, A., Rizzi, A., Frattale Mascioli, F.M.: Efficient approaches for solving the large-scale k-medoids problem. In: Proceedings of the 9th International Joint Conference on Computational Intelligence, IJCCI, INSTICC, vol. 1, pp. 338–347. SciTePress (2017)

16. Martino, A., Giuliani, A., Rizzi, A.: Granular computing techniques for bioinformatics pattern recognition problems in non-metric spaces. In: Chen, S.M., Pedrycz, W. (eds.) Computational Intelligence for Pattern Recognition. Springer, Accepted for Publication (2018). https://rd.springer.com/chapter/10.1007%2F978-3-319-89629-8_3
17. Ogston, E., Overeinder, B., Van Steen, M., Brazier, F.: A method for decentralized clustering in large multi-agent systems. In: Proceedings of the Second International Joint Conference on Autonomous Agents and Multiagent Systems, pp. 789–796. ACM (2003)
18. Pan, X., Chen, H.: Multi-agent evolutionary clustering algorithm based on manifold distance. In: 2012 Eighth International Conference on Computational Intelligence and Security (CIS), pp. 123–127. IEEE (2012)
19. Park, J., Oh, K.: Multi-agent systems for intelligent clustering. Proc. World Acad. Sci. Eng. Technol. **11**, 97–102 (2006)
20. Rizzi, A., Del Vescovo, G., Livi, L., Frattale Mascioli, F.M.: A new granular computing approach for sequences representation and classification. In: The 2012 International Joint Conference on Neural Networks (IJCNN), pp. 1–8. IEEE (2012)
21. Theodoridis, S., Koutroumbas, K.: Pattern Recognition, 4th edn. Academic Press (2008)
22. Zhang, T., Ramakrishnan, R., Livny, M.: Birch: an efficient data clustering method for very large databases. ACM Sigmod Rec. ACM **25**, 103–114 (1996)

Chapter 4
Neural Beamforming for Speech Enhancement: Preliminary Results

Stefano Tomassetti, Leonardo Gabrielli, Emanuele Principi, Daniele Ferretti
and Stefano Squartini

Abstract In the field of multi-channel speech quality enhancement, beamforming algorithms play a key role, being able to reduce noise and reverberation by spatial filtering. To that extent, an accurate knowledge of the Direction of Arrival (DOA) is crucial for the beamforming to be effective. This paper reports extremely improved DOA estimates with the use of a recently introduced neural DOA estimation technique, when compared to a reference algorithm such as Multiple Signal Classification (MUSIC). These findings motivated for the evaluation of beamforming with neural DOA estimation in the field of speech enhancement. By using the neural DOA estimation in conjunction with beamforming, speech signals affected by reverberation and noise improve their quality. These first findings are reported to be taken as a reference for further works related to beamforming for speech enhancement.

Keywords Beamforming · Speech enhancement · Artificial neural networks

4.1 Introduction

Algorithms for enhancing the quality and the intelligibility of speech signals find application in many important practical scenarios. In smart home scenarios [21, 22], automatic speech recognition (ASR) is widely employed and a cause of high word

S. Tomassetti · L. Gabrielli (✉) · E. Principi · D. Ferretti · S. Squartini
Dipartimento di Ingegneria dell'Informazione, Università Politecnica delle Marche,
Via Brecce Bianche, 60131 Ancona, Italy
e-mail: l.gabrielli@univpm.it

S. Tomassetti
e-mail: s.tomassetti@gmail.com

E. Principi
e-mail: e.principi@univpm.it

D. Ferretti
e-mail: d.ferretti@univpm.it

S. Squartini
e-mail: s.squartini@univpm.it

© Springer International Publishing AG, part of Springer Nature 2019
A. Esposito et al. (eds.), *Neural Advances in Processing Nonlinear Dynamic Signals*, Smart Innovation, Systems and Technologies 102,
https://doi.org/10.1007/978-3-319-95098-3_4

error rate is represented by the acoustic distortion due to additive noise and reverberation [18, 23]. Speech intelligibility in noisy environments is severely compromised for users of cochlear implants and during the communication with mobile phones [19]. Due to the importance of the problem, the literature of speech enhancement methods is vast and presents several different approaches. Hereby interest is given to multi-channel methods, i.e. those methods exploiting the information contained in signals acquired by multiple microphones.

Beamforming [4], i.e., the use of spatial filtering, is one of the most popular multi-channel approaches and it is ideally equivalent to steering the microphone polar pattern in the direction of the source. In delay-and-sum beamformer (DS), signals are aligned by taking into account the phase shifts among the microphones signals, while the filter-and-sum (FS) beamformer includes additional processing with linear filters. More advanced beamformers such as the Minimum Variance Distortionless Response (MVDR) [5] and the Generalized Sidelobe Canceller (GSC) [8, 9] adapt to the acoustic environment.

Nonlinear speech enhancement [12] based on deep neural networks (DNN) has also been recently proposed for enhancing speech from multiple microphone signals, in particular with the objective of improving the performance of ASRs. The advantage of this approach is that the parameters of the enhancement algorithm, i.e., the network weights, can be trained jointly with the ASR acoustic model, thus they are optimized under the same objective function. In [2], a DNN is employed as a speaker separation stage and the ASR employs bottle neck features as well as filter-bank coefficients extracted from multi-channel signals. In [24, 28], multi-channel MEL filter bank coefficients are employed as input to an ASR based on convolutional neural networks (CNNs). Hoshen et al. [11] developed a similar approach but using raw waveforms as input to the network. Actual DNN-based beamformers for ASR have been proposed in [17, 29]. In [29], the algorithm operates on multiple complex-valued short-time Fourier transforms (STFTs) to estimate a single enhanced signal. The network operates on generalized cross correlation (GCC) coefficients to estimate the weights of a FS beamformer that then processes the STFTs related to the microphone signals. The algorithm is employed as a preprocessing stage of an ASR and it can be trained jointly with the acoustic model in order to further optimize the process. A similar approach has been proposed in [17], where the spatial filter coefficients are estimated by a long short-term memory (LSTM) network [10] that takes raw waveform signals as input, instead of STFT coefficients. In [7, 32], the authors employ a DNN and a Bidirectional-LSTM for estimating the time-frequency mask in the MVDR beamformer.

Up to the authors' knowledge, few works propose DNN-based algorithms targeted at enhancing the quality and the intelligibility of the perceived speech. In an early work [16], a single-layer perceptron filter is introduced in the GSC beamformer framework to suppress noise. In [31], the work is extended by introducing alternative structures of the noise reduction algorithm. In a more recent work [3], the authors employ a denoising autoencoder with multi-channel features. In particular, they augment single-channel log-mel filter-bank features with information extracted from multiple channels. The authors evaluated the noise suppression performance with segmental signal-to-noise ratio (SSNR) and the cepstral distortion (CD) measures, and showed that employing pre-enhanced speech features their approach improves with respect to the single-channel autoencoder.

These works suggest that investigating on the capabilities of neural networks to improve current state-of-art speech enhancement algorithm may be fruitful. The paper first compares a well known algorithm for DOA estimation, Multiple Signal Classification (MUSIC) [26], with a recently introduced neural approach [30] with the intent of verifying the effectiveness of the experiments conducted in [30] and providing additional experimental information. The results are very promising and motivated for application to a multi-channel speech enhancement scenario where the estimated DOA angle feeds a FS beamforming algorithm to improve the intelligibility of speech signals affected by noise and reverberation, showing a satisfying improvement with respect to a state-of-the-art method. Adopting neural algorithms for DOA estimation removes some constraints, i.e. those related to the microphone array geometry.

4.2 Algorithm

In this work, the MUSIC method, is taken as a baseline and it is compared to a more recent and promising technique for neural DOA estimation based on machine learning reported in [30], from now on referred to as NDOA. NDOA follows a data-driven approach, where a corpus of data is provided during the training phase to estimate the algorithm parameters. A feature-set is first extracted and treated as input to a multi-layer perceptron (MLP). The output of the MLP is the DOA estimate required by subsequent stages (e.g. beamforming). Figure 4.1 resumes the complete algorithm, including beamforming and speech quality evaluation.

Among possible feature-sets, the generalized cross-correlation coefficients (GCC) are employed for their wide acceptance in the field of DOA Estimation and their ability in capturing phase related information [30]. Specifically, the GCC are more reliable compared to time difference of arrival (TDOA). In our implementation, the GCC-PHAT algorithm [15] is used to extract GCC vectors, based on the cross-correlation of spectral coefficients between all microphone pairs. For each microphone pair combination C, only a part of the GCC values are taken, depending on the microphones distance. Let D be the maximum distance between microphones in the array, the time delay is $\tau = D/c$ seconds, or $N = F_s \cdot \tau$ samples. Under such conditions only the center $2N + 1$ GCC values contain useful information, i.e. those values corresponding to delays in the range $\pm N$ samples. The rest of the GCC values can be discarded. In more rigorous terms, the cross correlation between the power spectra of any two microphone signals in the array, i.e. $S_{12}(f)$, is defined as:

$$S_{12}(f) = X_1(f) * \mathring{X}_2(f)^*, \tag{4.1}$$

Fig. 4.1 Flow diagram of
the dataset generation **a** and
neural DOA estimation **b**.
The clean speech is finally
compared to the processed
speech for the objective
evaluation

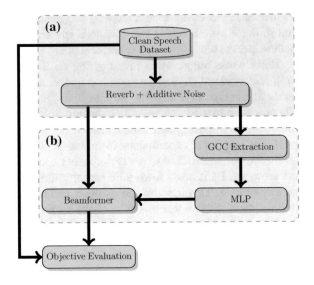

where $\mathring{X}_2(f)$ is a circular-shifted version of the FFT of $x_2(t)$ needed to operate
with coherent GCC vectors. The circular shift is of N samples, thus the only GCC
containing information are $S_{12}(0 \div 2N + 1)$.

To improve the robustness of the GCC, a further processing stage, histogram
equalization (HEQ) is undertaken. This improves reliability of the GCC by increasing
the spread between noisy and useful coefficients. Figure 4.2 shows a GCC feature
set before and after histogram equalization.

Finally, under stationary conditions for the sound source position, averaging of
the GCC can be done between successive frames. Under testing and training this is
implemented by averaging all coefficients from the same file in the dataset, which is
known to be stationary. In real-world conditions, under the reasonable hypothesis of
a slowly time-varying position, a moving average can be employed, with a suitable
averaging window.

Once feature extraction is completed, the features are given as input to a multilayer
perceptron (MLP). The MLP employed for DOA classification has an input layer
with nodes equal to the input feature dimensions $C(2N + 1)$. One hidden layer is
employed, with a sigmoid activation function. Differently from [30] the output layer
is fed to a nonlinear combination neuron which outputs a continuum estimate of
the DOA. In the original paper classes of $1°$ or more were used, making difficult to
compare different experiments with varying noise and reverberation.

A voice activity detection (VAD) algorithm [27] is employed to discard GCC from
audio frames not containing speech, that would harm learning.

(a)   **(b)**

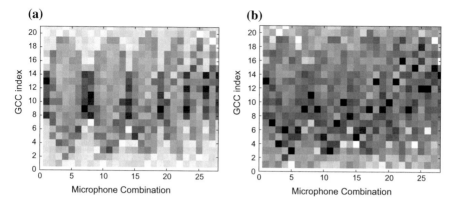

Fig. 4.2 GCC matrix extracted from a speech frame in the dataset before applying HEQ (**a**) and after (**b**)

4.3 Dataset

The generation of the dataset for training and testing follows previous works in the field. Specifically, the multichannel noisy speech is generated from a widely used mono clean speech dataset with American English sentences (WSJ0) [25]. All speech signals are sampled at 16 kHz. The speech signals are transformed into multichannel signals with noise and reverb, in order to test the robustness of the proposed approach.

The generation of the dataset assumes a microphone displacement following a uniform circular array (UCA) of 8 microphones, with a radius of 0.1 m. Different configurations of reverberation and additive noise are introduced to simulate different use cases, and they have been applied according to the following rules:

1. Reverberation was added employing the RIR Generator[1] tool for Matlab, based on [1].
2. Three reverberation schemes were used, simulating small, medium and large rooms, each one with the speaker in the far field or near field. One shortcoming of the work in [30] was that three different datasets were created depending on the room size, and each one was evaluated on its own.
3. The direction of arrival in the UCA were randomly selected.
4. Other randomly selected parameters were: speaker distance, room dimension, T60 reverberation time, SNR, noise type.
5. The added noise came from noise samples provided by the REVERB CHALLENGE dataset and its SNR was selected randomly from 0 to 20 dB. 40 different types of noise were used, divided in simulated rooms, according to the REVERB CHALLENGE [14].

---

[1] https://www.audiolabs-erlangen.de/fau/professor/habets/software/rir-generator.

Table 4.1 Training and Testing datasets details

	Training	Testing
Content	7768 sentences from the WSJ0 training set	507 sentences from the WSJ0 test set
Room size (m)	Small (7×5), medium (12×10), large (17×15)	Small (6×4), medium (10×8), large (14×12)
Distance (m)	Near (1) and far (2, 4, 6.5 for small, medium, large)	Near (1) and far (1.5, 3, 5 for small, medium, large)
T60 (s)	0.1–1.0 s with 0.1 s step	Three steps: 0.3, 0.6, 0.9 s
SNR (dB)	Randomly selected from 0 to 20 dB	Randomly selected from 0 to 20 dB

The microphone array displacement affects some of the algorithm parameters. Specifically, the maximum distance $D = 0.1\ m$ implies a delay of $N = 10$ samples, and the selection of 21 GCC coefficients. The array of 8 microphones allows $C = 28$ combinations of signals to compute the GCC. A total of 588 features are, thus, computed for each frame and are fed as input to the MLP. The signals sampling frequency also affects the frame size employed for FFT of the input signals. In our implementation the frame size chosen was 0.2 s with 50% overlap.

Training and Testing sets were organized as shown in Table 4.1.

4.4 Implementation and Results

Training and testing has been performed using Keras running on Theano as a backend, while all the audio preprocessing was done in Matlab. For each MLP parameter set, the training was done over 5000 epochs, interleaved with periodical validation every 1000 epochs. MLP weights were taken from the validation obtaining the lowest RMS error.

Experimenting all possible parameter sets is not feasible as it would require a large number of very time-consuming experiments. To reduce the number of trials, discrete steps have been used for all numerical parameters. Furthermore, a heuristic procedure has been employed to look for a sub-optimum. Its first step consists in conducting several experiments by varying a single parameter, with all other parameters fixed to a initial value. The value of the parameter under test yielding best results is taken and the procedure is repeated for another parameter until all parameters have been experimented with. Finally, a number of random experiments are conducted to gather more confidence that there are no other RMS error minima below the one previously found. Details regarding the experiments follow:

- MLP network size: from 80 to 512 in discrete steps;
- MLP update rule: stochastic gradient descent (SGD), Adam, AdaMax [13];
- activation functions: tanh, rectified linear unit (ReLU);

Table 4.2 Some of the MLP parameter sets employed during training and the RMS error obtained during testing of the related parameter set. The RMS error is expressed as the difference in angle with respect to the correct DOA. The last layer has dimension 1 and outputs a floating point value

Net size	Activations	Optimizer	RMS error
(588,100,80)	C	SGD	11.95
(588,100,80)	A	SGD	11.37
(588,100,80)	B	SGD	11.09
(588,100,80)	B	Adam	14.92
(588,100,80)	B	AdaMax	10.02
(588,250,200)	B	AdaMax	16.79
(588,160,80)	B	AdaMax	**4.48**

- mini-batch: 1–3000 in discrete steps;
- learning rate: 1e-9–1e-5 in discrete steps;
- momentum: 0.8, 0.9.

During preliminary tests, the latter three parameters were shown to yield improved performance with values, respectively, 3000, 1e-8, 0.9, notwithstanding the choice of the former three parameters. Furthermore, preliminary tests show that the first three combinations of the activation functions leading to good results are the following:

1. A: (tanh, tanh, tanh);
2. B: (tanh, ReLU, tanh);
3. C: (ReLU, ReLU, tanh).

Choice of MLP network size, weight optimization algorithm and activation functions has been done according to the heuristic procedure described above. Some results are reported in Table 4.2. The activation functions combination yielding the best results in first place is B. AdaMax, is found to be the best optimization algorithm and a network size of (588, 160, 80) largely improves performance.

After training and selection of the best parameter set, a first evaluation has been conducted on the DOA estimation algorithm with respect to the MUSIC algorithm. The results are extremely convincing as the error decrease is larger than one order of magnitude:

- MUSIC RMS Error: 122.8.
- NDOA RMS Error: 4.48.

The excellent capability of the NDOA algorithm to track the DOA is shown in Fig. 4.3, where DOA estimation errors are reported from an excerpt of 75 randomly selected sentences. Error values of NDOA are up two orders of magnitude below MUSIC. This motivates for application of NDOA to many scenarios, such as speech enhancement. To verify the effect of the improved accuracy of DOA estimation, both NDOA and MUSIC are first applied to a FS beamforming algorithm and the resulting speech quality is evaluated. Processed audio evaluation is carried on by employing

Fig. 4.3 DOA estimation RMS Error for MUSIC (dotted line) and NDOA (solid line)

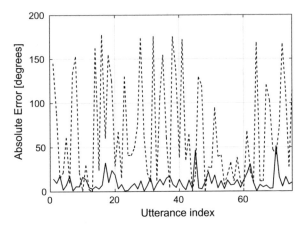

Table 4.3 Speech Quality comparison between unprocessed speech (NONE), beamformed speech with NDOA (NDOA), beamformed speech with MUSIC, speech enhanced with [6] (SE), speech enhanced with NDOA beamforming and [6] (NDOA + SE). Please note that with IS, lower values correspond to better performance

	NONE	MUSIC	NDOA	SE	NDOA + SE
PESQ	1.74	1.8	1.89	1.88	**1.95**
IS	3.4	3.55	3.49	3.28	**3.11**

two speech quality measures [20]: Perceptual Evaluation of Speech Quality (PESQ) and Itakura-Saito distance (IS). The former is defined as a standard for speech quality assessment for communication technologies, standardized as ITU-T recommendation P.862(02/01).[2] It provides off-line evaluation of speech signals quality by amplitude and time alignment, in order to provide a meaningful sample-by-sample comparison of the original signal and the processed signal. Furthermore it makes use of auditory transforms and cognitive models to predict a human Mean Opinion Score (MOS) speech quality assessment. The Itakura-Saito distance, on the other hand, is not a perceptual measure and it provides a measure of the difference between two spectra, in this case the original signal and its processed version.

The results, reported in Table 4.3 and compared to the original speech source (with noise and reverb applied), show that the higher DOA estimation accuracy achieved by the NDOA improves also the speech signal quality. Motivated by these findings, we tested whether the NDOA can further improve speech quality when applied in conjunction to an established speech enhancement technique by Ephraim et al. [6] (in short SE).

---

[2]http://www.itu.int/rec/T-REC-P.862/en.

These results are summarized in Table 4.3. The speech quality improvement obtained by the combination of both SE and NDOA, compared to SE only, is of 50% when evaluated with PESQ and of 70% when evaluated with IS. This confirms the validity of the approach.

4.5 Discussion and Future Directions

The experiments reported in this work show that neural DOA estimation exhibits excellent performance with respect to a reference technique such as MUSIC. When used in conjunction with a classic beamforming algorithm, its higher precision also improves its capability in enhancing the quality of speech affected by noise and reverberation with respect to a MUSIC DOA estimator in conjunction with the same beamforming algorithm. The performance, evaluated in terms of both PESQ and Itakura-Saito distance, is further increased in conjunction with a well-known speech algorithm by Ephraim et al.

These preliminary results in the field of speech enhancement motivate for further research in neural beamforming for speech enhancement. More recent machine learning algorithms can be applied to DOA estimation. For instance, while the computational cost of a MLP network is lower compared to most RNN techniques, these may potentially yield improved results in accuracy that are worth investigating. End-to-end learning could be applied, resulting in a whole beamforming architecture based solely on machine learning, with a deep neural network trained to cover both DOA estimation and beamforming. One of the advantages of this approach is the possibility to adopt irregular microphone array geometries. Preliminary experiments by the authors are ongoing in this challenging field and the exploitation of both classic and deep neural network architectures are under investigation.

Acknowledgements We acknowledge the CINECA award under the ISCRA initiative, for the availability of high performance computing resources and support.

References

1. Allen, J., Berkley, D.: Image method for efficiently simulating small-room acoustics. J. Acoust. Soc. Am., 943 (1979)
2. Araki, S., Hayashi, T., Delcroix, M., Fujimoto, M., Takeda, K., Nakatani, T.: Using neural network front-ends on far field multiple microphones based speech recognition. In: Proceedings of ICASSP, Florence, Italy, pp. 5542–5546, 4–9 May 2014
3. Araki, S., Hayashi, T., Delcroix, M., Fujimoto, M., Takeda, K., Nakatani, T.: Exploring multi-channel features for denoising-autoencoder-based speech enhancement. In: Proceedings of ICASSP, pp. 116–120 (2015)
4. Benesty, J., Chen, J., Huang, Y.: Microphone Array Signal Processing, vol. 1. Springer Science & Business Media (2008)

5. Capon, J.: High resolution frequency-wavenumber spectrum analysis. Proc. IEEE **57**(8), 1408–1418 (1969)
6. Ephraim, Y., Malah, D.: Speech enhancement using a minimum-mean square error short-time spectral amplitude estimator. IEEE Trans. Acoust. Speech Signal Process. **32**(6), 1109–1121 (1984)
7. Erdogan, H., Hayashi, T., Hershey, J.R., Hori, T., Hori, C., Hsu, W.n., Kim, S., Roux, J.L., Meng, Z., Watanabe, S.: Multi-channel speech recognition: LSTMs all the way through. In: Proceedings of the 4th CHiME Speech Separation and Recognition Challenge, San Francisco, CA, USA (2016)
8. Gannot, S., Cohen, I.: Speech enhancement based on the general transfer function gsc and postfiltering. IEEE Trans. Speech Audio Process. **12**(6), 561–571 (2004)
9. Griffiths, L., Jim, C.: An alternative approach to linearly constrained adaptive beamforming. IEEE Trans. Antennas Propag. **30**(1), 27–34 (1982)
10. Hochreiter, S., Schmidhuber, J.: Long short-term memory. Neural comput. **9**(8), 1735–1780 (1997)
11. Hoshen, Y., Weiss, R., Wilson, K.: Speech Acoustic Modeling from Raw Multichannel Waveforms, pp. 4624–4628 (2015)
12. Hussain, A., Chetouani, M., Squartini, S., Bastari, A., Piazza, F.: Nonlinear Speech Enhancement: An Overview, pp. 217–248. Springer Berlin (2007)
13. Kingma, D., Ba, J.: Adam: a method for stochastic optimization. In: International Conference on Learning Representations (2015)
14. Kinoshita, K., Delcroix, M., Yoshioka, T., Nakatani, T., Sehr, A., Kellermann, W., Maas, R.: The REVERB challenge: a common evaluation framework for dereverberation and recognition of reverberant speech. In: 2013 IEEE Workshop on Applications of Signal Processing to Audio and Acoustics, pp. 1–4. IEEE (2013)
15. Knapp, C., Carter, G.: The generalized correlation method for estimation of time delay. IEEE Trans. Acoust. Speech Signal Process. **24**(4), 320–327 (1976)
16. Knecht, W., Schenkel, M.E., Moschytz, G.S.: Neural network filters for speech enhancement. IEEE Trans. Speech Audio Process. **3**(6), 433–438 (1995)
17. Li, B., Sainath, T., Weiss, R., Wilson, K., Bacchiani, M.: Neural network adaptive beamforming for robust multichannel speech recognition. In: Proceedings of Interspeech, pp. 1976–1980, 8–12 Sept 2016
18. Li, J., Deng, L., Haeb-Umbach, R., Gong, Y.: Robust Automatic Speech Recognition: A Bridge to Practical Applications. Academic Press (2015)
19. Loizou, P.: Speech processing in vocoder-centric cochlear implants. In: Cochlear and Brainstem Implants, vol. 64, pp. 109–143. Karger Publishers (2006)
20. Philipos C. Loizou: Speech Enhancement: Theory and Practice. CRC Press (2013)
21. Principi, E., Fuselli, D., Squartini, S., Bonifazi, M., Piazza, F.: A speech-based system for in-home emergency detection and remote assistance. In: Proceedings of the 134th International AES Convention, Rome, Italy, pp. 560–569, 4–7 May 2013
22. Principi, E., Squartini, S., Bonfigli, R., Ferroni, G., Piazza, F.: An integrated system for voice command recognition and emergency detection based on audio signals. Expert Syst. Appl. **42**(13), 5668–5683 (2015)
23. Principi, E., Squartini, S., Piazza, F.: Power normalized cepstral coefficients based supervectors and i-vectors for small vocabulary speech recognition. In: Proceedings of the International Joint Conference on Neural Networks (IJCNN), Beijing, China, pp. 3562–3568, 6–11 July 2014
24. Renals, S., Swietojanski, P.: Neural networks for distant speech recognition. In: Proceedings of HSCMA, pp. 172–176 (2014)
25. Robinson, T., Fransen, J., Pye, D., Foote, J., Renals, S.: WSJ-CAM0: a british english corpus for large vocabulary continuous speech recognition. In: Proceedings of International Conference on Acoustics, Speech, and Signal Processing (1994)
26. Schmidt, R.: Multiple emitter location and signal parameter estimation. IEEE Trans. Antennas Propag. **34**(3), 276–280 (1986)

27. Sohn, J., Kim, N.S., Sung, W.: A statistical model-based voice activity detection. IEEE Signal Process. Lett. **6**(1), 1–3 (1999)
28. Swietojanski, P., Ghoshal, A., Renals, S.: Convolutional neural networks for distant speech recognition. IEEE Signal Process. Lett. **21**(9), 1120–1124 (2014)
29. Xiao, X., Watanabe, S., Erdogan, H., Lu, L., Hershey, J., Seltzer, M., Chen, G., Zhang, Y., Mandel, M., Yu, D.: Deep beamforming networks for multi-channel speech recognition. In: Proceedings of ICASSP, pp. 5745–5749 (2016)
30. Xiao, X., Zhao, S., Zhong, X., Jones, D.L., Chng, E.S., Li, H.: A learning-based approach to direction of arrival estimation in noisy and reverberant environments. In: 2015 IEEE International Conference on Acoustics, Speech and Signal Processing (ICASSP), pp. 2814–2818. IEEE (2015)
31. Yoganathan, V., Moir, T.: Multi-microphone adaptive neural switched Griffiths-Jim beamformer for noise reduction. In: Proceedings of the 10th International Conference on Signal Processing, pp. 299–302 (2010)
32. Zhang, H., Zhang, X., Gao, G.: Multi-channel speech enhancement based on deep stacking network. In: Proceedings of the 4th CHiME Speech Separation and Recognition Challenge, San Francisco, CA, USA (2016)

Chapter 5
Error Resilient Neural Networks on Low-Dimensional Manifolds

Alexander Petukhov and Inna Kozlov

Abstract We introduce an algorithm that improves Neural Network classification/registration of corrupted data belonging to low-dimensional manifolds. The algorithm combines ideas of the Orthogonal Greedy Algorithm with the standard gradient back-propagation engine incorporated in Neural Networks. Therefore, we call it *the Greedient algorithm*.

Keywords Neural networks · Greedient algorithm · Error correction

5.1 Introduction

We consider the problem of classification of corrupted data with neural networks. For the linear data model, this problem is mathematically equivalent to finding sparse representation of the data. To our knowledge, the recently designed methods of the data recovery/error elimination are based on "genuine" neural network ideas [6, 9, 10]. To reconstruct the encoded low-dimensional data approximately, the non-linear function is constructed with unsupervised training of an appropriate Neural Network (NN) which is known as an auto-encoder. Since the problem of the sparse reconstruction is known as a problem with high complexity, increasing the system's capability can be reached by making the auto-encoder more complex Another feature of the methods based on the auto-encoder consists in the necessity to create and train this (probably complex) auto-encoder [9, 10] even when the corrupted data has to be just classified and no correction of the data is required.

Our approach is based on an "old-fashioned" compressed sensing with the Orthogonal Greedy Algorithm (OGA) incorporated into a NN trained to register the fact

A. Petukhov (✉)
University of Georgia, Athens, GA 30602, USA
e-mail: petukhov@uga.edu

I. Kozlov
Algosoft-Tech USA, Bishop, GA, USA
e-mail: inna@algosoft-tech.com

© Springer International Publishing AG, part of Springer Nature 2019
A. Esposito et al. (eds.), *Neural Advances in Processing Nonlinear Dynamic Signals*, Smart Innovation, Systems and Technologies 102,
https://doi.org/10.1007/978-3-319-95098-3_5

that a vector belongs or does not belong to some low-dimensional manifold. It does not require any auto-encoder over an existing trained NN.

5.1.1 Compressive Sampling

The interest of researchers in representation/extraction of low dimensional data embedded into the spaces with high dimensions has been widely spread, especially for the last 10–15 years, which can be explained by their applicability to the real world problems and the discovery of the way to solve the non-convex problems of the sparse representations with the ℓ^1 convex relaxation [2, 3, 8]. The standard setting of the numerical part of the problem can be described in the form of the recovery of the data linearly representable in some redundant system with a sparse representation. A vector of data $\mathbf{x} \in \mathbb{R}^n$ has to be represented in the form $A\mathbf{y} = \mathbf{x}$, where A is a matrix of size $n \times N$, $\mathbf{y} \in \mathbb{R}^N$ with m entries non-equal to 0, $m < n < N$. If a sparse solution of the system above exists, then, for generic matrix, it is unique with probability 1. However, generally speaking, finding this solution is a problem with non-polynomial complexity. The convex relaxation allows to solve the convex problem $\|\mathbf{y}\|_1 \to \min$ instead of $\|\mathbf{y}\|_0 := \#\{y_i \mid y_i \neq 0\} \to \min$, which requires direct minimization of the sparsity.

The information theory interpretation of the problem above can be given in terms of the error correcting codes. If we assume that data from a linear space of dimension k are linearly encoded by applying the matrix $B \in \mathbb{R}^{N \times k}$ to the data vector $\mathbf{z} \in \mathbb{R}^k$:

$$\mathbf{w} = B\mathbf{z}, \quad N > k, \tag{5.1}$$

then (with the probability 1) the data will be protected from corruption of up to $N - k - 1$ entries of the "code vector" $\mathbf{w}$. When no corruption is applied to the vector $\mathbf{w}$, the data vector can be restored by the formula $\mathbf{z} = B^+\mathbf{w}$, where B^+ is the pseudo-inverse matrix.

If the vector $\mathbf{w}$ is corrupted by an unknown sparse error vector $\mathbf{y}$, the standard way for precise recovering $\mathbf{z}$ is to multiply the corrupted vector $\mathbf{w} + \mathbf{y}$ by the matrix $A := (B^\perp)^T$, where the columns of the matrix $B^\perp$ constitute an (orthonormal) basis in the space orthogonal to the column space of the matrix B. Then the problem of error correction is reduced to finding the sparse solution $\mathbf{y}$ of the system

$$A\mathbf{y} = \mathbf{x} := (B^\perp)^T (\mathbf{w} + \mathbf{y}), \tag{5.2}$$

i.e., the problem of error correction is equivalent to the recovery of sparse representations in a (tight) frame.

Since the time when the role of ℓ^1-minimization in compressed sensing was well understood, other methods solving the problems were investigated. Among those methods, greedy algorithms allowing to pick up the terms of sparse representation one-by-one or by small groups of terms relying on some specific were studied.

Greedy algorithms proved to be very efficient and, in most cases, outperformed ℓ^1-optimization.

The problem considered above can be interpreted in the following way. If the data set constitutes a low-dimensional linear manifold (a plane) S of dimension k embedded into an ambient space $\mathbb{R}^N$, $N > k$, then with overwhelming probability the data vectors are protected from corruption of up to $N - k - 1$ entries.

5.1.2 Orthogonal Greedy Algorithm

The OGA has many nice features. It is most transparent among the group of algorithms applied in compressive sampling settings. When implemented in an appropriate way, it is extremely efficient from the point of view of computational complexity.

We will describe its fast version given in [7]. Without loss of generality, we assume that the rows of the matrix A in (5.2) constitute an orthonormal basis of its span. Then its columns constitute a tight frame.

We denote by Λ the set of indexes of non-zero entries of the vector $\mathbf{y}$. If we know Λ, the problem of finding a vector $\mathbf{y}$ "loses" the non-polynomial complexity and becomes a "simple" problem of finding the decomposition of $\mathbf{x}$ as a linear combination of those columns of A. However, if N is very large, the problem of solving a large system of linear equations may become hard. At the same time, if operations of multiplication by A and by A^T are very computationally efficient, the iterative decomposition algorithm can also be implemented efficiently. Any compactly supported tight wavelet frame is an example of such transform.

In what follows, the matrix P_Λ is a diagonal matrix whose diagonal elements are 1s for indexes from Λ and 0s otherwise; S_Λ is a span of the columns of A with indexes from Λ. The algorithm of the orthogonal projection on a span of the vectors with index set Λ is as follows.

The Algorithm A:

1. *Initialization.* Set $i = 0$, $\mathbf{x}_0 := \mathbf{x}$, $\mathbf{y}_0 = \mathbf{0}$.
2. *Iteration* $\mathbf{x}_{i+1} := \mathbf{x}_i - A P_\Lambda A^T \mathbf{x}_i$; $\mathbf{y}_{i+1} := \mathbf{y}_i + P_\Lambda A^T \mathbf{x}_i$; $i := i + 1$.

For Algorithm A, $A\mathbf{y}_i$ converges (see [7]) to the orthogonal projection of $\mathbf{x}$ onto S_Λ, $\mathbf{x}^* = \mathrm{Pr}_{S_\Lambda} \mathbf{x}$. In particular, if (5.2) takes place, $\mathbf{y}_i$ converges to its solution and $\mathbf{x}_i \to \mathbf{0}$. Generally, the solution is not unique when $\dim S_\Lambda < \#\Lambda$.

The case, when the set Λ is known, corresponds to the case of data erasures. To fight errors in data, we need to identify their location. Then Algorithm A can be applied for this estimated set Λ. Let $\hat{A}$ be the matrix A after normalization of its columns. The following algorithm is a version of OGA:

Algorithm B.

1. **Initialization.** Set $\Lambda_0 := \emptyset$; $\mathbf{x}_0 := \mathbf{x}$; $\mathbf{y}_0 := \mathbf{0}$; $\delta, 0 < \delta \leq 1$; $i := 0$.
2. Find

$$M_i := \max_{1 \leq j \leq N} \left\{ \left\| (\hat{A}^T \mathbf{x}_i)_j \right\| \right\},$$

3. **Greedy Selection Step**.

$$\Lambda_{i+1} := \Lambda_i \cup \{j \mid |(\hat{A}^T \mathbf{x}_k)_j| \geq \delta M_i\}.$$

4. **Orthogonal Projection Step** Apply Algorithm A for $\mathbf{x} := \mathbf{x}_i$, $\Lambda := \Lambda_k$. Using its output $\mathbf{y}^*$, $\mathbf{x}^*$, update $\mathbf{y}_{i+1} := \mathbf{y}_i + \mathbf{y}^*$, $\mathbf{x}_{i+1} := \mathbf{x}_i - \mathbf{x}^*$.
5. If $\|\mathbf{x}_{i+1}\| > \epsilon$, $i := i + 1$, go to Step 2.

While the Greedy Selection Step has a common sense justification as a selection of directions most probable to be directions of the error vector, it may have a different interpretation. Formula (5.1) gives a parametric description of a k-plane S, whereas $A\mathbf{y}$ (when A is a tight frame) defines the vector of orthogonal projection of $\mathbf{y}$ to the orthogonal complement of S. Then the square of the magnitude of that projection is $r(\mathbf{x}) = \|\mathbf{x}\|^2 = \mathbf{y}^T A^T A\mathbf{y}$ (the square of the distance from the k-plane) serves as a measure of deviation of $\mathbf{y}$ from the k-plane. The gradient of r with respect to $\mathbf{y}$ can be found as

$$\nabla_y r(\mathbf{x}) = 2A^T A\mathbf{y} = 2A^T \mathbf{x}.$$

Thus, the update formulas in Algorithm A are just updates along the partial gradient of the functional r, where the entries to be updated were selected as the most probable (from the point of view of OGA) corrupted entries. If the normalization is not used, the greedy selection is implemented by comparing the magnitude of the gradient entries. Unfortunately, the frames with equal lengths of all vectors are rare. Therefore, the Greedy Selection Step in Algorithm B is different from the selection according to the maximum of gradient entries.

Thus, in reconstruction iteration formulas we have

$$\mathbf{z}_{i+1} := \mathbf{z}_i - P_\Lambda A^T A\mathbf{y}_i = \mathbf{z}_i - \frac{1}{2}P_\Lambda \nabla_y r(\mathbf{x}_i),$$

$$\mathbf{x}_{i+1} := \mathbf{x}_i - A P_\Lambda A^T \mathbf{x}_i,$$

where $\mathbf{z}_0 := \mathbf{z} + \mathbf{y}$, $\mathbf{x}_0 := \mathbf{x}$. This algorithm converges to the "clean data" $\mathbf{z}$ when the projector P_Λ is defined appropriately, i.e., when we know coordinates of corrupted entries.

5.1.3 Non-linear Models and Neural Networks

In many real world problems the model of the data sparsity is much more complicated than a linear manifold. Typically, the data redundancy is provided either by nature or by attempting to measure and to describe the data in detail in overcomplete systems. Two typical examples of such representations are:

(A) The results of medical tests of a patient may be redundant because they are collected in different time without a goal to identify a specific type of illness.

The redundant information may have errors and missing entries which still can be recovered or at least will not lead to wrong conclusions.

(B) A celluloid film or any video footage contains information represented at 18–30 frames (images) per second. Due to the redundancy, the survived extremely corrupted old films can be almost perfectly restored.

In the applied problems mentioned above, the linearity of the data model is questionable since the sum of two data vectors can be beyond of the data model. At the same time, some small change of the information vectors probably keeps the result within the data model. Therefore, it seems that the smooth manifold model is a good extension of the classical linear error correction settings.

The goal of this paper is to conduct the feasibility study of the greedy algorithms for registration of corrupted data belonging to some smooth manifold by means of a NN trained on clean data.

Suppose that a low-dimensional manifold does not have any analytical description available at the processing time and it is learned by a NN with the standard gradient back-propagation algorithm.

We assume that the data vectors belonging to the class of our interest constitute a k-dimensional manifold S in an N-dimensional linear space. The precise analytical description is unknown, while significant amount of data from S as well as from its complement $\bar{S} := \mathbb{R}^N \backslash S$ is available. We will not discuss here the size of this data set from S available. We just assume that it is significant for training a NN for making a reliable decision: either a tested vector belongs to S or to $\bar{S}$. This decision is made by means of the functional $\Phi(\mathbf{x})$ which has the range $[0, 1]$. It is obtained by minimization of the loss function of a NN ending with the SoftMax layer. If $\Phi(\mathbf{x}) < 1/2$, we make decision $\mathbf{x} \in S$. Otherwise, $\mathbf{x} \in \bar{S}$.

The notion of *an erasure* is understood as missing information when the entry value is unknown, whereas the coordinate of the missed information is available for the tester. The notion of *an error* means that the data entry is corrupted and its coordinate is unavailable. We will consider both types of corruption.

In this paper, we assume that the training set is free of errors. We want to test an arbitrary vector for the property of belonging to a given smooth manifold if some of vector entries may be corrupted or unavailable.

In particular, this means that if our algorithm introduces the "correction" of a vector $\mathbf{x} + \mathbf{e} \to \tilde{\mathbf{x}}$ and $\Phi(\tilde{\mathbf{x}}) < 1/2$, we consider such experiment as "success" when $\mathbf{x} \in S$, even if $\tilde{\mathbf{x}}$ is far from S.

5.2 Neural Network Architecture Description

In our experiments, we consider a few "toy" data models, which suffer from many drawbacks, and a very simple network architecture. At the same time, the design of a functional for the registration of vectors of a manifold learned by a NN is a very hot research area (cf. [11]) in the framework of unsupervised NN training.

We consider a simple NN consisting of 4 layers accepting data from the space $\mathbb{R}^N$, $N = 10$. This is the network architecture:

1. Fully connected layer with 100 neurons.
2. Activation layer x_+^2.
3. Fully connected layer with 2 neurons.
4. SoftMax layer.

The role of network training is to adapt the parameters of fully connected layers to provide the minimum of the loss function on the set of input (labeled) data belonging to either S or $\bar{S}$.

When the parameters are optimized and fixed, the trained network provides the (non-linear) functional $\Phi(\mathbf{y})$ as output of the channel corresponding to $\bar{S}$ of the last (SoftMax) network layer. Then Φ is defined on the entire $\mathbb{R}^N$ and has the range $[0, 1]$. In what follows,

$$U_\lambda := \{\mathbf{y} \mid \Phi(\mathbf{y}) < \lambda\}, \quad S_\lambda := \{\mathbf{y} \mid \Phi(\mathbf{y}) = \lambda\}, \ 0 \leq \lambda \leq 1.$$

The decision $\mathbf{y} \in S$ is made when $\mathbf{y} \in U_{0.5}$. Otherwise, $\mathbf{y} \in \bar{S}$.

5.3 Training and Testing Data

Obviously, linear or non-linear manifolds are potentially protected from corruption of arbitrary k coordinates if and only if they do have at most one intersection point with any k-plane parallel to some coordinate k-plane. At the same time, this property can be omitted if we just want to register the fact that a (corrupted) point belongs to the manifold.

Some difficulties are expected. Assume that our non-linear k-manifold can be embedded in some linear m-plane, $k \leq m < N$. Then, having a comparable number of training points from S and $\bar{S}$, the obvious simplification available for a trained network is to set 2 parallel very close hyperplanes (or their minor deformation) as the level surface $S_{0.5}$ around S. Since the size of $U_{0.5}$ between the hyperplanes can be insignificant, a number of the test points from $\bar{S}$ hitting this set (as well as their contribution into average loss function) is insignificant. Thus, despite the seemingly huge source of the false identification, i.e., identification of points from $\bar{S}$ as points from S, the rate of erroneous decisions is not suffer from this drawback, provided that the models of training and testing sets are coordinated.

5.3.1 Models for S

For the manifold S, we use one of 3 models:

Model 1. The unit circumference in a random 2-plane of $\mathbb{R}^N$.
Model 2. The unit sphere in a random 3-plane of $\mathbb{R}^N$.

Model 3. A spiral curve whose parametric representation coordinates are generated as $\{\sin \alpha_i t\}_{i=1}^{N}$, where $\alpha_i = 1 + 3\eta_i$, η_i are independent random variables uniformly distributed on $[0, 1]$.

The first two models refer to 2- and 3-planes. As we mentioned above, this, in general, allows the NN not to follow a manifold shape but just to place it within a thin 2D or 3D stripe. This stripe may have tiny N-measure. Hence, the points inside (or even outside) the circle (or the sphere) will be classified as a point of $\mathcal{S}$, i.e., the classifier would actually work on linear subspaces of $\mathbb{R}^N$ containing $\mathcal{S}$ rather than on the non-linear 1D or 2D nonlinear manifolds. Model 3 is a 1D manifold that does not have this potential drawback, provided that we select incomparable frequencies $\{\alpha_i\}$.

5.3.2 Models for $\bar{\mathcal{S}}$

To get meaningful results, the set $\bar{\mathcal{S}}$ has to be selected according to the principles which are different for training and testing sets.

For a training set, on the one hand, it should be close enough to $\mathcal{S}$ to obtain the final functional having a sharp minimum at the points of $\mathcal{S}$. On the other hand, it should not be too close to allow a NN with a fixed number of neurons to separate $\mathcal{S}$ and $\bar{\mathcal{S}}$.

For the test, the set $\bar{\mathcal{S}}$ does not have to be as close to $\mathcal{S}$ as for the training set. A point close to $\mathcal{S}$ finds itself in the "zone of influence" of $\mathcal{S}$. This phenomenon can be described as follows. While $\mathcal{S}$ is a low-dimensional manifold, the set $U_{0.5}$ is a set of full dimension N. Then, if a point of $\bar{\mathcal{S}}$ is close to $U_{0.5}$, any small change of its one or a few coordinates with a high probability results in getting that point into $U_{0.5}$. This may look like a wrong error correction and may reduce the success rate of the algorithm, while this point is "almost" a point of $\mathcal{S}$ from the point of view of the Euclidean distance (or at least because of proximity of $\Phi(\mathbf{y})$ to 0.5). In fact, its deviation from $U_{0.5}$ is within the radius of data noise. Hence, it should not be judged as an algorithm failure.

We generate $\bar{\mathcal{S}}$ distributed over $\mathbb{R}^N$ as a random variable

$$\xi = \nu + \sigma\eta, \tag{5.3}$$

where ν is a point uniformly distributed over $\mathcal{S}$ and η has a standard normal distribution. We use $\sigma = 1$ for the training set and $\sigma = 2$ for the test. In all our experiments, the train set has 10^4 point of $\mathbb{R}^N$, while the test set has 1000 points. In both cases, we generate an equal number of points for $\mathcal{S}$ and $\bar{\mathcal{S}}$.

5.3.3 Model for Errors

The set of errors in our experiments is an additive corruption introduced into randomly selected data entries with probability p_{err}. The value of corruption is generated by a Gaussian distribution with the variance $\sigma_{err} = 0.5$. We consider data corruption with random erasures with probability p_{ers}. In the classification algorithm, the missed data are initialized with 0.

5.3.4 Greedient Algorithm Implementation

We train the NN described in Sect. 5.2 with the standard Stochastic Gradient Back-Propagation algorithm. We apply 10^5 iterations to ensure the maximum accurate separation in the training set consisting of clean (uncorrupted) data.

Due to back-propagation procedure, the trained NN generates a functional $\Phi(\mathbf{y})$ which has to have minimal values at the points of S.

For a test data vector $\Phi(\mathbf{y}) < 0.5$, we make the decision $\mathbf{y} \in S$. Otherwise, we run the greedy gradient algorithm, checking the opportunity to change a few coordinates of the vector $\mathbf{y}$ obtaining the vector $\mathbf{y}^*$ such that $\Phi(\mathbf{y}^*) < 0.5$. If such change is found, we make a decision that $\mathbf{y}$ is a corrupted version of some vector from S. Otherwise, we consider $\mathbf{y}$ as (maybe corrupted version of) a vector from $\bar{S}$.

Using the back-propagation algorithm with a trained network we can compute the values $\frac{\partial \Phi}{\partial y_j}$ which allow to introduce corrections of the values of $\mathbf{y}$ toward smaller values of Φ. We will apply the greedy approach to select coordinates to be used for gradient descent iterations.

Algorithm C. (Greedient Algorithm)

1. **Initialization.** Set $\Lambda_0 := \emptyset$; $\mathbf{y}^0 := \mathbf{y}$; $i := 0$; $n = 0$; $e := 0$; γ, $Imax$, $Kmax$, $Emax$.
2. If $e \geq Emax$, then $\Lambda_{i+1} := \Lambda_i$, go to Step 5
3. **Greedy selection step.** Find

$$m := \arg\max_{1 \leq j \leq N, \, j \notin \Lambda_i} \left\{ \left| \left| \frac{\partial \Phi}{\partial y_j}(\mathbf{y}^i) \right| \right| \right\},$$

4. $\Lambda_{i+1} := \Lambda_i \cup \{m\}$.
5. $k := 0$.
6. **Orthogonal Projection Step.** $k := k + 1$;
   Update $y_j^{i+1} := y_j^i - \gamma \frac{\partial \Phi}{\partial y_j}(\mathbf{y}^i)$, $j \in \Lambda_{i+1}$; $i := i + 1$.
7. If $k < Kmax$, then go to 6;
8. If $i < Imax$, then go to Step 2.

We apply the greedient algorithm with the parameters $\gamma = 1$, $Emax = 4$, $Kmax := 4$, $Imax = 20$ to all test vectors corrupted by errors. We also use the same iterations for the test vectors corrupted by erasures. However, in that case the set of erasures Λ is known in advance and does not need not be updated. The Greedy selection steps 3 and 4 should be skipped as well as the partial iteration loops controlled by Step 7. For erasures, we use $Imax := 100$ iterations of the gradient descent algorithm along the coordinates corresponding to erasures.

Generally, like in [7], a group of the largest entries can be updated in Step 4.

5.4 Numerical Results

The results of numerical experiments are given in Tables 5.1, 5.2 and 5.3. The last two columns of each row of the table contain the misclassification rate for the raw data (with no correction) followed by the rate for the data processed by the Greedient Algorithm.

The first 3 rows of each table give results for data corrupted with erasures with probabilities of missed entries $p_{ers} = 0.1, 0.3, 0.5$.

The remaining 3 rows present the results for data corrupted by sparse errors with the same probabilities.

Table 5.1 Misclassification Rate (%). Model 1

p_{err}	p_{ers}	No correction	Corrected
0	0.1	17.4	<0.1
0	0.3	32.4	0.3
0	0.5	40.0	4.5
0.1	0	19.7	2.5
0.3	0	43.0	4.2
0.5	0	47.5	14.7

Table 5.2 Misclassification Rate (%). Model 2

p_{err}	p_{ers}	No correction	Corrected
0	0.1	12.4	0.1
0	0.3	31.0	1.2
0	0.5	33.0	7.0
0.1	0	17.5	3.0
0.3	0	41.3	4.0
0.5	0	48.3	12.2

Table 5.3 Misclassification Rate (%). Model 3

p_{err}	p_{ers}	No correction	Corrected
0	0.1	27.5	0.1
0	0.3	46.7	0.3
0	0.5	49.0	2.0
0.1	0	17.6	1.9
0.3	0	38.9	4.3
0.5	0	48.1	12.6

5.4.1 Result Discussion

The numerical results in Tables 5.1, 5.2 and 5.3 show that the suggested algorithm has a very good potential capability for better data classification.

For corruption in the form of erasures, gradient descent algorithm gives almost perfect results when up to 30% of entries are missed. The improvement due to correction is 2 orders of the magnitude less than for the raw data.

While the case $p_{ers} = 0.5$ does not look very impressive, the larger amount ($2 \div 7\%$) of misclassified data can be explained by reasons not related to the algorithm capacity. In order to have at least theoretical opportunity to identify whether a vector belongs to a K-dimensional manifold when P entries are missing or corrupted, the inequality $K + P < N$ has to be satisfied. It is easy to compute that, when $p_{ers} = 0.5$, a big portion (more than half) of misclassification rate is due to the high probability of the data not satisfying that condition. Therefore, the higher misclassification rates can be completely explained by natural factors, those cannot be overcome by any algorithm, and by inaccurate training rather than by a drawback of the error correction procedure.

The experimental results show that the misclassification rate for data with errors is much higher than for data with erasures. Let us identify what factors have influence on this rate. It turns out that if we run the test on clean data (no such line in Tables 5.1, 5.2 and 5.3), for all three data models, the result is almost identical to the rows with $p_{err} = 0.1$. At the same time, if we do not run the error correcting block, the misclassification rate is 0.

This result tells us that our decision, based on the descent gradient along the always fixed number $Emax = 4$ of the "suspicious" coordinates, is too simplified. It leads to reduction of the system capability. This is especially visible on the clean data.

Nevertheless, turns out that the cost of those wrong decisions in classification ($\mathbf{y} \in \mathcal{S}$ instead of $\mathbf{y} \in \bar{\mathcal{S}}$) is not too high. The misclassification occurs only for points which are close to $U_{0.5}$ with respect to the Euclidean distance. While more accurate implementation of the greedy selector deserves an additional study, the current algorithm version does not have any serious drawbacks even in the presented form.

5.5 Conclusion

In this paper, we introduced a greedy algorithm to improve the capability of a NN in processing corrupted data from low-dimensional manifolds. The algorithm does not require any additional computational environment and is based on the standard back-propagation engine incorporated in a NN. The Greedient Algorithm shows the high performance on the corrupted synthetic data.

References

1. Becker, S., Le Cun, Y.: Improving the convergence of backpropagation learning with second order methods. In: Proceedings of the 1988 Connectionist Models Summer School, pp. 29–37. Morgan Kaufmann, San Matteo (1988)
2. Candès, E.J., Tao, T.: Decoding by linear programming. IEEE Trans. Inf. Theory **51**, 4203–4215 (2005)
3. Donoho, D.: Compressed sensing. IEEE Trans. Inf. Theory **52**, 1289–1306 (2006)
4. Martens, J., Sutskever, I., Swersky, K.: Estimating the Hessian by back-propagating curvature (2012). arXiv:1206.6464
5. Martens, J.: Second-order optimization for neural networks. Ph.D. thesis, University of Toronto (2016)
6. Mousavi, A., Baraniuk, R. G.: Learning to invert: signal recovery via deep convolutional networks (2017). arXiv:1701.03891v1 [stat.ML]
7. Petukhov, A.: Fast implementation of orthogonal greedy algorithm for tight wavelet frames. Eur. J. Signal Process. **86**, 471–479 (2006)
8. Rudelson, M., Vershynin, R.: Geometric approach to error correcting codes and reconstruction of signals. Int. Math. Res. Not. **64**, 4019–4041 (2005)
9. Simpson, A.: Deep Transform: Error Correction via Probabilistic Re-Synthesis (2015). arXiv:1502.04617 [cs.LG]
10. Simpson, A.: Deep Transform: Time-Domain Audio Error Correction via Probabilistic Re-Synthesis (2015). arXiv:1503.05849 [cs.SD]
11. Zhao, J., Mathieu, M., LeCun, Y.: Energy-Based Generative Adversarial Network (2017). arXiv:1609.03126 [cs.LG]

Chapter 6
FIS Synthesis by Clustering for Microgrid Energy Management Systems

Stefano Leonori, Maurizio Paschero, Antonello Rizzi and Fabio Massimo Frattale Mascioli

Abstract Microgrids (MGs) are the most affordable solution for the development of smart grid infrastructures. They are conceived to intelligently integrate the generation from Distributed Energy Resources (DERs), to improve Demand Response (DR) services, to reduce pollutant emissions and curtail power losses, assuring the continuity of services to the loads as well. In this work it is proposed a novel Fuzzy Inference System (FIS) synthesis procedure as the core inference engine of an Energy Management System (EMS) for a grid-connected MG equipped with a photovoltaic power plant, an aggregated load and an Energy Storage System (ESS). The EMS is designed to operate in real time by defining the ESS energy flow in order to maximize the revenues generated by the energy trade with the distribution grid considering a Time Of Use (TOU) energy prices policy. The FIS adopted is a first order Tagaki-Sugeno type, designed through a data driven approach. In particular, multidimensional Membership Functions (MFs) are modelled by a K-Means clustering algorithm. Successively, each cluster is used to define both the antecedent and the consequent parts of a tailored fuzzy rule, by estimating a multivariate Gaussian MF and the related interpolating hyperplane. Results have been compared with benchmark references obtained by a Linear Programming (LP) optimization. The best solution found is characterized by a small number of MFs, namely a limited number of fuzzy rules. Its performances are close to the optimum solution in terms of profit generated and, moreover, it shows a smooth exploitation of the ESS.

S. Leonori (✉) · M. Paschero (✉) · A. Rizzi · F. M. F. Mascioli
DIET Department, University of Rome "La Sapienza", Via Eudossiana 18, 00184 Rome, Italy
e-mail: stefano.leonori@uniroma1.it

M. Paschero
e-mail: maurizio.paschero@uniroma1.it

A. Rizzi
e-mail: antonello.rizzi@uniroma1.it

F. M. F. Mascioli
e-mail: fabiomassimo.frattalemascioli@uniroma1.it

© Springer International Publishing AG, part of Springer Nature 2019
A. Esposito et al. (eds.), *Neural Advances in Processing Nonlinear Dynamic Signals*, Smart Innovation, Systems and Technologies 102,
https://doi.org/10.1007/978-3-319-95098-3_6

61

6.1   Introduction

A Microgrid (MG) is an electric grid able to intelligently manage local electric power systems affected by stochastic and intermittent behaviours such as electric generation from renewable energy sources, electric vehicles charging and, controllable, deferrable and shiftable loads. The MG infrastructure relies on power converters able to connect the MG components in order to locally manage the MG power flows and power exchange with the connected grid. It must be equipped with a communication infrastructure able to monitor and supervise the state of all the MG components in real time.

Usually the MG is supported by an Energy Storage System (ESS) able to guarantee the quality of service, the electric stability and to give a certain autonomy to the system when it is disconnected to the grid (i.e. islanded mode).

In [1] authors discuss the advantages in terms of power control and power efficiency of a MG with the main bus in DC rather than AC, and how it will affect the modernization of the distribution grid towards a smart grid [2].

The implementation of a suitable Demand Side Management (DSM) into the Energy Management System (EMS) allows to apply Demand Response (DR) services to the costumer, which is referred to as prosumer, in case it is also equipped with a power generation system.

In [3] are summarized the DR main services (i.e. valley filling, load shifting, peak shaving operations). These services, together with Vehicle-2-Grid (V2G) operations and the intelligent use of the ESS, allow to reduce the stress caused by the MG to the connected distribution grid in order to get incentives, avoid penalties, reduce both the consumptions and the operational costs, which strictly depend on the energy price policies adopted by the distribution grid. Concerning this topic, in [4, 5] it is discussed the development of new energy policies which will involve the costumer to assume an active role in the energy market by means of the application of DR services. In this work it is proposed an EMS able to define in real time the energy flow exchanged with the grid in order to maximize the MG profit by considering a Time Of Use (TOU) energy policy. The EMS synthesis is based on K-Means clustering algorithm used to identify a set of Fuzzy Rules in a Fuzzy Inference System (FIS) optimized off-line through a data driven approach.

The paper is organized in sections. The MG problem formulation is given in Sect. 6.2. The EMS design and the considered benchmark solutions are illustrated in Sect. 6.3. In Sect. 6.4 are reported the simulation settings, whereas the achieved results are illustrated in Sect. 6.5. Finally, conclusion and future works are discussed in Sect. 6.6.

6.2 MG Problem Formulation

In this paper it is considered a prosumer grid-connected MG equipped with a DSM EMS. It is in charge of efficiently manage the MG components represented as aggregated systems grouped in renewable sources power generation systems, aggregated loads and ESSs. Their energy flows are managed in real time by an EMS that acts as a decision making system. It must efficiently redistribute the prosumer energy balance between the grid and the ESS by maximizing the profit of the energy trade with the main grid. This work is based on several hypotheses that define the correct level of abstraction to properly place the problem under analysis. The power value of the MG components has been considered constant within each 15 min time slot. Low level operations such as voltage and reactive power control are not considered. The power transmission losses within the MG are considered negligible. The on-line control module ensure that the power balance is achieved during the real-time operation. The EMS has a sample time equal to the time slot duration which is considerably greater than the characteristic time of the ESS power control, therefore the ESS inner loop has been neglected. The power converters which connects the MG sub-components to each other, included the one allowing the MG-grid connection, are neglected in terms of power losses and characteristic time of control.

The MG aggregated energy generation, aggregated load request, energy exchanged with the ESS and energy exchanged with the grid during the n-th time slot are denoted with E_n^L, E_n^G, E_n^S and E_n^N, respectively. In Fig. 6.1 it is represented a schematic diagram of the MG, where the power lines are drawn in black and the signal wires in red. The Battery Management System (BMS) monitors the ESS and estimates its State Of Charge (*SoC*) which is used as an input of the EMS [6, 7].

By assuming that the prosumer energy production E_n^G has the priority to meet the prosumer energy demand E_n^L, the prosumer energy balance E_n^{GL} can be defined as

$$E_n^{GL} = E_n^G + E_n^L, \qquad n = 1, 2, \ldots \tag{6.1}$$

Moreover, in this work it is assumed that the prosumer energy balance E_n^{GL} is a known quantity read in real time by an electric meter. In each time slot, E_n^{GL} must

Fig. 6.1 MG architecture. Signal wires in red, power lines in black

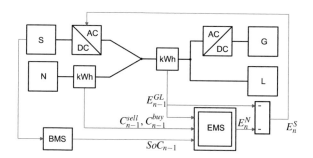

be exchanged with both the main grid and the ESS by fulfilling the following energy balance relation

$$E_n^S + E_n^N + E_n^{GL} = 0, \qquad n = 1, 2, \ldots \tag{6.2}$$

The energy E_n^S is assumed positive or negative when the ESS is discharged or recharged, respectively. Similarly, the energy E_n^N is considered positive (negative) when the network is selling (buying) energy to the MG. Considering a TOU price policy, as made in previous works [8–10], it is possible to formulate the profit P generated by the energy trade with the main grid in a time period composed by N_{slot} time slots as

$$P = \sum_{n=1}^{N_{slot}} P_n \quad \text{where} \quad P_n = \begin{cases} E_n^N \cdot C_n^{buy} & \text{if } E_n^N > 0 \\ E_n^N \cdot C_n^{sell} & \text{if } E_n^N \leq 0 \end{cases} \tag{6.3}$$

where C_n^{buy} and C_n^{sell} define the energy prices in purchase and sale during the n-th time slot. According with [11], it is assumed that during the n-th time slot the MG cannot exchange with the grid an amount of energy greater than the current energy balance E_n^{GL}. In other words, in case of over-production (over-demand) (i.e. $E^{GL} > 0$ ($E^{GL} < 0$)) the ESS can be only charged (discharged).

6.3 FIS EMS Design

In the literature EMSs are designed following several different methods. In [12] the MG EMS is formulated as a rolling horizon Markov decision process used for the optimization of the MG power flow. In [13], it is considered an EMS for a V2G service in a charge station. By assuming to know the future energy demand profile, the authors formulated the problem in a closed form which allows to schedule the energy flows through a Linear Programming (LP) formulation aimed to the maximization of the profit generated by the energy exchange with the grid. LP and stochastic approach problem formulations are often difficult to be employed. In fact they need an accurate MG state prediction model (i.e. MG generation and load prediction), a huge amount of data samples, several simplifications that could compromise the accuracy of the results, and high computational costs. Other works are rather based on the use of FIS for EMS decision making process or other soft computing techniques such as Neural Networks and clustering algorithms. FISs can be efficiently optimized through heuristics which consider Single Objective or Multi Objective problem formulation. Moreover, soft computing techniques can provide efficient and robust solutions even when only a small dataset is available. In [14] a FIS is used to minimize the power peaks and the fluctuations of the energy exchange with the connected grid while keeping the battery SoC evolution within certain security limits. It is known that, when the FIS has a large number of inputs, it can be affected by a huge Rule Base system, especially if it is based on a grid partition approach. This fact, beside a huge computational cost, could compromise an efficient design of the FIS.

In this paper it has been studied the behaviour of a EMS FIS able to efficiently estimate in real time the energy E_n^N and E_n^S to be exchanged during the n-th time slot with the connected grid and the ESS, respectively (see Fig. 6.1). The EMS is supposed to be fed by the input vector $\mathbf{u}$ constituted by 4 variables, namely, the current energy balance E_{n-1}^{GL}, the current energy prices in sale and purchase, C_{n-1}^{sell} and C_{n-1}^{buy} and the SoC value SoC_{n-1}. It should be noted that whereas E_{n-1}^{GL}, C_{n-1}^{sell} and C_{n-1}^{buy} are instantaneous quantities read by the meters, Soc_{n-1} is a status variable depending on the previous history of the ESS.

Before entering the EMS, the E^{GL}, C^{sell}, C^{buy} inputs, must be normalized in the range [0, 1], whereas the SoC belongs to [0, 1] by its own definition. In the following, the normalized input vector will be referred to as $\bar{\mathbf{u}}$.

In order to measure and compare the performance of different FIS EMSs, two benchmark solutions have been considered representing the upper and lower bound of the profit (6.3), respectively.

For the lower benchmark solution it has been chosen to consider the MG without ESS, which is obtained by considering $E_n^S = 0$ and hence $E_n^N = -E_n^{GL}$ for each n (see (6.2)). The profit achieved in this situation will be referred to as P_{lower}.

The upper benchmark solution referred to as P_{upper} corresponds to the optimal solution achieved through a LP formulation relative to the maximization of the profit defined in (6.3). The value of P_{upper} is evaluated by assuming the a priori knowledge of the overall data set used in the MG simulation, namely the E^{GL} and the C^{buy} C^{sell} energy price profiles. The estimation of P_{upper} and P_{lower} allows to define a normalized performance index $\bar{P}$ which will be used to fairly compare different EMS synthesis.

$$\bar{P} = \frac{P_{upper} - P}{P_{upper} - P_{lower}} \tag{6.4}$$

Moreover, it should be noted that the LP simulation allows to evaluate the SoC_{opt} and the E_{opt}^N time sequences associated with the optimal solution. These sequences, together with the measurable quantities E^{GL}, C^{buy} and C^{sell}, will be used at training stage in order to synthesize the FIS by means of K-Means algorithm [15, 16].

The FIS MFs are modeled by means of multivariate Gaussian functions which assure the coverage of the entire fuzzy domain regardless of the number of employed MFs. The generic MF $\Phi(\bar{\mathbf{u}})$ is defined as follows:

$$\Phi(\bar{\mathbf{u}}) = e^{-\frac{1}{2}(\bar{\mathbf{u}} - \boldsymbol{\mu}) \cdot \mathbb{C}^{-1} \cdot (\bar{\mathbf{u}}^T - \boldsymbol{\mu}^T)} \tag{6.5}$$

where $\boldsymbol{\mu}$ and $\mathbb{C}$ are the mean value vector and the covariance matrix of the multivariate Gaussian function which assumes here the role of rule antecedent set. The consequent fuzzy rule is modeled adopting a first order Takagi-Sugeno method. The Rule Consequent (RC) E^N, which represent the energy exchanged with the grid, is represented by means of a suitable hyperplane defined as follows:

$$E^N = a_0 + a_1 E^{GL} + a_2 C^{sell} + a_3 C^{buy} + a_4 SoC \qquad (6.6)$$

where the coefficient set $A = \{a_0, \ldots, a_4\}$ completely define the hyperplane. The overall output of the FIS is computed by a Winner Takes All (WTA) strategy. The synthesis of the FIS EMS is made through a data driven approach that rely on a given dataset composed by an E^{GL}, C^{buy} and C^{sell} profiles. The partitioning of the dataset in Test Set (TsS), Validation Set (VlS) and Training Set (TrS) is made on a daily base, i.e. all time slots associated with the same day will belong to a single set, namely TsS or TrS or VlS. More precisely, the whole dataset is firstly divided in two subsets having the same cardinality. The first subset constitutes the TsS whereas the second one is partitioned in 5 different ways in order to constitute 5 different couples of TrSs and VlSs.

The LP optimization is ran on the j-th TrS TrS_j in order to evaluate the corresponding values of $(SoC_{opt})_j$ and $(E^N_{opt})_j$.

The construction of the FIS MFs is based on a clustering procedure performed through an appropriately designed K-Means algorithm based on the Euclidean distance function. The K-Means algorithm is ran for all values of k ranging from $k = 2$ to $k = 30$ on the j-th sets Γ_j constituted by the 4-dimensional patterns obtained by joining the E^{GL}, C^{buy} and C^{sell} samples belonging to the j-th TrS TrS_j with the corresponding j-th SoC time sequence $(SoC_{opt})_j$ estimated by means of the LP optimization. The K-Means algorithm, executed on the j-th set Γ_j for a given k value, is in charge of partitioning Γ_j into k clusters and outputs the respective k centroids and k covariance matrices that are then used to define the rule antecedent set according with (6.5). Moreover, all the patterns of the j-th set Γ_j associated with the k-th cluster are collected in the Γ_{jk} array. Similarly the values of $(E^N_{opt})_j$ associated with the Γ_{jk} are collected in the $(E^N_{opt})_{jk}$ set.

For the k-th cluster of the j-th TrS a suitable hyperplane can be defined. Each hyperplane is represented by means of the coefficients set A_{jk} estimated through the Linear Least Square Regression (LLSR) on the set of input-output pairs $\{\Gamma^{in}_{jk}, (E^N_{opt})_{jk}\}$ (see (6.6)).

The previously described procedure results in 29×5 different FIS synthesis. More precisely, for each k value there exist 5 different FIS corresponding to the 5 different TrSs. Each FIS in now simulated on the relative VlS and the achieved normalized performance indexes are evaluated according with (6.4). For each value of k, the best FIS FIS^{best}_k among the available 5 is selected whereas the remaining 4 are discarded. Finally, for each value of k, the MG is simulated on the TsS data on the corresponding FIS^{best}_k and the corresponding profit P is estimated according whit (6.3). The FIS FIS^{best}, which produces the higher profit on the TsS, is selected to design the final EMS. The overall procedure of the EMS FIS design and optimization is illustrated in Algorithm 6.1.

6.4 Simulation Setting

In this work it has been considered a MG composed by the following energy systems: a PV generator of 19 kW, an aggregated load with a peak of power around 8 kW, and ESS with an energy capacity of 24 kWh. For the ESS modeling it has been taken into consideration the Toshiba ESS SCiB module having a rated voltage of 300 [V], a current rate of 8 [C-Rate] and a capacity of about 80 [Ah].

The dataset used in this work has been provided by *AReti S.p.A.*, the electricity distribution company of Rome. The energy prices are the same used in previous works, [9, 10].

Algorithm 6.1 EMS Design

1: **procedure** EMS DESIGN
2: Partition of the dataset in TsS and 5 pairs of TrS_j and VlS_j
3: **for** $j = 1$ to 5 **do**   ▷ for each TrS_j
4:   $\{E_{opt}^N, SoC_{opt}\}_j := \mathrm{LP}(TrS_j)$ ▷ evaluation of the optimal solution on the 5 TrS_j
5:   $\Gamma_j^{in} := \{TrS_j, SoC_{opt}\}$ ▷ join TrS and SoC_{opt}
6: **end for**
7: **for** $k = 2$ to 30 **do**   ▷ for each value of k
8:   **for** $j = 1$ to 5 **do**   ▷ for each TrS_j
9:    $\{\mu_{kj}, \mathbb{C}_{kj}, \Gamma_{kj}\} := \mathrm{KMeans}(k, \Gamma_j^{in})$ ▷ K-Means execution
10:    $\Phi_{kj} := \{\mu_{kj}, \mathbb{C}_{kj}\}$ ▷ MF evaluation
11:    select the points $(E_{opt}^N)_{kj}$ corresponding to Γ_{kj}
12:    $A_{kj} := \mathrm{LLSR}(\{\Gamma_{kj}, (E_{opt}^N)_{kj}\})$ ▷ hyperplanes evaluation
13:    $FIS_{kj} := \{\Phi_{kj}, A_{kj}\}$ ▷ FIS synthesis
14:    simulation of $FIS_{kj}(VlS_j)$ on the j-th VlS and evaluation of $\bar{P}_{kj}$
15:   **end for**
16:   selection of FIS_k^{best} among FIS_{kj} according with $\bar{P}_{kj}$
17:   simulation of $FIS_k^{best}(TsS)$ on TsS and evaluation of the profit P_k
18: **end for**
19: selection of FIS^{best} among FIS_k^{best} according with P_k
20: **end procedure**

The considered dataset covers an overall period of 20 days sampled with a 15 min frequency. The overall dataset is shown in Fig. 6.2 together with the energy prices both in sale (positive) and purchase (negative).

The even days are assigned to the TsS whereas the odd days are partitioned with a random selection between the TrS and VlS in order to form 5 different TrS-VlS partitions. Being the FISs designed on the TrS information it has been chosen to assign the 70% of the odd days to the TrS and the reaming 30% to the VlS. Regarding the K-Means, it is defined by a number of iterations equal to 20 and every execution of the algorithm is repeated 30 times (i.e. number of replicates) with a new pattern initialization which consists in a random selection of the centroids. The solution chosen by the K-Means algorithm is that one which presents the minimum sum of the Euclidean distances between the patterns and their respective centroids, i.e. the one which maximizes the average of clusters compactness.

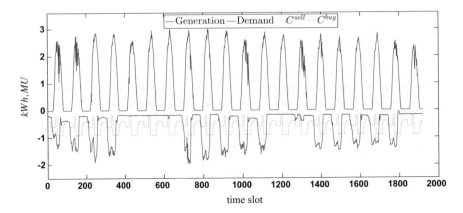

Fig. 6.2 Dataset of the MG energy production and demand and energy prices

6.5 Results

In Table 6.1 are reported the results of the best ten solutions sorted according with the profit P achieved by the corresponding FIS simulated on the TsS values. Moreover, for each solution it is reported the number of clusters, the Davies-Bouldin (DB) index [17], the actual upper and lower profit expressed in [*Monetary Units Per Day*] and the performance index $\bar{P}$ calculated on the VlS. The best solution is achieved for $k = 9$. In order to evaluate its performances, the upper and the lower limit of the profit have been calculated for the TsS, obtaining $P_{TsS}^{upper} = 8.84$ and $P_{TsS}^{lowper} = 2.45$,

Table 6.1 Best solutions results

FIS	TrS	VlS				TsS
	BD index	P	P^{upper}	P^{lower}	$\bar{P}$	P
	Adim	MU/day	MU/day	MU/day	P.U.	MU/day
FIS_9^{best}	0,75	10,00	11,33	5,62	0,23	7,66
FIS_4^{best}	0,32	10,18	11,33	5,62	0,20	7,49
FIS_{11}^{best}	1,04	9,99	11,33	5,62	0,23	7,46
FIS_7^{best}	0,49	10,05	11,33	5,62	0,22	7,42
FIS_5^{best}	0,34	10,09	11,33	5,62	0,22	7,41
FIS_3^{best}	0,19	9,85	11,33	5,62	0,26	7,39
FIS_6^{best}	0,39	10,08	11,33	5,62	0,22	7,37
FIS_{12}^{best}	1,12	12,78	14,26	9,71	0,32	7,35
FIS_{15}^{best}	1,67	13,45	14,82	10,25	0,30	7,34
FIS_{17}^{best}	2,27	13,34	14,82	10,25	0,32	7,34

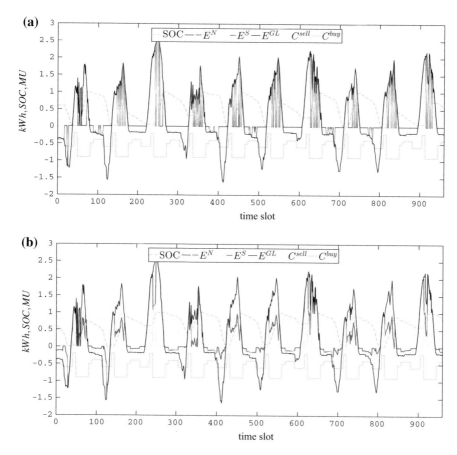

Fig. 6.3 MG energy flow profiles on the TsS: **a** Optimal solution **b** Proposed solution

respectively. By comparing these values with the profit achieved by each EMS it can be stated that their performances are quite close to the optimal one. More precisely, made 100 the P^{upper} profit and 0 the P^{lower} one, the FIS based synthesized solution is placed at 82. Moreover, by comparing the time evolution of the optimal solution with the FIS based one, shown in Fig. 6.3a and b, respectively, it can be seen that the optimal solution presents a strongly intermittent behaviour for both E^S and E^N, whereas the proposed FIS based one has much flatter energy profiles. The better smoothness of the FIS based solution produces a lower degradation of the ESS, whereas the optimal solution stresses much more the ESS due to the high variations of power exchanged in short periods.

6.6   Conclusion

In this work it is proposed a novel procedure to design a first order Takagi-Sugeno FIS EMS with the aim to maximize the profit generated by energy trading with the grid, assuming a TOU energy price policy. It is characterized by Gaussian multi-dimensional MFs defined through a K-Means clustering algorithm on a given TrS. Each of them is associated to a different RC hyperplane. By assuming to know a priori the considered dataset, these are evaluated relying on the optimum MG energy flows profiles by means of a LP problem formulation. In order to optimized the EMS, different FISs have been synthesized, by considering different TrS-VIS pairs and changing the number of clusters (i.e. MFs and RCs). For a given k value, the best FIS has been selected testing the FISs performance on their respective VISs. The proposed synthesis procedure allows to considerably reduce the number of MFs with respect to a grid partition based synthesis approach adopted in previous works. The FIS selected gives a profit on the TsS greater than 80% of the maximum performance yielded by the LP solution. This benchmark solution shows irregular high oscillations of the ESS SoC that makes more tricky an efficient FIS EMS design, avoiding solutions characterized by an high number of MFs (i.e. characterized by an higher granularity). Therefore, for future works it will be investigated the implementation of a smoother reference surface solution in order to give more regularity to it. In particular, it will be taken into consideration a Multi Objective function formulation in order to consider the degradation of the ESS as well.

References

1. Dragicevic, T., Vasquez, J.C., Guerrero, J.M., Skrlec, D.: Advanced lvdc electrical power architectures and microgrids: a step toward a new generation of power distribution networks. IEEE Electr. Mag. **2**(1), 54–65 (2014)
2. Patterson, B.T.: Dc, come home: Dc microgrids and the birth of the "enernet". IEEE Power Energy Mag. **10**(6), 60–69 (2012)
3. Deng, R., Yang, Z., Chow, M.Y., Chen, J.: A survey on demand response in smart grids: mathematical models and approaches. IEEE Trans. Ind. Inform. **11**(3), 570–582 (2015)
4. Amer, M., Naaman, A., M'Sirdi, N.K., El-Zonkoly, A.M.: Smart home energy management systems survey. International Conference on Renewable Energies for Developing Countries **2014**, 167–173 (2014)
5. Kirschen, D.S.: Demand-side view of electricity markets. IEEE Trans. Power Syst. **18**(2), 520–527 (2003)
6. Paschero, M., Storti, G.L., Rizzi, A., Frattale Mascioli, F.M., Rizzoni, G.: A novel mechanical analogy based battery model for soc estimation using a multi-cell ekf. IEEE Trans. Sustain. Energy **7**(4), 1695–1702 (2016)
7. Luzi, M., Paschero, M., Rossini, A., Rizzi, A., Mascioli, F.M.F.: Comparison between two non-linear kalman filters for reliable soc estimation on a prototypal bms. In: Industrial Electronics Society, IECON 2016-42nd Annual Conference of the IEEE, pp. 5501–5506. IEEE (2016)
8. De Santis, E., Rizzi, A., Sadeghiany, A., Mascioli, F.M.F.: Genetic optimization of a fuzzy control system for energy flow management in micro-grids. In: IFSA World Congress and NAFIPS Annual Meeting (IFSA/NAFIPS), 2013 Joint, pp. 418–423 (2013)

9. Leonori, S., De Santis, E., Rizzi, A., Mascioli, F.M.F.: Optimization of a microgrid energy management system based on a fuzzy logic controller. In: IECON 2016-42nd Annual Conference of the IEEE Industrial Electronics Society, pp. 6615–6620 (2016)
10. Leonori, S., De Santis, E., Rizzi, A., Mascioli, F.M.F.: Multi objective optimization of a fuzzy logic controller for energy management in microgrids. In: 2016 IEEE Congress on Evolutionary Computation (CEC), pp. 319–326 (2016)
11. Leonori, S., Paschero, M., Rizzi, A., Mascioli, F.M.F.: An optimized microgrid energy management system based on fis-mo-ga paradigm. In: 2017 IEEE International Conference on Fuzzy Systems (FUZZ-IEEE), pp. 1–6 (2017)
12. Lan, Y., Guan, X., Wu, J.: Rollout strategies for real-time multi-energy scheduling in microgrid with storage system. IET Gener. Transm. Distrib. **10**(3), 688–696 (2016)
13. Farzin, H., Fotuhi-Firuzabad, M., Moeini-Aghtaie, M.: A practical scheme to involve degradation cost of lithium-ion batteries in vehicle-to-grid applications. IEEE Trans. Sustain. Energy **7**(4), 1730–1738 (2016)
14. Arcos-Aviles, D., Pascual, J., Marroyo, L., Sanchis, P., Guinjoan, F.: Fuzzy logic-based energy management system design for residential grid-connected microgrids. IEEE Trans. Smart Grid **PP**(99), 1 (2016)
15. Rizzi, A., Mascioli, F.M.F., Martinelli, G.: Automatic training of anfis networks. In: 1999 IEEE International Fuzzy Systems Conference Proceedings, 1999, FUZZ-IEEE '99, vol. 3, pp. 1655–1660 (1999)
16. Leonori, S., Martino, A., Rizzi, A., Mascioli, F.M.F.: Anfis synthesis by clustering for microgrids ems design. In: Proceedings of the 9th International Joint Conference on Computational Intelligence, IJCCI, vol. 1, pp. 328–337. INSTICC, SciTePress (2017)
17. Davies, D.L., Bouldin, D.W.: A cluster separation measure. IEEE Trans. Pattern Anal. Mach. Intell. **PAMI-1**(2), 224–227 (1979)

Chapter 7
Learning Activation Functions from Data Using Cubic Spline Interpolation

Simone Scardapane, Michele Scarpiniti, Danilo Comminiello
and Aurelio Uncini

Abstract Neural networks require a careful design in order to perform properly on a given task. In particular, selecting a good activation function (possibly in a data-dependent fashion) is a crucial step, which remains an open problem in the research community. Despite a large amount of investigations, most current implementations simply select one *fixed* function from a small set of candidates, which is not adapted during training, and is shared among all neurons throughout the different layers. However, neither two of these assumptions can be supposed optimal in practice. In this paper, we present a principled way to have data-dependent adaptation of the activation functions, which is performed *independently* for each neuron. This is achieved by leveraging over past and present advances on cubic spline interpolation, allowing for local adaptation of the functions around their regions of use. The resulting algorithm is relatively cheap to implement, and overfitting is counterbalanced by the inclusion of a novel damping criterion, which penalizes unwanted oscillations from a predefined shape. Preliminary experimental results validate the proposal.

Keywords Neural network · Activation function · Spline interpolation

S. Scardapane (✉) · M. Scarpiniti · D. Comminiello · A. Uncini
Department of Information Engineering, Electronics and Telecommunications (DIET),
"Sapienza" University of Rome, Via Eudossiana 18, 00184 Rome, Italy
e-mail: simone.scardapane@uniroma1.it

M. Scarpiniti
e-mail: michele.scarpiniti@uniroma1.it

D. Comminiello
e-mail: danilo.comminiello@uniroma1.it

A. Uncini
e-mail: aurel@ieee.org

73

7.1 Introduction

Neural networks (NNs) are extremely powerful tools for approximating complex
nonlinear functions [7]. The nonlinear behavior is introduced in the NN architecture
by the elementwise application of a given nonlinearity, called the activation function
(AF), at every layer. Since AFs are crucial to the dynamics and computational power
of NNs, the history of the two over the last decades is deeply connected [15]. As
an example, the use of *differentiable* AFs was one of the major breakthroughs in
NNs, leading directly to the back-propagation algorithm. More recently, progress on
piecewise linear functions was shown to facilitate backward flow of information for
training very deep networks [4]. At the same time, it is somewhat surprising that the
vast majority of NNs only use a small handful of fixed functions, to be hand-chosen
by the practitioner before the learning process. Worse, there is no principled reason to
believe that a 'good' nonlinearity might be the same across all layers of the network,
or even across neurons in the same layer.

This is shown clearly in a recent work by Agostinelli et al. [1], where every
neuron in a deep network was endowed with an adaptable piecewise linear function
with possibly different parameters, concluding that *"the standard one-activation-
function-fits-all approach may be suboptimal"* in current practice. Experiments in
AF adaptation have a long history, but they have never met a wide applicability in the
field. The simplest approach is to parameterize each sigmoid function in the network
by one or more 'shape' parameters to be optimized, such as in the seminal 1996 paper
by Chen and Chang [3] or the later work by Trentin [16]. Along a similar line, one
may consider the use of polynomial AFs, wherein each coefficient of the polynomial
is adapted by gradient descent [11]. Additional investigations can be found in [2, 5,
9, 10, 19]. One strong drawback of these approaches is that the parameters involved
affect the AF *globally*, such that a change in one region of the function may be
counterproductive on a different, possibly faraway, region.

Several years ago, an alternative approach was introduced by using spline interpolating functions as AFs [6, 17], resulting in what was called a spline AF (SAF).
Splines are an attractive choice for interpolating unknown functions, since they can
be described by a small amount of parameters, yet each parameter has a local effect,
and only a fixed number of them is involved every time an output value is computed
[18]. The original works in [6, 17] had two main drawbacks that prevented a wider
use of the underlying theory. First, SAFs were only investigated in an online setting,
where updates are computed one sample at a time. Whether an efficient implementation is possible (and feasible) also for batch (or mini-batch) settings was not shown.
Secondly, the obtained SAFs had a tendency to overfit training data, resulting in
oscillatory behaviors which hindered performance. Inspired by recent successes in
the field of nonlinear adaptive filtering [13, 14], our aim in this paper is two-fold.
On one hand, we provide a modern introduction to the use of SAFs in neural networks, with a particular emphasis on their efficient implementation in the case of
batch (or mini-batch) training. Our treatment clearly shows that the major problem in
their implementation, which is evident from the discussion above, is the design of an

efficient way to regularize their control points. In this sense, as a second contribution we provide a simple (yet effective) 'damping' criterion to prevent unwanted oscillations in the testing phase, which penalizes deviations from the original points in terms of ℓ_2 norm. A restricted set of experiments shows that the resulting formulation is able to achieve a lower test error than a standard NN with fixed AFs, while at the same time learning non-trivial activations with different shapes across different neurons.

The rest of the paper is organized as follows. Section 7.2 presents the basic theory of SAFs for the case of a single neuron. Section 7.3 extends the treatment to the case of a NN with one hidden layer, by deriving the gradient equations for the SAFs parameters in the internal layer. Then, Sect. 7.4 goes over the experimental results, while we conclude with some final remarks in Sect. 7.5.

7.2 The Spline Activation Function

We begin our treatment of SAFs with the simplest case of a single neuron endowed with a flexible AF (see [13, 17] for additional details). Given a generic input $\mathbf{x} \in \mathbb{R}^D$, the output of the SAF is computed as:

$$s = \mathbf{w}^T \mathbf{x}, \tag{7.1}$$

$$y = \varphi(s; \mathbf{q}), \tag{7.2}$$

where $\mathbf{w} \in \mathbb{R}^D$ (we suppose that an eventual bias term is added directly to the input vector), and the AF $\varphi(\cdot)$ is parameterized by a vector $\mathbf{q} \in \mathbb{R}^Q$ of internal parameters, called *knots*. The knots are a sampling of the AF values over Q representative points spanning the overall function. In particular, we suppose the knots to be uniformly spaced, i.e. $q_{i+1} = q_i + \Delta x$, for a fixed $\Delta x \in \mathbb{R}$, and symmetrically spaced around the origin. Given s, the output is computed by spline interpolation over the closest knot and its P rightmost neighbors. The common choice $P = 3$, which we adopt in this paper, corresponds to *cubic* interpolation, and it is generally a good trade-off between locality of the output and interpolating precision.

Given the index i of the closest knot, we can define the normalized abscissa value between q_i and q_{i+1} as:

$$u = \frac{s}{\Delta x} - \left\lfloor \frac{s}{\Delta x} \right\rfloor, \tag{7.3}$$

where $\lfloor \cdot \rfloor$ is the floor operator. From u we can compute the normalized reference vector $\mathbf{u} = \left[u^P \ u^{P-1} \dots u \ 1\right]^T$, while from i we can extract the relevant control points $\mathbf{q}_i = \left[q_i \ q_{i+1} \dots q_{i+P}\right]^T$. We refer to the vector $\mathbf{q}_i$ as the ith *span*. The output (7.2) is then computed as:

$$y = \varphi(s) = \mathbf{u}^T \mathbf{B} \mathbf{q}_i, \tag{7.4}$$

where $\mathbf{B} \in \mathbb{R}^{(P+1) \times (P+1)}$ is called the spline basis matrix. In this work, we use the Catmull-Rom (CR) spline with $P = 3$, given by:

$$\mathbf{B} = \frac{1}{2} \begin{bmatrix} -1 & 3 & -3 & 1 \\ 2 & -5 & 4 & -1 \\ -1 & 0 & 1 & 0 \\ 0 & 2 & 0 & 0 \end{bmatrix}. \tag{7.5}$$

Different bases give rise to alternative interpolation schemes, e.g. a spline defined by a CR basis passes through all the control points, but its second derivative is not continuous.

Apart from the locality of the output, SAFs have two additional interesting properties. First, the output in (7.4) is extremely efficient to compute, involving only vector-matrix products of very small dimensionality. Secondly, derivatives with respect to the internal parameters are equivalently simple and can be written down in closed form. In particular, the derivative of the nonlinearity $\varphi(s)$ with respect to the input s is given by:

$$\frac{\partial \varphi(s)}{\partial s} = \varphi'(s) = \frac{\partial \varphi(s)}{\partial u} \cdot \frac{\partial u}{\partial s} = \left(\frac{1}{\Delta x} \right) \dot{\mathbf{u}} \mathbf{B} \mathbf{q}_i , \tag{7.6}$$

where:

$$\dot{\mathbf{u}} = \frac{\partial \mathbf{u}}{\partial u} = \left[P u^{P-1} \ (P-1) u^{P-2} \dots 1 \ 0 \right]^T . \tag{7.7}$$

Given this, the derivative of the SAF output y with respect to $\mathbf{w}$ is straightforward:

$$\frac{\partial \varphi(s)}{\partial \mathbf{w}} = \varphi'(s) \cdot \frac{\partial s}{\partial \mathbf{w}} = \varphi'(s) \mathbf{x} , \tag{7.8}$$

Similarly, for $\mathbf{q}_i$ we obtain:

$$\frac{\partial \varphi(s)}{\partial \mathbf{q}_i} = \mathbf{B}^T \mathbf{u} . \tag{7.9}$$

while we have $\frac{\partial \varphi(s)}{\partial q_k} = 0$ for any element q_k outside the current span $\mathbf{q}_i$.

7.3 Designing Networks with SAF Neurons

7.3.1 Computing Outputs and Inner Derivatives

Now we consider the more elaborate case of a single hidden layer NN, with a D-dimensional input, H neurons in the hidden layer, and O output neurons.[1] Every neuron in the network uses a SAF with possibly different adaptive control points, which are set independently during the training process. For easiness of computation, we suppose that the sampling set of the splines is the same for every neuron (i.e., each neuron has Q points equispaced according to the same Δx), and we also have a single shared basis matrix $\mathbf{B}$. The forward phase of the network is similar to that of a standard network. In particular, given the input $\mathbf{x}$, we first compute the output of the ith hidden neuron, $i = 1, \ldots, H$, as:

$$h_i = \varphi(\mathbf{w}_{h_i}^T \mathbf{x}; \mathbf{q}_{h_i}) . \tag{7.10}$$

These are concatenated in a single vector $\mathbf{h} = [h_1, \ldots, h_H, 1]^T$, and the ith output of the network, $i = 1, \ldots, O$, is given by:

$$f_i(\mathbf{x}) = y_i = \varphi(\mathbf{w}_{y_i}^T \mathbf{h}; \mathbf{q}_{y_i}) . \tag{7.11}$$

The derivatives with respect to the parameters $\{\mathbf{w}_{y_i}, \mathbf{q}_{y_i}\}$, $i = 1, \ldots, O$ can be computed directly with (7.8)–(7.9), substituting $\mathbf{x}$ with $\mathbf{h}$. By back-propagation, the derivative of the ith output with respect to the jth (inner) weight vector $\mathbf{w}_{h_j}$ is similar to a standard NN:

$$\frac{\partial y_i}{\partial \mathbf{w}_{h_j}} = \varphi'(s_{y_i}) \cdot \varphi'(s_{h_j}) \cdot \lfloor \mathbf{w}_{h_j} \rfloor_i \cdot \mathbf{x} , \tag{7.12}$$

where with a slight abuse of notation we let s_{y_i} denote the activation of the ith output (and similarly for s_{h_j}), $\lfloor \cdot \rfloor_i$ extracts the ith element of its input vector, and the two $\varphi'(\cdot)$ are given by (7.6). For the derivative of the control points of the jth hidden neuron, denote by $\mathbf{q}_{h_j,k}$ the currently active span, and by $\mathbf{u}_{h_j}$ the corresponding reference vector. The derivative with respect to the ith output is then given by:

$$\frac{\partial y_i}{\partial \mathbf{q}_{h_j,k}} = \varphi'(s_{y_i}) \cdot \lfloor \mathbf{w}_{h_j} \rfloor_i \cdot \mathbf{B}^T \mathbf{u}_{h_j} . \tag{7.13}$$

---

[1]We note that the following treatment can be extended easily to the case of a network with more than one hidden layer. However, restricting it to a single layer allow us to keep the discussion focused on the problems/advantages arising in the use of SAFs. We leave this extension to a future work.

7.3.2   Initialization of the Control Points

An important aspect that we have not discussed yet is how to properly initialize the control points. One immediate choice is to sample their values from an AF which is known to work well on the given problem, e.g. a hyperbolic tangent. In this way, the network is guaranteed to work similarly to a standard NN in the initial phase of learning. Additionally, we have found good improvements in error by adding Gaussian noise $\mathcal{N}(0, \sigma^2)$ with small variance σ^2 to a randomly chosen subset of control points (around 5% in our experiments). This provides a good variability in the beginning, similarly to how connections are set close to (but not identically equal to) zero during initialization.

7.3.3   Choosing a Training Criterion

Suppose we are provided with a training set of N input/output pairs in the form $\{\mathbf{x}_i, \mathbf{d}_i\}_{i=1}^N$. For simplicity of notation, we denote by $\mathbf{w}$ the concatenation of all weight vectors $\{\mathbf{w}_{h_i}\}$ and $\{\mathbf{w}_{y_i}\}$, and by $\mathbf{q}$ a similar concatenation of all control points. Training can be formulated as the minimization of the following cost function:

$$J(\mathbf{w}, \mathbf{q}) = \frac{1}{N} \sum_{i=1}^N L(\mathbf{d}_i, \mathbf{f}(\mathbf{x}_i)) + \lambda_w R_w(\mathbf{w}) + \lambda_q R_q(\mathbf{q}), \qquad (7.14)$$

where $L(\cdot, \cdot)$ is an error function, while $R_w(\cdot)$ and $R_q(\cdot)$ provide meaningful regularization on the two set of parameters. The first two terms are well-known in the neural network literature [7], and they can be set accordingly. Particularly, in our experiments we consider a squared error term $L(\mathbf{d}_i, \mathbf{f}(\mathbf{x}_i)) = \|\mathbf{d}_i - \mathbf{f}(\mathbf{x}_i)\|_2^2$, and ℓ_2 regularization on the weights $R_w(\mathbf{w}) = \|\mathbf{w}\|_2^2$. The derivatives of $L(\cdot, \cdot)$ can be computed straightforwardly with the formulas presented in Sect. 7.3.1.

The term $R_q(\mathbf{q})$ is used to avoid overfitted solutions for the control points. In fact, its presence is the major difference with respect to previous attempts at implementing SAFs in neural networks [17], wherein overfitting was counterbalanced by choosing a large value for Δx, which in a way goes outside the philosophy of spline interpolation itself. At the same time, choosing a proper form for the regularization term is nontrivial, as the term should be cheap to compute, and it should introduce just as much *a priori* information as needed, without hindering the training process. Most of the literature on regularizing $\mathbf{w}$ cannot be used here, as the corresponding formulations do not make sense in the context of spline interpolation. As an example, simply penalizing the ℓ_2 norm of $\mathbf{q}$ leads to functions close to the zero function, while imposing sparsity is also meaningless.

For the purpose of this paper, we consider the following 'damping' criterion:

$$R_q(\mathbf{q}) = \|\mathbf{q} - \mathbf{q}_o\|_2^2, \qquad (7.15)$$

where $\mathbf{q}_o$ represents the initial values for the control points, as discussed in the previous section (without considering additional noise). The criterion makes intuitive sense as follows: while for $\mathbf{w}$ we wish to penalize unwanted deviations from very small weights (which can be justified with arguments from learning theory), in the case of $\mathbf{q}$ we are interested in penalizing changes with respect to a 'good' function parameterized by the initial control points $\mathbf{q}_o$, namely one of the standard AFs used in NN training. In fact, setting a value for λ_q very high essentially deactivates the adaptation of the control points. Clearly, other choices are possible, and in this sense this paper serves as a starting point for further investigations towards this objective. As an example, we may wish to penalize first (or second) order derivatives of the splines in order to force a desired level of smoothness [18].

7.3.4 Remarks on the Implementation

In order to be usable in practice, SAFs require an efficient implementation to compute outputs and derivatives concurrently for the entire training dataset or, alternatively, for a properly chosen mini-batch (in the case of stochastic optimization algorithms). To begin with, we underline that the equations for the reference vector (see 7.3) do not depend on the specific neuron, and for this reason they can easily be vectorized layer-wise on most numerical algebra libraries to obtain all vectors concurrently. Additionally, the indexes and relative terms $\mathbf{Bq}_i$ in (7.4) can be cached during the forward pass, to be reused during the computation of the derivatives. In this sense, the outputs of a layer *and* its derivatives can be computed by one 4×4 matrix-vector computation, and three 4-dimensional inner products, which have to be repeated for every pair input/neuron. In our experience, the cost of a relatively well-optimized implementation does not exceed twice that of a standard network for medium-sized batches, where the most onerous operation is the reshaping of the gradients in (7.9) and (7.13) into a single vector of gradients relative to the global vector $\mathbf{q}$.

7.4 Experimental Results

7.4.1 Experimental Setup

To evaluate the preliminary proposal, we consider two simple regression benchmarks for neural networks, the 'chemical' dataset (included among MATLAB's testbeds for function fitting), and the 'California Housing'.[2] They have respectively 498 and 20640 examples, and 8 numerical features. Inputs are normalized in the $[-1, +1]$ range, while outputs are normalized in the $[-0.5, +0.5]$ range. We compare a NN

---

[2]http://www.dcc.fc.up.pt/~ltorgo/Regression/cal_housing.html.

with 5 hidden neurons and $\tanh(\cdot)$ AFs (denoted as 'Standard' in the results), and a NN with the same number of neurons and SAF nonlinearities. The weight vector $\mathbf{w}$ is initialized with the method described in [4]. Each SAF is initialized from a $\tanh(\cdot)$ nonlinearity, and control points are defined in the $[-2, +2]$ range with $\Delta x = 0.2$, which is a good compromise between locality of the SAFs and the overall number of adaptable parameters. For the first scenario, λ_q is kept to a small value of 10^{-5}. For each experiment, a random 30% of the dataset is kept for testing, and results are averaged over 15 different splits to average out statistical effects. Error is computed with the Normalized Root Mean-Squared Error (NRMSE). The optimization problems are solved using a freely available MATLAB implementation of the Polack-Ribiere variant of the nonlinear conjugate gradient optimization algorithm by C. E. Rasmussen [12].[3] The optimization process is allowed 1500 maximum iterations. MATLAB code for the experiments is also available on the web.[4] We briefly remark that the MATLAB library, apart from repeating the experiments presented here, is also designed to handle networks with more than a single hidden layer, and implements the ADAM algorithm [8] for stochastic training in case of a larger dataset.

7.4.2 Scenario 1: Strong Underfitting

As a first example, we consider a scenario of strong underfitting, wherein the user has misleadingly selected a very large value of $\lambda_w = 1$, leading in turn to extremely small values for the elements of $\mathbf{w}$ after training. Results in terms of training and test RMSE are provided in Table 7.1. Since the activations of the NN tend to be very close to 0 (where the hyperbolic tangent operates in an almost-linear regime), standard NNs have a constant zero output, leading to a RMSE of 1. Nonetheless, SAF networks are able to reach a very satisfactory level of performance, which in the first case is almost comparable to that of a fully optimized network (see the following section).

To show the reasons for this, we have plotted four representative nonlinearities after training in Fig. 7.1. It is easy to see that the nonlinearities have adapted to act as 'amplifiers' for the activations in their operating regime, with mild and strong peaks around 0. Of particular interest is the fact that the resulting SAFs need not be perfectly centered around 0 (e.g. Fig. 7.1c), or even symmetrical around the y-axis (e.g. Fig. 7.1d). In fact, the splines are able to efficiently counterbalance a bad setting for the weights, with behaviors which would be very hard (or close to impossible) using standard setups with fixed, shared, mild nonlinearities.

---

[3]http://learning.eng.cam.ac.uk/carl/code/minimize/.

[4]https://bitbucket.org/ispamm/spline-nn.

Table 7.1 Average results for scenario 1 ($\lambda_w = 1$), together with one standard deviation

Dataset	Nonlinearity	Tr. RMSE	T.st NRMSE
Chemical	Standard	1.00 ± 0.00	1.00 ± 0.01
	SAF	$\mathbf{0.29 \pm 0.02}$	$\mathbf{0.31 \pm 0.02}$
Calhousing	Standard	1.02 ± 0.00	1.01 ± 0.01
	SAF	$\mathbf{0.56 \pm 0.01}$	$\mathbf{0.57 \pm 0.02}$

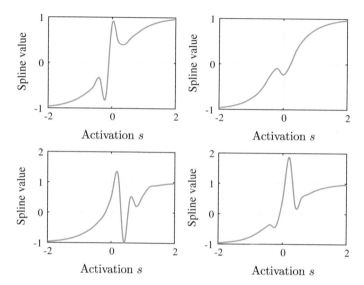

Fig. 7.1 Non-trivial representative SAFs after training for scenario 1

7.4.3 Scenario 2: Well-Optimized Parameters

In our second scenario, we consider a similar comparison with respect to before, but we fine-tune the parameters of the two methods using a grid search with a 3-fold cross-validation on the training data as performance measure. Both λ_w and λ_q (only for the proposed algorithm) are searched in an exponential interval 2^j, with $j = -10, \ldots, 5$. Optimal parameters found by the grid-search are listed in Table 7.2, while results in terms of training and test NRMSE are collected in Table 7.3.

Overall, we see that the NNs endowed with the SAF nonlinearities are able to surpass by a large margin a standard NN, and the results from the previous scenario. The only minor drawback evidenced in Table 7.3 is that the SAF network has some overfitting occurring in the 'chemical' dataset (around 2 points of NRMSE), showing that there is still some room for improvement in terms of spline optimal regularization.

Also in this case, we plot some representatives SAFs after training (taken among those which are not trivially identical to the tanh nonlinearity) in Fig. 7.2. As before, in general SAFs tend to provide an amplification (with a possible change of sign)

Table 7.2 Optimal parameters (averaged over the runs) found by the grid-search procedure for scenario 2

Dataset	Nonlinearity	λ_w	λ_q
Chemical	Standard	10^{-3}	–
	SAF	10^{-2}	10^{-4}
Calhousing	Standard	10^{-4}	–
	SAF	10^{-3}	10^{-4}

Table 7.3 Average results for scenario 2 (fine-tuning for parameters), together with one standard deviation

Dataset	Nonlinearity	Tr. RMSE	T.st NRMSE
Chemical	Standard	0.32 ± 0.01	0.32 ± 0.02
	SAF	$\mathbf{0.26 \pm 0.01}$	$\mathbf{0.28 \pm 0.02}$
Calhousing	Standard	0.55 ± 0.01	0.55 ± 0.01
	SAF	$\mathbf{0.51 \pm 0.02}$	$\mathbf{0.51 \pm 0.02}$

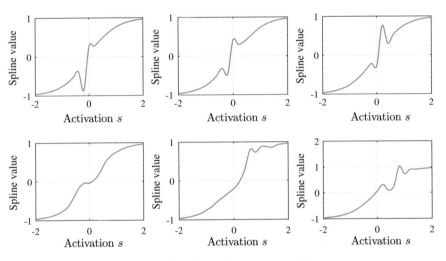

Fig. 7.2 Non-trivial representative SAFs after training for scenario 2

of their activation around some region of operation. It is interesting to observe that, also in this case, the optimal shape need not be symmetric (e.g. Fig. 7.2a), and might even be far from centered around 0 (e.g. Fig. 7.2c). Resulting nonlinearities can also present some additional non-trivial behaviors, such as a small region of insensibility around 0 (e.g. Fig. 7.2d), or a region of pre-saturation before the actual tanh saturation (e.g. Fig. 7.2e and f).

7.5 Conclusion

In this paper, we have presented a principled way to adapt the activation functions of a neural network from training data, locally and independently for each neuron. Particularly, each nonlinearity is implemented with cubic spline interpolation, whose control points are adapted in the optimization phase. Overfitting is controlled by a novel ℓ_2 regularization criterion avoiding unwanted oscillations. Albeit efficient, this criterion does constrain the shapes of the resulting functions by a certain degree. In this sense, the design of more advanced regularization terms is a promising line of research. Additionally, we plan on exploring the application of SAFs to deeper networks, where it is expected that the statistics of the neurons' activations can change significantly layer-wise [4].

References

1. Agostinelli, F., Hoffman, M., Sadowski, P., Baldi, P.: Learning activation functions to improve deep neural networks (2014). arXiv preprint arXiv:1412.6830
2. Chandra, P., Singh, Y.: An activation function adapting training algorithm for sigmoidal feed-forward networks. Neurocomputing **61**, 429–437 (2004)
3. Chen, C.T., Chang, W.D.: A feedforward neural network with function shape autotuning. Neural Netw. **9**(4), 627–641 (1996)
4. Glorot, X., Bengio, Y.: Understanding the difficulty of training deep feedforward neural networks. In: International Conference on Artificial Intelligence and Statistics, pp. 249–256 (2010)
5. Goh, S., Mandic, D.: Recurrent neural networks with trainable amplitude of activation functions. Neural Netw. **16**(8), 1095–1100 (2003)
6. Guarnieri, S., Piazza, F., Uncini, A.: Multilayer feedforward networks with adaptive spline activation function. IEEE Trans. Neural Netw. **10**(3), 672–683 (1999)
7. Haykin, S.: Neural Networks and Learning Machines, 3rd edn. Pearson Education (2009)
8. Kingma, D., Ba, J.: Adam: A method for stochastic optimization. In: 3rd International Conference for Learning Representations (2015). arXiv preprint arXiv:1412.6980
9. Lin, M., Chen, Q., Yan, S.: Network in Network (2013). arXiv preprint arXiv:1312.4400
10. Ma, L., Khorasani, K.: Constructive feedforward neural networks using hermite polynomial activation functions. IEEE Trans. Neural Netw. **16**(4), 821–833 (2005)
11. Piazza, F., Uncini, A., Zenobi, M.: Artificial neural networks with adaptive polynomial activation function. In: International Joint Conference on Neural Networks, vol. 2, pp. II–343. IEEE/INNS (1992)
12. Rasmussen, C.: Gaussian Processes for Machine Learning. MIT Press (2006)
13. Scarpiniti, M., Comminiello, D., Parisi, R., Uncini, A.: Nonlinear spline adaptive filtering. Signal Process. **93**(4), 772–783 (2013)
14. Scarpiniti, M., Comminiello, D., Scarano, G., Parisi, R., Uncini, A.: Steady-state performance of spline adaptive filters. IEEE Trans. Signal Process. **64**(4), 816–828 (2016)
15. Schmidhuber, J.: Deep learning in neural networks: an overview. Neural Netw. **61**, 85–117 (2015)
16. Trentin, E.: Networks with trainable amplitude of activation functions. Neural Netw. **14**(4), 471–493 (2001)
17. Vecci, L., Piazza, F., Uncini, A.: Learning and approximation capabilities of adaptive spline activation function neural networks. Neural Netw. **11**(2), 259–270 (1998)
18. Wahba, G.: Spline Models for Observational Data. SIAM (1990)
19. Zhang, M., Xu, S., Fulcher, J.: Neuron-adaptive higher order neural-network models for automated financial data modeling. IEEE Trans. Neural Netw. **13**(1), 188–204 (2002)

Chapter 8
Context Analysis Using a Bayesian Normal Graph

Amedeo Buonanno, Paola Iadicicco, Giovanni Di Gennaro and Francesco A. N. Palmieri

Abstract Contextual information can be used to help object detection in video and images, or to categorize text. In this work we demonstrate how the Latent Variable Model, expressed as a Factor Graph in Reduced Normal Form, can manage contextual information to support a scene understanding task. In an unsupervised scenario our model learns how various objects can coexist, by associating object variables to a latent Bayesian cluster. The model, that is implemented using probabilistic message propagation, can be used to correct or to assign labels to new images.

Keywords Object detection · Contextual information · Belief propagation
Bayesian network · Factor graph

8.1 Introduction

Contextual information can be very useful for improving object detection and for approaching scene understanding tasks [1, 2]. The fact that, in a particular scene, some objects may be more likely to occur than others, may improve our confidence on a pattern recognition task, or on a smart image captioning procedure [3–7]. The challenge is to build a probabilistic model that, as in the human visual system, can store the statistical relationships among objects and between scene and objects to support high-level image handling.

A. Buonanno (✉) · P. Iadicicco · G. Di Gennaro · F. A. N. Palmieri
Dipartimento di Ingegneria Industriale e dell'Informazione,
Università della Campania "Luigi Vanvitelli" (ex SUN), via Roma 29,
81031 Aversa, CE, Italy
e-mail: amedeo.buonanno@unicampania.it

P. Iadicicco
e-mail: paolaiadicicco@libero.it

G. Di Gennaro
e-mail: giovannidigennaro@virgilio.it

F. A. N. Palmieri
e-mail: francesco.palmieri@unicampania.it

© Springer International Publishing AG, part of Springer Nature 2019
A. Esposito et al. (eds.), *Neural Advances in Processing Nonlinear Dynamic Signals*, Smart Innovation, Systems and Technologies 102,
https://doi.org/10.1007/978-3-319-95098-3_8

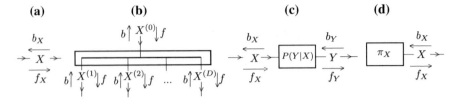

Fig. 8.1 FGrn components: **a** Variable; **b** Replicator; **c** SISO block (factor); **d** Source block. The forward and backward messages are shown for each variable

Fig. 8.2 The LVM model as a Bayesian graph (**a**) and as a Factor Graph in Reduced Normal Form (**b**)

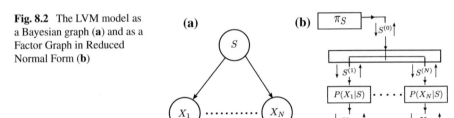

The relationship among objects and between object and context can be modeled in a probabilistic way. We focus here on a *Latent Variable Model* (LVM) [8] (Fig. 8.2) where the N variables $\{X_j\}_{j=1..N}$, representing absence, presence, or multiple presence of the objects, are related to the unobserved context variable S.

Probabilistic modeling for context analysis has been proposed before [4, 9] but, to our knowledge, this is the first work that uses the Factor Graph in Reduced Normal Form (FGrn) paradigm to tackle this type of problem.

The aim of this work is to demonstrate that the simple LVM model, together with the FGrn framework, is a powerful and flexible tool to perform object detection tasks using contextual information. Moreover the LVM, trained in an unsupervised way, is able to recognize the context and, in inference, can use contextual information to help correct and/or enrich scene description.

In Sects. 8.2 and 8.3 we briefly review the FGrn paradigm and how it is implemented. In Sect. 8.4 we present a set of simulations on the standard COCO Dataset [10]. Conclusions are reported in Sect. 8.5.

8.2 Factor Graphs in Reduced Normal Form

Using the FGrn framework, a Bayesian graph can be designed using only four building blocks: *Variables*, *Replicators* (or *Diverters*), *Single-Input/Single-Output (SISO) blocks* and *Source blocks* (Fig. 8.1). The FGrn paradigm allows to handle the Bayesian Graph architectures [11] in an easier way, using a unique learning equation [12], distributed among several SISO blocks of the network. The great flexibility and modularity of this paradigm brought us to define a Simulink library for the rapid prototyping of architectures based on this paradigm [13].

The parameters in the SISO and source blocks are learned from messages available locally. The learning process follows an Expectation Maximization [EM] algorithm to maximize global likelihood of observed data (see [12] for derivations and comparisons). A complete review of the paradigm is beyond the scope of this work and the interested reader can find more details in our recent works [12–15] or in the classical papers [11, 16].

8.3 Latent Variable Model

The *Latent-Variable Model* (LVM) [8], that is the focus of this paper, is shown in Fig. 8.2 with its N variables $\{X_j\}_{j=1:N}$, that take values from the discrete alphabet $\mathcal{X} = \{\xi_1, \xi_2, \ldots, \xi_{|\mathcal{X}|}\}$.

The N variables are connected to one Hidden (Latent) Variable S that takes values from the discrete alphabet $\mathcal{S} = \{\sigma_1, \sigma_2, \ldots, \sigma_{|\mathcal{S}|}\}$. The SISO blocks represent $|\mathcal{S}| \times |\mathcal{X}|$ row-stochastic probability conditional matrices: $P(X_n|S) = [Pr\{X_n = \xi_j|S = \sigma_i\}]_{i=1:|\mathcal{S}|}^{j=1:|\mathcal{X}|}$.

This generative model can be seen also as a Mixture of Categorical Distributions [17] where each element of the alphabet $\mathcal{S}$ is a "Bayesian cluster" for the N-dimensional stochastic vector, $\mathbf{X} = (X_1, X_2, ..., X_N)$. More information on LVM described using FGrn can be found in [14, 15, 18, 19].

Note that in the FGrn formulation [12] each SISO block of Fig. 8.2 uses its own backward message $\mathbf{b}_{X_j}$ at the bottom, and a different incoming message from the diverter $\mathbf{f}_{S(j)}$. Similarly the sum rule produces specific forward messages $\mathbf{f}_{X_j}$ at the bottom and different backward messages $\mathbf{b}_{S(j)}$ toward the diverter.

In this work the N observed variables are the object categories in the current scene and the hidden variable S represents the context variable. More specifically, the label in $\mathcal{X} = 0, 1, 2, .., D$, describes how many times that object category is present in a realization and S groups the occurrences in a context by similarity to previous outcomes. Generally the complexity of the whole system increases with the cardinality of the defined alphabets. In our experiments we will use $D = 2$, where we assign the value 2 when the number of occurrences is greater or equal to 2.

After learning, the network in Fig. 8.2 represents the joint probability distribution:

$$p_{X_1, X_2, \ldots, X_N, S}(x_1 x_2 \ldots x_N, s) = p_{X_1|S}(x_1|s) p_{X_2|S}(x_2|s) \ldots p_{X_N|S}(x_N|s) \pi_S(s).$$

When an example $(x_1[n]x_2[n]\ldots x_N[n])$ at time n is observed, backward messages $b_{X_j} = \delta(x_j - x_j[n])$, $j = 1, \ldots, N$, are injected into the network, where δ is the Kronecker delta. After message propagation, the forward distribution at the ith variable will be proportional to the posterior probability of X_i given all the other observed variables. For example, for X_1 we have:

$$f_{X_1}(x_1) = \sum_{s \in S} p_{X_1|S}(x_1|s) p_{X_2|S}(x_2[n]|s) \dots p_{X_N|S}(x_N[n]|s) \pi_S(s)$$
$$= p_{X_1,X_2,\dots,X_N}(x_1 x_2[n]\dots x_N[n]), \tag{8.1}$$

which is proportional to the posterior distribution of the X_1 given all the other observed variables ($p_{X_1|X_2,\dots,X_N}(x_1|x_2[n]\dots x_N[n])$). Note that variable S is never observed and there is no interaction between forward and backward messages at each X_i because the graph has no loops.

8.3.1 Model Evaluation

Once the model parameters have been learned via probability propagation and the EM algorithm [12], in this unsupervised scenario we need to define an appropriate metric to evaluate the system in performing inference. In this paper we use empirical average likelihood, average entropy and error probability, as described in the following.

8.3.1.1 Likelihood

Note that, as in any cycle-free Bayesian network, the likelihood for each example is available anywhere in the network. For example at time n at X_1 we have:

$$L[n] = p_{X_1,X_2,\dots,X_N}(x_1[n], x_2[n], \dots, x_N[n])$$
$$= p_{X_1,X_2,\dots,X_N}(x_1, x_2[n], \dots, x_N[n])\delta(x_1 - x_1[n]) = f_{X_1}(x_1) \cdot b_{X_1}(x_1[n]), \tag{8.2}$$

and the product $f_{X_j}(x_j) \cdot b_{X_j}(x_j[n])$ at any other j would give the same result. To get the exact values, we should not normalize messages during propagation but normalization is important for numerical stability and we usually do so. Therefore to estimate the likelihood for the nth example, we average among the different values obtained from different variables $L[n] = \frac{1}{N}\sum_{j=1}^{N} f_{X_j}(x_j[n])$. Since we have T_{tr} training examples and T_{va} validation examples, we compute the Negative Log Likelihood of the entire sets as $\mathcal{L}^{tr} = -\frac{1}{T_{tr}}\sum_{n=1}^{T_{tr}} \log L[n]$ and $\mathcal{L}^{va} = -\frac{1}{T_{va}}\sum_{n=1}^{T_{va}} \log L[n]$.

8.3.1.2 Entropy

To have a more detailed understanding of the capability of the system to provide sharp responses for each variable, given all the others, we also compute the conditional entropy or each variable X_i given all the others [20]. For example, for variable X_1:

$$\mathcal{H}(X_1|X_2\dots X_N) =$$
$$= -\sum_{x_2\dots x_N} p(x_2\dots x_N)\sum_{x_1} p(x_1|x_2, \dots, x_N) \log p(x_1|x_2, \dots, x_N) \tag{8.3}$$
$$= -\sum_{x_1,x_2,\dots,x_N} p(x_1, x_2, \dots, x_N) \log p(x_1|x_2, \dots, x_N),$$

that can be computed from the messages as:

$$\mathcal{H}(X_1 | X_2 = x_2[n], ..., X_N = x_N[n])$$
$$= -\sum_{x_1} p(x_1, x_2[n], ..., x_N[n]) \log p(x_1 | x_2[n], ..., x_N[n]) \qquad (8.4)$$
$$\sim -\sum_{x_1} f_{X_1}(x_1) \cdot \log f_{X_1}(x_1)$$

This computation can be performed for each $\{X_j\}_{j=1:N}$ and possibly be averaged over the Training or the Validation Set.

8.3.1.3 Error Probability

The marginal probability of each observed variable $p(X_j)$ is proportional to the product between the two messages $f_{X_j}(x_j) \cdot b_{X_j}(x_j)$. However, during testing, when X_j is available, $b_{X_j}(x_j)$ is a sharp distribution $\delta(x_j - x_j[n])$ with $x_j[n] = \delta_k = [0, .., 1, .., 0]$ (a vector with all zero except for the k-th position), and the forward distribution $f_{X_j[k]}$ is the posterior distribution for that variable given all the others. Therefore to evaluate the accordance between the *ground truth*, described by the backward message $\mathbf{b}_{X_j}$, and the posterior inferred by the network carried by the forward message $\mathbf{f}_{X_j}$, we can use a Maximum A Posteriori (MAP) criterion: given that h is the position of the maximum of the forward message, if h is different from k, the position of the ground truth value, we declare an error. The error count is then averaged over the entire dataset and variables in order to provide a synthetic measure of performance. Clearly some variables may be more accurately predicted than others. This depends on the intrinsic dependence with the other variables and with the cluster variable.

8.4 Simulations on the COCO Dataset

COCO Dataset [10] is a large collection of natural images with segmentation and annotations provided for all objects in the scene. The complete Dataset (Release 2014) is composed by $123,287$ images representing 80 common objects resulting in a total of $886,284$ labeled object instances. We have trained our model using a subset of 3000 images and annotations randomly extracted from the complete COCO Dataset.

Since our objective is to evaluate the presence of the 80 categories in the image, we have represented each image with a vector of 80 observed category variables $\mathbf{X} = \{X_j\}_{j=1:80}$. Each observed variable X_j is related to a forward message $\mathbf{f}_{X_j}$ and to a backward message $\mathbf{b}_{X_j}$.

The Training Set of 3000 images, can be represented as a matrix $\mathbf{X}$ of dimension 3000×80 with the generic element, $\{X_{ij}\}_{i=1:3000}^{j=1:80}$, that can assume three possible values:

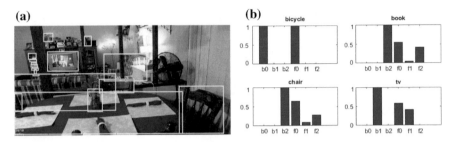

Fig. 8.3 **a** A scene from the Training Set; **b** Backward messages (b_0, b_1, b_2) injected in the network and forward messages (f_0, f_1, f_2) collected at the end of the inference process. Other 76 variables have not been shown for space limitations but most of the objects they represent are not present in the scene

$$\{X_{ij}\}_{i=1:3000}^{j=1:80} = \begin{cases} 0 & \text{object of category } j \text{ is absent in image } i, \\ 1 & \text{object of category } j \text{ is present only once in image } i, \\ 2 & \text{object of category } j \text{ is present more than once in image } i. \end{cases}$$

To define a Validation Set we have randomly extracted from the COCO Dataset 750 images not included in the Training Set and without duplicated patterns.

In probabilistic terms the generic observed variable X_j is a categorical variable that can assume values in a vocabulary $\mathcal{X} = \{\xi_0 = 0, \xi_1 = 1, \xi_2 = 2\}$. Through the injection of the backward messages into the network, we describe the evidence related to each variable, with sharp one-hot representation (i.e. if the j-th category is present once in the current image, we represent this situation as $\mathbf{b}_{X_j} = [b_{X_j}(\xi_0) = 0, b_{X_j}(\xi_1) = 1, b_{X_j}(\xi_2) = 0]$).

For example, considering the scene in Fig. 8.3a, since the object *chair* is present more than once, the backward message for the *chair* variable will be: $\mathbf{b}_{X_{chair}} = [0, 0, 1]$.

In the learning phase we have 3000 backward messages. All the other messages at the beginning of the learning process are initialized to uniform distributions. Also the forward message (prior) for the latent variable is set to be uniform. For N_e epochs these messages flow in the network and are used to learn the 80 SISO blocks in the system, following the rules described in [12, 15].

In Figs. 8.4a, b and c we show, respectively, the Likelihood, the Entropy and the Probability Error on the observed data when we vary the embedding space size on the Training and the Validation Set. These trends are useful to choose the embedding space size.

From Fig. 8.4a we can see that the Negative Log Likelihood on the Training Set and on the Validation Set decreases until the embedding space size of about 85. After that both curves become substantially stable even though on the Validation Set we note a slow increase.

The Entropy (Fig. 8.4b) gives us the information on the capability of the network to respond sharply for each variable, given all the others. It may be interpreted as

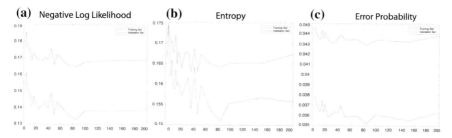

Fig. 8.4 **a** Likelihood, **b** Entropy, **c** Error Probability of the LVM for various embedding space dimension

the average uncertainty for all variables: low values should correspond to higher model accuracy. In our simulations, both on the Training Set and on Validation Set, when the embedding space increases, the Entropy decreases indicating that a higher Embedding Space improves the model performance. At the Embedding Space of 25 the Entropy on the Validation Set starts to oscillate and after 85 it increases indicating a reduction of the generalization power of the network.

The Error Probability (Fig. 8.4c) is estimated using the decision rule described before applied to backward and forward messages. On the Training Set, and the Validation Set the Error Probability decreases until the embedding space has size approximately 85.

8.4.1 Inference

In the following we set the embedding space size $|\mathcal{S}|$ to 30 and, after learning, we have inspected the information provided by the network. In Fig. 8.3b we can see the backward and forward messages related to some observed variables for the image of Fig. 8.3a (from the Training Set). For example, if we consider the variable *bicycle*, the backward message, representing the ground truth, tells us that the bicycle is absent in the scene. The forward message, that merges the information learned by the network, confirms this evidence. The same behavior is shown by other vehicle objects (not shown in the figure for space limitations) confirming that it is very unlikely to find these type of objects in an internal environment as a kitchen.

As another example, consider the variable *chair*. The backward message indicates that this object is present various times in the current scene. The forward message returned by our model says that it is very likely either find the chair more times, or not at all in this type of scene. Instead it seems to be less unlikely that this object is present only once. Another interesting situation is represented by the variable *tv* that can be present in this type of scene, according to the forward message, once or not at all, but it is completely unlikely that this object is present more than once. Finally, if we consider the variable *book*, the backward message indicates that this

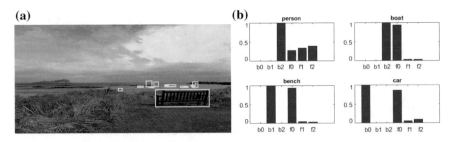

Fig. 8.5 **a** A scene from the Validation Set; **b** Some backward (b_0, b_1, b_2) and forward (f_0, f_1, f_2) messages after the inference process. Other 76 variables have not been shown for space limitations but most of the objects they represent are not present in the scene

object is present more than once in the current image. The model instead, that has seen all images of the Training Set, has built an internal representation of the reality that believes that, in this type of scene, the book can be absent or present more than once. It has learned that it is less likely that a book is present only once.

In Fig. 8.5 an image from the Validation Set is shown with some backward and forward messages. In particular, in this image there are two main objects: a bench in the foreground and some boats in the background, as we can see from the backward input messages. The small values for forward messages f_1 and f_2 for variables *bench* and *boat* suggest it is unlikely that these two objects coexist in the same context because in the Training Set this coexistence is rare.

8.4.2 Recall

If the backward messages for all observed variables are set to uniform except for some variables (evidence), we can perform a Bayesian Recall task, obtaining the prediction of the presence of other objects in the scene according to the provided evidence (fill-in). For example, for image in Fig. 8.6a we set all observed variables in our system to uniform except for variables: *chair*, *dining table* and *bottle* (for space limitations only the *dining table* is represented in Fig. 8.6b), obtaining the forward messages in Fig. 8.6b. In the context with the evidence injected, our model predicts that it is more likely to have more instances of a *person*, than having one or none. At the same time the model predicts the possible presence of forks and knives. Spoons and bowls are also predicted (not shown for space limitations).

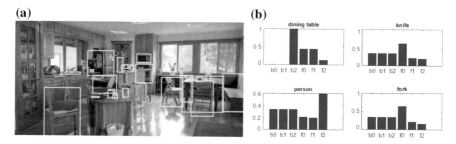

Fig. 8.6 a A scene from the Training Set; **b** Some backward messages (b_0, b_1, b_2) injected in the network and some forward messages (f_0, f_1, f_2) collected at the end of the inference process. Other 76 variables have not been shown for space limitations but most of the objects they represent are not present in the scene

8.4.3 Inspection of Learned Clusters

In order to inspect the learned representation and express the characteristics of the clusters present in the observed data, we set the backward messages for all observed variables to uniform and inject a delta distribution as forward message at the latent variable S.

In this experiment we show results for an embedding space dimension equal to 10. Figure 8.7 shows the forward messages for three values of S (three of the ten learned contexts). The first context (Fig. 8.7a) is clearly related to an indoor scene where there are one *person* or more, at least one *dinner table*, *fork* and *knife*. The second context (Fig. 8.7b) seems to be related to an outdoor sport activity with one *person* or more, *sport ball*, *tennis rackets* and *baseball ball*. The third context (Fig. 8.7c) is clearly related to a road scene, in fact there are more than one *person*, *cars* and the probability of the presence of *bus* and *traffic light* is not negligible.

8.4.4 Error Correction

The context analysis can be useful also for error correction in a pattern recognition task. Figure 8.8a shows an image where we simulate a wrong detection of a *dining table* and of a *car* in the considered scene. The LVM responds with a forward message that is proportional to the posterior probability on a detection given all the others (Fig. 8.8b). These values can confirm or modify the injected evidence based on the data observed during the Training Phase.

We can note that the network "corrects" the observation of the *dining table* because this object is not coherent with the context inferred using the other observations. Differently the observation of one *car* is a plausible detection: in fact the probability

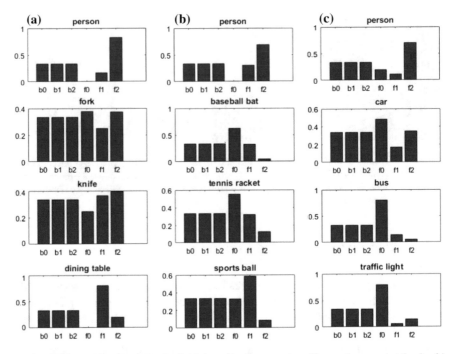

Fig. 8.7 Backward messages (b_0, b_1, b_2) injected in the network and forward messages (f_0, f_1, f_2) collected at the end of the inference process for three different context learned by the network

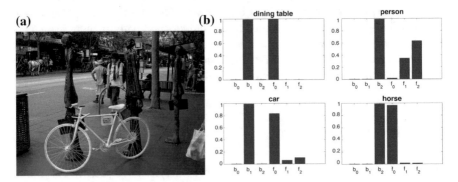

Fig. 8.8 **a** A scene from the Training Set; **b** Some backward (b_0, b_1, b_2) and posterior (f_0, f_1, f_2) messages after the inference process. Other 76 variables have not been shown for space limitations

of the presence of a car (or more than one), in similar context, is greater than zero. Most observations of the other variables are substantially confirmed by the posterior (not shown for space limitations) except for some observations that are not likely (e.g. more than one *horse*).

8.5 Conclusion

In this work we have used a Latent Variable Model in the framework of Factor Graphs in Reduced Normal Form for context analysis with experiments on a subset of the COCO Dataset. The experiments conducted here demonstrate that the LVM is a very flexible model that can manage the context variables in order to understand the scene depicted in an image. The extreme flexibility of the FGrn paradigm permits the fusion of the information coming from different situations and provides an embedded code for a scene understanding task. In future work more complex structures for automated scene captioning systems based on the contextual information will be investigated.

References

1. Oliva, A., Torralba, A.: The role of context in object recognition. Trends Cogn. Sci. **11**, 520–527 (2007)
2. Mottaghi, R., Chen, X., Liu, X., Cho, N.G., Lee, S.W., Fidler, S., Urtasun, R., Yuille, A.: The role of context for object detection and semantic segmentation in the wild. In: 2014 IEEE Conference on Computer Vision and Pattern Recognition, pp. 891–898 (2014)
3. Torralba, A.: Contextual priming for object detection. Int. J. Comput. Vis. **53**(2), 169–191 (2003)
4. Jin, Y., Geman, S.: Context and hierarchy in a probabilistic image model. In: 2006 IEEE Computer Society Conference on Computer Vision and Pattern Recognition (CVPR 2006), vol. 2, pp. 2145–2152 (2006)
5. Galleguillos, C., Rabinovich, A., Belongie, S.: Object categorization using co-occurrence, location and appearance. In: IEEE Conference on Computer Vision and Pattern Recognition, 2008. CVPR 2008. pp. 1–8. IEEE (2008)
6. Heitz, G., Koller, D.: Learning Spatial Context: Using Stuff to Find Things, pp. 30–43. Springer, Berlin Heidelberg (2008)
7. Yao, B., Li, F.: Recognizing human-object interactions in still images by modeling the mutual context of objects and human poses. IEEE Trans. Pattern Anal. Mach. Intell. **34**(9), 1691–1703 (2012)
8. Murphy, K.P.: Machine Learning: A Probabilistic Perspective (Adaptive Computation and Machine Learning series). The MIT Press (2012)
9. Choi, J.M., Torralba, A., Willsky, A.S.: A tree.based context model for object recognition. IEEE Trans **34**, 240–252 (2012)
10. Lin, T., Maire, M., Belongie, S.J., Bourdev, L.D., Girshick, R.B., Hays, J., Perona, P., Ramanan, D., Dollár, P., Zitnick, C.L.: Microsoft COCO: common objects in context. In: CoRR, vol. abs/1405.0312 (2014)
11. Loeliger, H.A.: An introduction to factor graphs. IEEE Signal Process. Mag. vol. 21, pp. 28–41 (2004)
12. Palmieri, F.A.N.: A comparison of algorithms for learning hidden variables in bayesian factor graphs in reduced normal form. IEEE Trans. Neural Netw. Learn. Syst. **27**, 2242–2255 (2016)
13. Buonanno, A., Palmieri, F.A.N.: Simulink implementation of belief propagation in normal factor graphs. In: Proceedings of the 24th Workshop on Neural Networks, WIRN 2014, May 15–16, Vietri sul Mare, Salerno, Italy (2014)
14. Buonanno, A., Palmieri, F.A.N.: Towards building deep networks with bayesian factor graphs (2015). http://arxiv.org/abs/1502.04492
15. Buonanno, A., Palmieri, F.A.N.: Two-dimensional multi-layer factor graphs in reduced normal form. In: Proceedings of the International Joint Conference on Neural Networks, IJCNN2015, July 12–17, 2015, Killarney, Ireland (2015)

16. Kschischang, F.R., Frey, B., Loeliger, H.: Factor graphs and the sum-product algorithm. IEEE Trans. Inform. Theory **47**, 498–519 (2001)
17. Koller, D., Friedman, N.: Probabilistic Graphical Models: Principles and Techniques. MIT Press (2009)
18. Palmieri, F.A.N., Buonanno, A.: Discrete independent component analysis (dica) with belief propagation. In: Proceedings of IEEE Machine Learning for Signal Procesing Conference, MLSP2015, Sept. 17–20, Boston, US (2015)
19. Buonanno, A., di Grazia, L., Palmieri, F.A.N.: Bayesian clustering on images with factor graphs in reduced normal form. In: Proceedings of the 25th Workshop on Neural Networks, WIRN 2015, May 20–22, Vietri sul Mare, Salerno, Italy (2015)
20. Cover, T.M., Thomas, J.A.: Elements of Information Theory. Wiley (2006)

Chapter 9
A Classification Approach to Modeling Financial Time Series

Rosa Altilio, Giorgio Andreasi and Massimo Panella

Abstract In this paper, several classification methods are applied for modeling financial time series with the aim to predict the trend of successive prices. By using a suitable embedding technique, a pattern of past prices is assigned a class if the variation of the next price is over, under or stable with respect to a given threshold. Furthermore, a sensitivity analysis is performed in order to verify if the value of such a threshold influences the prediction accuracy. The experimental results on the case study of WTI crude oil commodity show a good classification accuracy of the next (predicted) trend, and the best performance is achieved by the K-Nearest Neighbors classification strategy.

9.1 Introduction

Stock market prediction is one of the most challenging fields for time series forecasting. In Random Walk Theory [1], prices do not follow a trend and therefore their prediction is no worth of being considered. Nevertheless, in recent years a lot of forecasting methods have been developed. Two approaches are commonly used to analyze the movement of prices (i.e., up, down, hold, etc.) in order to predict future trends [2]. The first one is the Fundamental Analysis, which utilizes economic factors to estimate the intrinsic values of securities. The second one, known as Technical Analysis, is based on the principles of the Dow Theory [3] and it uses the history of prices to predict future movements, by considering the historical graph of prices and looking for recurrent structures. Recently, other methods based on computational

R. Altilio · G. Andreasi · M. Panella (✉)
Department of Information Engineering, Electronics and Telecommunications (DIET),
University of Rome "La Sapienza", Via Eudossiana 18, 00184 Rome, Italy
e-mail: massimo.panella@uniroma1.it

R. Altilio
e-mail: rosa.altilio@uniroma1.it

G. Andreasi
e-mail: giorgio.andreasi@gmail.com

© Springer International Publishing AG, part of Springer Nature 2019   97
A. Esposito et al. (eds.), *Neural Advances in Processing Nonlinear
Dynamic Signals*, Smart Innovation, Systems and Technologies 102,
https://doi.org/10.1007/978-3-319-95098-3_9

intelligence have been developed. These approaches have shown that it is possible to achieve a high accuracy for prediction in spite of the early results obtained by Random Walk Theory.

The main problem for financial forecasting is that the majority of economic behaviors are nonlinear [4] and hence, if we use a linear model we cannot capture all the information within the time series. Traditional models based on statistical assumptions, as Autoregressive Integrated Moving Average (ARIMA) [5] and Generalized Autoregressive Conditional Heteroskedasticity (GARCH) [6], have limitations in estimating the underlying structure of data due to the complexity that a real system presents. Therefore, in recent years many machine learning models [7], as Artificial Neural Networks (ANNs) [8], Gaussian Mixture Models [9], Support Vector Machine (SVM) [10], and Genetic Algorithms (GAs) [11], have been proposed for modeling financial time series.

Generally speaking, the aim of financial forecasting is to make a punctual prediction, that is to look for the next day's closing value. For this purpose, the proposed methods can achieve good results in terms, for example, of Mean Squared Error (MSE), Normalized Mean Squared Error (NMSE), Noise-to-Signal Ratio (NSR), Mean Absolute Percentage Error (MAPE) [12–14], but they have a problem: the correspondence on the price movement up or down of the predicted time series with respect to the actual one is not assured.

Instead, in a lot of real applications it is more important to know the trend of stock prices rather than the punctual value. This approach can be formulated as a classification problem [15]. Some introductory papers in this regard have been presented many years ago, with no effective results. In [16], four features of the NASDAQ-100 Technology Sector Index are considered (i.e., stock price volatility, stock momentum, index volatility and index momentum), in order to predict a label that represents if the stock's price increases or decreases. The SVM technique is used for classification with the best accuracy of 61% approximately. In [17], the Dow Jones Industrial Average Index is predicted and it is found out that some periods are more predictable than others. In particular, the Hurst exponent is used to select the most predictability period. ANNs, K-Nearest Neighbor (KNN), decision trees, and ensemble methods are used to solve the classification problem. The best result achieves 65% in terms of accuracy. In [18], twelve technical indicators are used to predict the label of Korea composite stock price index (KOSPI) with a binary classification for next day's direction. An SVM classifier is adopted with an accuracy in classification of 57%.

In this paper, a more general classification approach is proposed by using enhanced models for time series forecasting based on embedding theory. The aim is to predict if the stock's price increases, decreases or remains stationary in a certain threshold for the next trading day. The proposed approach is applied to a case study pertaining to the WTI crude oil energy commodity. In fact, energy is the main factor for every economic market [19]. Energy price dynamics is affected by complex risk factors, such as political events, extreme weather conditions, and financial market behavior. Crude oil is the key component for the economic development as well as the growth of industrialized and developing countries. In this market, predictions are fundamental for covering commodity's risk because the commodity price impacts on company and country profits.

This contribution is organized as follows. In Sect. 9.2, the methodology of the proposed predictive classification is introduced, while in Sect. 9.3 the experiments and the empirical results are summarized and discussed together with a sensitivity analysis on the algorithm parameters. Finally, our conclusions are drawn in Sect. 9.4.

9.2 Proposed Methodology

The main goal of forecasting financial time series is to estimate the closing price of the next trading day. In this work, we will focus on a trend classification rather than on a punctual, numerical prediction; thus, we will consider only if the stock price increases, decreases or it holds. A preliminary step, which is mandatory to follow the said approach, is to build a dataset suited to a classification problem and hence, to associate past observations of a time series with a symbolic label describing the next trend to be predicted.

Let C_k, $k \geq 1$, be the sample of a time series representing the closing price of the k-th day. The trend of each day is associated with a symbolic label in the set $\mathcal{L} = \{up,\ hold,\ down\}$, which is determined by comparing the current closing price to the one of the previous day. The class L_k of the k-th day is obtained by the following comparisons:

- Class *up*:

$$\frac{C_k - C_{k-1}}{|C_{k-1}|} > \varepsilon \tag{9.1}$$

- Class *hold*:

$$-\varepsilon \leq \frac{C_k - C_{k-1}}{|C_{k-1}|} \leq \varepsilon \tag{9.2}$$

- Class *down*:

$$\frac{C_k - C_{k-1}}{|C_{k-1}|} < -\varepsilon \tag{9.3}$$

The trend's prediction of the k-th day can be based on past observations of the closing prices. Usually, this is obtained by the so-called 'embedding technique' by which the vector $\underline{C}_k$ of past samples is defined as:

$$\underline{C}_k = \begin{bmatrix} C_{k-(D-1)T} & C_{k-(D-2)T} & \cdots & C_{k-T} & C_k \end{bmatrix}, \tag{9.4}$$

where D is the embedding dimension and T is the time lag between samples [20]. In the following, D and T will be chosen by a rule-of-thumb as successively specified.

Given the embedded vector of past samples, to complete the predictive model of the trend behavior it is mandatory to determine a classifier $f(\cdot)$ that represents the link between the embedded vector at day k and the successive trend label at day $k + 1$, that is $L_{k+1} = f(\underline{C}_k)$. As financial time series usually have a complex behavior, which is often noisy and even chaotic, the classifier model must be based on machine learning techniques by which the non-linear and/or non-stationary nature of time series is captured by learning the model parameters using data driven techniques. To this end, a training set $\mathscr{D}$ based on available observations is arranged. Let K be the last available day for which the closing price is available before starting the training procedure, then we have:

$$\mathscr{D} = \begin{pmatrix} C_{K-n-(D-1)T} & \cdots & C_{K-n-T} & C_{K-n} & L_{K-n+1} \\ C_{K-n+1-(D-1)T} & \cdots & C_{K-n+1-T} & C_{K-n+1} & L_{K-n+2} \\ \vdots & & \vdots & \vdots & \vdots \\ C_{K-2-(D-1)T} & \cdots & C_{K-2-T} & C_{K-2} & L_{K-1} \\ C_{K-1-(D-1)T} & \cdots & C_{K-1-T} & C_{K-1} & L_{K} \end{pmatrix} \tag{9.5}$$

Such a training set is made of n observations arranged by rows. In order to determine how many rows are available (or may be used) at day K, we need to satisfy the following boundary condition:

$$K - n - (D - 1)T \geq 1 \quad \Rightarrow \quad n \leq K - 1 - (D - 1)T . \tag{9.6}$$

An illustrative example regarding the embedding procedure and label association is shown in Fig. 9.1, by using $D = 3$, $T = 1$, and $\varepsilon = 0.01$. Considering at day 6 the embedded vector:

$$\underline{C}_6 = [77.23 \quad 76.98 \quad 73.14] ,$$

the relative variation from day 6 to day 7 is:

$$\frac{71.19 - 73.14}{73.14} = -0.0267 < -\varepsilon ,$$

and so the trend label of day 7 is $L_7 = down$.

9.3 Experimental Results

In this paper, we consider data of the WTI time series relevant to the year 2010 and available at www.investing.com. A visual representation of the behavior is shown in Fig. 9.2, where 254 trading days are considered. In particular, closing prices are used as inputs to predict the label for the next day.

The numerical simulations are implemented using MATLAB R2015b on a Macbook Pro 2.7 GHz, Intel core i5 and 8 GB of RAM. Data are preprocessed on the basis of the embedding parameters and tests are done with different values of

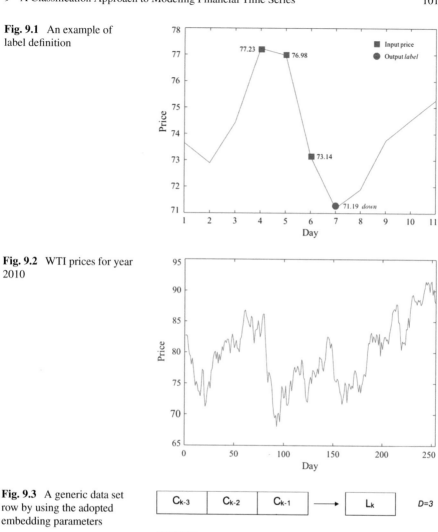

Fig. 9.1 An example of label definition

Fig. 9.2 WTI prices for year 2010

Fig. 9.3 A generic data set row by using the adopted embedding parameters

$D = \{3, 5, 10\}$, while keeping fixed $T = 1$ as a rule-of-thumb. The chosen values of D represent three days, one week and two weeks as input, respectively. An idea of how input data are organized to predict the label for the next day is shown in Fig. 9.3.

As previously explained, our aim is to predict the trend of the time series and this is formulated as a classification problem. Six classification models are adopted in this regard: Linear Discriminant Analysis (LDA); Quadratic Discriminant Analysis

Table 9.1 Classification accuracy (%) for LDA, QDA, KNN

ε (%)	LDA			QDA			KNN		
	D = 3	D = 5	D = 10	D = 3	D = 5	D = 10	D = 3	D = 5	D = 10
0.5	37.40	35.04	41.73	37.01	34.65	34.65	35.43	35.43	40.16
0.8	37.01	35.04	37.40	38.58	34.25	35.83	31.50	35.43	35.43
1.0	37.40	36.61	35.83	35.83	34.25	37.80	44.09	39.76	39.76
1.2	41.54	39.76	38.98	38.98	36.61	39.37	51.18	50.39	43.31
1.5	44.88	42.13	42.91	42.13	40.55	47.24	50.79	55.51	51.97
2.0	48.43	49.61	48.03	50.39	48.03	57.87	72.05	68.50	70.87
2.5	49.61	51.97	53.54	53.94	56.69	67.72	83.07	82.68	85.43

Table 9.2 Classification accuracy (%) for Naive Bayes, SVM, CART

ε (%)	Naive Bayes			SVM			CART		
	D = 3	D = 5	D = 10	D = 3	D = 5	D = 10	D = 3	D = 5	D = 10
0.5	40.55	37.80	36.22	39.37	34.25	36.61	38.98	36.22	41.74
0.8	36.22	35.43	37.40	36.61	32.28	34.25	34.25	34.65	29.13
1.0	42.52	31.89	36.61	36.61	33.46	37.40	38.58	36.22	37.01
1.2	51.18	44.88	39.19	41.74	39.76	35.83	45.67	41.73	38.58
1.5	57.48	56.69	48.82	40.16	38.58	42.91	46.85	45.67	48.43
2.0	74.41	69.69	62.20	44.09	46.85	54.33	63.78	64.17	59.84
2.5	84.25	71.65	62.20	48.43	49.61	61.02	81.89	74.02	79.53

(QDA); KNN; Naive Bayes; SVM; Classification and Regression Tree (CART). The results are evaluated through a confusion matrix and the classification accuracy. The confusion matrix is a matrix where rows represent the real labels while columns the predicted ones; the correct predictions are on the main diagonal. Numerical results are evaluated also in terms of classification accuracy (A) that, in respect to the elements a_{ij} of the confusion matrix, is given by the percentage ratio between the number of correct predictions and the total observations:

$$A = \frac{a_{11} + a_{22} + a_{33}}{\sum_{i=1}^{3} \sum_{j=1}^{3} a_{ij}} \times 100. \tag{9.7}$$

The numerical results are summarized in Tables 9.1 and 9.2 and a sensitivity analysis has been realized in order to ascertain how the value of ε influences the accuracy of prediction. It is evident that all of the classifiers obtain better performances as ε increases. A visual representation of this concept is given in Fig. 9.4, where we report the results for $D = 3$ only, as a similar behavior is obtained for other values of the embedding parameters. As it can be seen, all of the classifiers report an upward trend with ε increasing, with a knee above $\varepsilon = 1.5\%$.

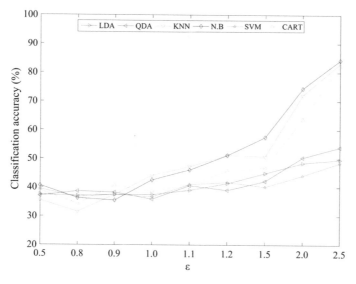

Fig. 9.4 Trend accuracy versus ε using $D = 3$

	$\tilde{u}$	$\tilde{h}$	$\tilde{d}$
u	0	17	1
h	1	216	0
d	2	16	1

(a) $\varepsilon = 2.5$

	$\tilde{u}$	$\tilde{h}$	$\tilde{d}$
u	38	10	8
h	14	108	12
d	6	15	43

(b) $\varepsilon = 2.0$

	$\tilde{u}$	$\tilde{h}$	$\tilde{d}$
u	34	17	12
h	20	71	23
d	11	25	41

(c) $\varepsilon = 1.5$

Fig. 9.5 Best confusion matrix obtained at different values of ε

The best prediction accuracy of 85.43% is obtained by the KNN classifier with $D = 10$ and $\varepsilon - 2.5\%$, the related confusion matrix is shown in Fig. 9.5a: the class predicted more accurately is *hold*, with 216 predicted labels on 217 real *hold* labels. However, this performance is obtained on an unbalanced dataset, mostly containing the class *hold* because of the relatively high range associated with the largest value of ε in (9.2). As previously discussed, the performances are adequate also for $\varepsilon = 2.0\%$, for which the best classifier is Naive Bayes with $D = 3$ and an accuracy of 74.41%. In such a case, considering the confusion matrix shown in Fig. 9.5b, the dataset is much more balanced with respect to the previous one and still achieving a good accuracy. The performances get worse rapidly for smaller values of the considered threshold; at $\varepsilon = 1.5\%$, the best classifier is still Naive Bayes, with $D = 3$ but scoring a classification accuracy of 57.48%. The related confusion matrix, reported in

Fig. 9.6 Box plot sensitivity
analysis

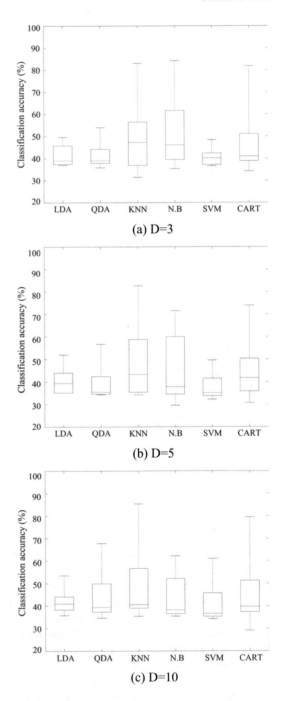

(a) D=3

(b) D=5

(c) D=10

Table 9.3 WTI classification with $\varepsilon = 2.5\%$

D	LSE (%)	LDA (%)	QDA (%)	KNN (%)	N-B (%)	SVM (%)	CART (%)
3	85.43	49.61	53.94	83.07	84.25	48.43	81.89
5	85.43	51.97	56.69	82.68	71.65	49.61	74.02
10	85.43	53.54	67.72	85.43	62.20	61.02	79.53

Fig. 9.5c, evidences an even more spread dataset. Nonetheless, the number of 'false positives' (i.e., the worst situation where a real *down* is predicted as *up*) remains limited in all almost all of the considered situations.

Finally, the results are summarized in terms of box plots using different values of D. As shown in Fig. 9.6, in all cases the classifier with the best accuracy is the KNN one. It achieves the highest box plot, but also the more spread due to the influence of ε. In order to compare these results with a reference baseline, a linear predictor whose parameters are determined by standard Least Squares Estimation (LSE) is considered. The threshold ε is set to 2.5% and the results are summarized in Table 9.3. Overall, LSE has a better performance for a reduced information taken into account within the embedded dataset, that is for $D = 3$ and $D = 5$, while for $D = 10$ the KNN classifier is worthy of being considered.

9.4 Conclusion

In this paper, a novel approach has been proposed to modeling the behavior of financial time series, with the aim to predict a label for the next trading day. It is formulated as a classification problem, where class *up* occurs when the price increases over a threshold, *down* when the price decreases under the same threshold and *hold* when the price remains stationary.

The results show that there are not significant differences among different choices of the embedding parameters, but the sensitivity analysis shows that the threshold's value influences the prediction accuracy. All of the classifiers perform better with a high value of the threshold. An accuracy of 74.41% on a balanced dataset is achieved at $\varepsilon = 2.0\%$ by using the Naive Bayes classifier with $D = 3$ and $T = 1$.

The use of complex classification models based on machine learning techniques, and taking into account the distributed nature of financial data sources across the markets worldwide [21, 22], should be further investigated in future research works. It is possible to improve the desired performance also by using more suited embedding techniques. Actually, the performances are similar to the ones obtained by a linear classification model, although they are fully adequate in respect to the goal of the approach proposed in this paper.

References

1. Fama, E.F.: Random walks in stock market prices. Financ. Anal. J. **51**(1), 75–80 (1995)
2. Teixeira, L.A., De Oliveira, A.L.I.: A method for automatic stock trading combining technical analysis and nearest neighbor classification. Expert Syst. Appl. **37**(10), 6885–6890 (2010)
3. Schannep, J.: Dow Theory for the 21st Century. Wiley, USA (2008)
4. Clements, M.P., Franses, P.H., Swanson, N.R.: Forecasting economic and financial time-series with non-linear models. Int. J. Forecast. **20**(2), 169–183 (2004)
5. Zhang, G.P.: Time series forecasting using a hybrid ARIMA and neural network model. Neurocomputing **50**, 159–175 (2003)
6. Garcia, R.C., Contreras, J., Van Akkeren, M., Garcia, J.B.C.: A GARCH forecasting model to predict day-ahead electricity prices. IEEE Trans. Power Syst. **20**(2), 867–874 (2005)
7. Zhang, G., Patuwo, B.E., Hu, M.Y.: Forecasting with artificial neural networks: the state of the art. Int. J. Forecast. **14**(1), 35–62 (1998)
8. Kaastra, I., Boyd, M.: Designing a neural network for forecasting financial and economic time series. Neurocomputing **10**(3), 215–236 (1996)
9. Panella, M., Rizzi, A., Martinelli, G.: Refining accuracy of environmental data prediction by MoG neural networks. Neurocomputing **55**(3–4), 521–549 (2003)
10. Tay, F.E., Cao, L.: Application of support vector machines in financial time series forecasting. Omega **29**(4), 309–317 (2001)
11. Haupt, R.L., Haupt, S.E.: Practical Genetic Algorithms. Wiley (2004)
12. van de Geer, S.: Least squares estimation with complexity penalties. Math. Methods Stat. **10**(3), 355–374 (2001)
13. Rosato, A., Altilio, R., Araneo, R., Panella, M.: Embedding of time series for the prediction in photovoltaic power plants. In: Proceedings of International Conference on Environment and Electrical Engineering (EEEIC 2016), pp. 1–4. IEEE (2016)
14. Proietti, A., Panella, M., Leccese, F., Svezia, E.: Dust detection and analysis in museum environment based on pattern recognition. Measurement **66**, 62–72 (2015)
15. Rizzi, A., Buccino, N.M., Panella, M., Uncini, A.: Genre classification of compressed audio data. In: Proceedings of Workshop on Multimedia Signal Processing (MMSP 2008), pp. 654–659. IEEE (2008)
16. Madge, S.: Predicting stock price direction using support vector machines. Independent Work Report Spring 2015 (2015)
17. Qian, B., Rasheed, K.: Stock market prediction with multiple classifiers. Appl. Intell. **26**(1), 25–33 (2007)
18. Kim, K.J.: Financial time series forecasting using support vector machines. Neurocomputing **55**(1), 307–319 (2003)
19. Panella, M., Barcellona, F., D'ecclesia, R.L.: Forecasting energy commodity prices using neural networks. Adv. Decis. Sci. **2012**, 1–26 (2012)
20. Abarbanel, H.: Analysis of Observed Chaotic Data. Springer Science & Business Media (2012)
21. Scardapane, S., Fierimonte, R., Di Lorenzo, P., Panella, M., Uncini, A.: Distributed semi-supervised support vector machines. Neural Netw. **80**, 43–52 (2016)
22. Rosato A., Altilio R., Panella M.: Recent advances on distributed unsupervised learning. In: Bassis S., Esposito A., Morabito F., Pasero E. (eds.) Advances in Neural Networks. Smart Innovation, Systems and Technologies, vol. 54, pp. 77–86 (2016)

Chapter 10
A Low-Complexity Linear-in-the-Parameters Nonlinear Filter for Distorted Speech Signals

Danilo Comminiello, Michele Scarpiniti, Simone Scardapane, Raffaele Parisi and Aurelio Uncini

Abstract In this paper, the problem of the online modeling of nonlinear speech signals is addressed. In particular, the goal of this work is to provide a nonlinear model yielding the best tradeoff between performance results and required computational resources. Functional link adaptive filters were proved to be an effective model for this problem, providing the best performance when trigonometric expansion is used as a nonlinear transformation. Here, a different functional expansion is adopted based on the Chebyshev polynomials in order to reduce the overall computational complexity of the model, while achieving good results in terms of perceived quality of processed speech. The proposed model is assessed in the presence of nonlinearities for both simulated and real speech signals.

Keywords Nonlinear modeling · Functional links · Chebyshev polynomials
Loudspeaker distortions · Nonlinear system identification

10.1 Introduction

In the recent years, a widespread availability of commercial hands-free speech communication systems has occurred, also due to the development of immersive speech communication techniques [4, 7]. However, such devices often mount low-cost components, which may affect the quality of the perceived speech. In particular, poor-quality loudspeakers, vibrations of plastic shells, D/A converters and power amplifiers may introduce a significant amount of nonlinearity in speech signals, especially during large signal peaks.

In online learning applications related to hands-free speech communications, such as nonlinear acoustic echo cancellation (NAEC) and active noise control (ANC), linear-in-the-parameters (LIP) nonlinear filters represent an effective and flexible

D. Comminiello (✉) · M. Scarpiniti · S. Scardapane · R. Parisi · A. Uncini
Department of Information Engineering, Electronics and Telecommunications (DIET),
"Sapienza" University of Rome, Via Eudossiana 18, 00184 Rome, Italy
e-mail: danilo.comminiello@uniroma1.it

© Springer International Publishing AG, part of Springer Nature 2019
A. Esposito et al. (eds.), *Neural Advances in Processing Nonlinear Dynamic Signals*, Smart Innovation, Systems and Technologies 102,
https://doi.org/10.1007/978-3-319-95098-3_10

solution [6, 8–10, 12, 18]. However, the modeling and compensations by LIP non-linear filters may require a large number of nonlinear elements, which involve a high computational load that may represent a problem of real-time applications like NAEC.

In order to address this problem, in this paper, we propose a LIP nonlinear filter that provides the best tradeoff between performance results and required computational resources. In particular, we take into account the nonlinear *functional link adaptive filters* (FLAFs) [6], which is based on a nonlinear expansion of the input signal by the so-called *functional links* [11, 13, 16], and an adaptive filtering of the transformed signal in cascade.

One of the most important advantages of the FLAF is its flexibility, since it is possible to set the different parameters of the FLAF individually in order to fit the model at best for a specific application. In the design of an FLAF, an important choice is the number of functional links to be adopted in the model. This choice is strictly related to the nonlinearity degree introduced by the unknown system and with the chosen type of functional expansion. Therefore, in order to reduce the computational complexity we directly aim at designing a suitable and efficient *functional expansion block* to be used for the modeling of nonlinear speech signals.

In particular, *Chebyshev functional links* are assessed within NAEC problems and compared with other classic functional expansions. Performance is evaluated in terms of both error-based criteria and speech quality measures, while considering the minimum possible computational load. Results are achieved over both simulated and real data and show the effectiveness of Chebyshev functional links to be used for a low-complexity FLAF model.

The paper is organized as follows: the FLAF-based model for the modeling of speech signals is introduced in Sect. 10.2, while Chebyshev functional links and their properties are described in Sect. 10.3. Results are discussed in Sect. 10.4 and, finally, in Sect. 10.5 our conclusions are drawn.

10.2 A Functional Link-Based Nonlinear Model for NAEC

The FLAF model is purely nonlinear, since the adaptive filter receives as input a trans-formed nonlinear signal. However, very often in acoustic speech signal processing, there is also a linear component to be modeled, as in the case of the presence of an acoustic impulse response in NAEC. To this end, we adopt a filtering scheme based on the FLAF that includes both linear and nonlinear filtering, called *split functional link adaptive filter* (SFLAF) [6].

The SFLAF architecture, depicted in Fig. 10.1, involves a linear branch and a nonlinear branch in parallel. The former is nothing but a linear adaptive filter totally aiming at modeling the linear components of the system to be identified. On the other hand, the nonlinear branch is a nonlinear FLAF. The output signal of the SFLAF is obtained from the sum of the outputs of the two parallel branches:

Fig. 10.1 The split
functional link adaptive filter

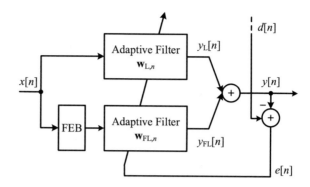

$$y[n] = y_L[n] + y_{FL}[n] = \mathbf{x}_n^T \mathbf{w}_{L,n-1} + \mathbf{g}_n^T \mathbf{w}_{FL,n-1}. \tag{10.1}$$

The error signal is then obtained as:

$$e[n] = d[n] - y[n] \tag{10.2}$$

where $d[n]$ is the desired signal that may includes any background noise. The
error signal (10.2) is used to adapt both the adaptive filters. In (10.1), $\mathbf{x}_n \in \mathbb{R}^M = \left[x[n]\ x[n-1] \ldots x[n-M+1] \right]^T$ is the input to the filter on the linear branch,
with M being the length of the input vector. Also, in (10.1), the vector $\mathbf{g}_n \in \mathbb{R}^{M_e} = \left[g_0[n]\ g_1[n] \ldots g_{M_e-1}[n] \right]^T$ is the *expanded buffer*, i.e., the output of the *functional
expansion block* (FEB) whose length is $M_e \geq M_i$.

Both the adaptive filters $\mathbf{w}_{L,n}$ and $\mathbf{w}_{FL,n}$ in (10.1) can be updated by using any
linear adaptive algorithm. Here, we use a *normalized least-mean square* (NLMS)
algorithm [17], so that:

$$\mathbf{w}_{L,n} = \mathbf{w}_{L,n-1} + \mu_L \frac{\mathbf{x}_n e[n]}{\mathbf{x}_n^T \mathbf{x}_n + \delta} \tag{10.3}$$

$$\mathbf{w}_{FL,n} = \mathbf{w}_{FL,n-1} + \mu_{FL} \frac{\mathbf{g}_n e[n]}{\mathbf{g}_n^T \mathbf{g}_n + \delta} \tag{10.4}$$

where μ_L and μ_{FL} are the step-size parameters and δ is a regularization factor.

10.3 Chebyshev Functional Link Expansion

10.3.1 Functional Expansion Block

One of the most important part in the FLAF-based model is the FEB, which contains a series of functions satisfying universal approximation properties. Such functions, called "functional links", are collected in a chosen set $\Phi = \{\varphi_0(\cdot),$ $\varphi_1(\cdot), \ldots, \varphi_{Q-1}(\cdot)\}$, with Q being the number of functional links. The FEB receives the first $M_i \leq M$ samples of $\mathbf{x}_n$, which are transformed and expanded in a higher-dimensional space by the chosen set of functional links, thus yielding the nonlinear expanded buffer $\mathbf{g}_n$:

$$g[n] = \varphi_0(x[n])$$

$$\vdots$$

$$g[n - Q + 1] = \varphi_{Q-1}(x[n])$$

$$\vdots$$

$$g[n - M_e + 1] = \varphi_{Q-1}(x[n - M_i + 1])$$

where P is the *order* of the functional link.

10.3.2 Chebyshev Polynomial Expansion

The chosen set of functional links must satisfy the universal approximation properties, and it can be a subset of orthogonal polynomials, such as Chebyshev, Legendre, Laguerre and trigonometric polynomials [2, 5, 13, 19] or just approximating functions, such as sigmoid functions [11, 15]. Among such functional expansions, *trigonometric polynomials* represent one of the most popular expansions, especially for applications involving audio and speech input signals [6, 16], since at best of their capabilities they provide the best performance results [5]. However, in this paper we focus on Chebyshev polynomial expansion to reduce the computational load.

Chebyshev polynomials are widely used in different fields of application due to their powerful nonlinear approximation capabilities. These properties were proved in [13, 19] within an artificial neural network (ANN), which also shows faster convergence than a multi-layer perceptron (MLP) network. Chebyshev polynomials involve functions of previously computed functions, thus increasing their effectiveness in dynamic problems. Moreover, being derived from a power series expansion, Chebyshev functional links may approximate a nonlinear function with a very small error near the point of expansion. On the other hand, the drawback is that, far from the point of expansion, the error often increases rapidly. Compared with other power series, Chebyshev polynomials show lower computational complexity and higher efficiency, when the polynomial order is rather low.

Considering the i-th input $x[n-i]$ of the nonlinear buffer, with $i = 0, \ldots, M_i$, the Chebyshev polynomial expansion can be expressed as:

$$\varphi_j(x[n-i]) = 2x[n-i]\varphi_{j-1}(x[n-i]) - \varphi_{j-2}(x[n-i]) \qquad (10.5)$$

for $j = 0, \ldots, P-1$. It can be noted from (10.5) that the number of Chebyshev functional links is equal to the expansion order, i.e., $Q = P$. Initial values in (10.5) (i.e., for $j = 0$) are:

$$\varphi_{-1}(x[n-i]) = x[n-i]$$
$$\varphi_{-2}(x[n-i]) = 1. \qquad (10.6)$$

10.3.3 Properties of Chebyshev Polynomials

Chebyshev polynomials are endowed with some interesting properties [3]. They are orthogonal in $\mathbb{R}_1$ with respect to the a weighting function $1/\pi\sqrt{1-x^2[n-i]}$:

$$\int_{-1}^{1} \varphi_j(x[n-i])\,\varphi_k(x[n-i]) \frac{1}{\pi\sqrt{1-x^2[n-i]}} dx = \begin{cases} 0, & j \neq k \\ 1, & j = k = 0 \\ 1/2, & j = k \neq 0 \end{cases}. \qquad (10.7)$$

For any $x[n-i] \in \mathbb{R}_1$, also $\varphi_j(x[n-i]) \in \mathbb{R}_1$, with values comprises in the range $[-1, 1]$. Therefore, $\varphi_j(x[n-i])$ are *equiripple functions* in $\mathbb{R}_1$.

Moreover, any polynomial with order P, $p(x[n-i])$, can be derived as a *linear combination* of Chebyshev polynomials [3]:

$$p(x[n-i]) = \sum_{i=0}^{P-1} c_j\varphi_j(x[n-i]). \qquad (10.8)$$

The last property is important since a linear combination of Chebyshev polynomials can arbitrarily well approximate any real continuous function $f(x[n-i])$. This can be proved via the *Stone-Weierstrass theorem* [14], as shown in [3].

Moreover, the approximation of a continuous function $f(x[n-i])$ with a linear combination of Chebyshev polynomials, $p(x[n-i])$, up to a degree P is very close to a min-max approximation [3]. Indeed, the approximation error is:

$$\epsilon[n-i] = f(x[n-i]) - p(x[n-i]) = \sum_{j=P}^{+\inf} c_j\varphi_j(x[n-i]). \qquad (10.9)$$

Table 10.1 Computational cost comparison of different functional link expansions in terms of multiplications

Expansion type	No. multiplications
Chebyshev Polynomial Expansion	$2M_e$
Legendre Polynomial Expansion	$4M_e$
Trigonometric Series Expansion	$M_e/2 + P$

For a continuous and differentiable function, the coefficients c_j converge to 0 rapidly, and therefore, $\epsilon[n-i] \approx c_P \varphi_P (x[n-i])$, which corresponds to an equiripple function.

Some of the above properties are proved in [3] for Chebyshev polynomials.

10.3.4 Analysis of the Computational Complexity

We briefly report the computational complexity of the Chebyshev functional link expansions with respect to other standard expansions, like trigonometric and Legendre series expansions. In order to provide a fair view of the computational resources required by the expansions, we do not consider additional cost of the SFLAF structure but we focus only on the operations made by the FEB.

The Chebyshev functional link expansion in (10.5) involves for each iteration $2PM_i$ multiplications and PM_i additions. Similarly, the complexity of Legendre and trigonometric functional link expansions is derived in [5]. In terms of the expanded buffer length, we can consider that $M_e = PM_i$ for Chebyshev and Legendre functional link expansions and $M_e = 2PM_i$ for trigonometric expansion. A comparison of the computational complexity, in terms of multiplications only, is summarized in Table 10.1. If we fix the expanded buffer length M_e and we consider that $P << M_e$, then it is easy to note that the trigonometric expansion involves the smallest number of multiplications. Therefore, in order to achieve the best tradeoff between performance and complexity using Chebyshev functional link, we necessary need to obtain superior performance than trigonometric functional links. As an alternative, we should try to obtain the same performance of trigonometric functional links but with a smaller number of nonlinear elements.

10.4 Experimental Results

We assess the proposed Chebyshev SFLAF within NAEC scenarios, comparing results with those obtained by trigonometric and Legendre series expansions. For each experiment we show the best possible SFLAF configuration, in terms of the chosen parameters, yielding the optimal tradeoff between performance and computational complexity.

We evaluate the performance results in terms of the *echo return loss enhancement* (ERLE), which describes the amount of echo canceled by the microphone signal, and it is defined as:

$$\text{ERLE}[n] = 10 \log_{10} \left(\frac{\text{E}\{d^2[n]\}}{\text{E}\{e^2[n]\}} \right) \tag{10.10}$$

The ERLE denotes how much echo signal is canceled, but this does not always correspond to a real signal enhancement in terms of perceived quality. Therefore, besides using the ERLE, we also consider another quality measure suitably designed for speech signals that may denote how "well" an SFLAF model produces a reliable estimate of the echo signal. In particular, we consider one of the most used objective measures for the speech quality evaluation that is the *perceptual evaluation of speech quality* (PESQ) [1], which estimates the overall loudness difference between the original signal and its estimation. Such signals are equalized to a reference listening level and then processed by a filter having a similar response to a standard telephone handset. An auditory transform is then applied to obtain the loudness spectra. The loudness difference between the two signals is averaged over time and frequency in order to achieve a prediction of subjective quality rating [5]. The PESQ score may be comprise in the range [1.0, 4.5], where 4.5 indicates the best possible quality.

10.4.1 Simulated NAEC Scenario

The first experiment is conducted in a simulated teleconferencing environment with reverberation time of $T_{60} \approx 150$ ms, in which the acoustic impulse response between the loudspeaker and the microphone is measured at 8 kHz sampling rate. A desktop computer equipped with an i3 CPU at 3.07 GHz is used for simulations. Female speech signal is used as input. Additive Gaussian noise is considered at the microphone signal, with 20 dB of *signal-to-noise ratio* (SNR). The simulated distortion applied to the female speech is a symmetrical soft-clipping nonlinearity, aiming at simulating a classic loudspeaker saturation effect, described by [5]:

$$\bar{x}[n] = \begin{cases} \frac{2}{3\zeta}x[n] & \text{for } 0 \le |x[n]| \le \zeta \\ \text{sign}(x[n]) \frac{3-(2-|x[n]|/\zeta)^2}{3} & \text{for } \zeta \le |x[n]| \le 2\zeta \\ \text{sign}(x[n]) & \text{for } 2\zeta \le |x[n]| \le 1 \end{cases} \tag{10.11}$$

where the clipping threshold $0 < \zeta \le 0.5$ determines the nonlinearity level. Here, we consider a strong distortion, provided by using $\zeta = 0.1$ in (10.11).

We set the step sizes at $\mu_L = \mu_{FL} = 0.5$, and $M_{ri} = M$ for all the SFLAF but we use the minimum possible expansion order for Chebyshev, trigonometric and Legendre series such that results can be comparable in terms of the ERLE.

Fig. 10.2 Performance comparison in terms of ERLE between SFLAFs with different expansions for speech input affected by a soft-clipping nonlinearity

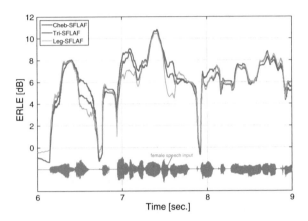

Table 10.2 Performance comparison in terms of PESQ and processing time between SFLAFs with different expansions for speech input affected by a soft-clipping nonlinearity

SFLAF type	M_e	PESQ	Sec.
Chebyshev SFLAF	$PM_i = 600$	3.765	5.339
Legendre SFLAF	$PM_i = 1200$	2.868	7.876
Trigonometric SFLAF	$PM_i = 1200$	3.582	5.672

In particular, we have chosen $P = 2$ for both Chebyshev and trigonometric SFLAF and $P = 4$ for Legendre SFLAF. Such results are shown in Fig. 10.2 where it is possible to see that Chebyshev SFLAF provides the best performance keeping the complexity contained. For better readability of the figures, we show a window of 3 out of 10 s of the ERLE behavior. This result is more evident by evaluating the quality measures in terms of the PESQ, which are reported in Table 10.2, where it can be also seen that Chebyshev SFLAF achieves the best PESQ score, while adapting the lowest number of nonlinear elements and, thus, involving the lowest computational time.

10.4.2   Real NAEC Scenario

In a second experiment, we evaluate the performance of the proposed method on real data from a classic scenario of acoustic echo cancellation, i.e. a hands-free desktop teleconference. For this experiment we consider a typical office room with a relatively low level of background noise, which guarantees sufficiently high signal-to-noise ratio (SNR). In this way it is possible to evaluate the proposed canceller fairly, thus avoiding external interferences that could require further processing modules (e.g., double-talk detectors). For the same reason, we used a high-quality microphone (AKG C562 CM), so that the most significant nonlinearities in the system are those

Fig. 10.3 Performance comparison in terms of ERLE between SFLAFs with different expansions for speech input in a real NAEC scenario

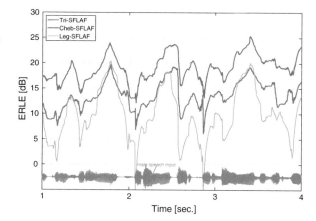

Table 10.3 Performance comparison in terms of PESQ and processing time between SFLAFs with different expansions for speech input in a real NAEC scenario

SFLAF type	M_e	PESQ	Sec.
Chebyshev SFLAF	$PM_i = 100$	3.857	2.132
Legendre SFLAF	$PM_i = 200$	2.742	3.626
Trigonometric SFLAF	$PM_i = 200$	3.124	2.951

produced by the loudspeaker. To this end, 40 cm far from the microphone, we placed a low-cost commercial loudspeaker, capable of introducing significant distortions. The input signal is male speech recorded at 16 kHz sampling frequency. The length of the experiments is 20 s. We consider a typical volume level of a quiet speech conversation, when usually loudspeaker distortions are mild and they cannot be perceived by the user. However, they do affect the echo cancellation, thus degrading the performance in the absence of a nonlinear modeling.

For this experiment, we use the same setting of the previous one, but with different filter lengths. In particular, we use $M = 200$ for the filter on the linear path, and $M_i = 50$ for the functional expansion. Therefore, the number of parameters for the filter on the nonlinear path is $M_e = PM_i = 100$ for the Chebyshev SFAF, and $M_e = 200$ for both the trigonometric and Legendre SFLAFs. Results in terms of the ERLE are shown in Fig. 10.3, showing a window of 3 out of 20 s, in which Chebyshev SFLAF clearly shows the best performance with respect to the other methods. This result is confirmed by the PESQ scores in Table 10.3, where Chebyshev SFLAF outperforms the other methods, while showing the lowest computational complexity.

10.5   Conclusion

In this paper, a Chebyshev functional link adaptive filter has been introduced as a low complexity model for the modeling of distorted speech signals. The proposed model exploits the properties of Chebyshev polynomial expansions and takes advantage of the fact that it achieves acceptable performance even with low expansion order, thus resulting the best possible functional link-based model when a low computational complexity is required by a specific problem, like NAEC. Performance are evaluated in terms of an error-based measure, i.e., the ERLE, but also in terms of a speech quality measure, i.e., the PESQ. Overall results proved that Chebyshev SFLAF is the best performing method when the minimum possible computational resources are available for the nonlinear modeling.

References

1. Perceptual evaluation of speech quality (PESQ): An objective method for end-to-end speech quality assessment of narrowband telephone networks and speech codecs (2000)
2. Carini, A., Cecchi, S., Romoli, L., Sicuranza, G.L.: Legendre nonlinear filters. Sig. Process. **109**, 84–94 (2015)
3. Carini, A., Sicuranza, G.L.: A study about Chebyshev nonlinear filters. Sig. Process. **122**, 24–32 (2016)
4. Comminiello, D., Cecchi, S., Scarpiniti, M., Gasparini, M., Romoli, L., Piazza, F., Uncini, A.: Intelligent acoustic interfaces with multisensor acquisition for immersive reproduction. IEEE Trans. Multimedia **17**(8), 1262–1272 (2015)
5. Comminiello, D., Scardapane, S., Scarpiniti, M., Parisi, R., Uncini, A.: Functional link expansions for nonlinear modeling of audio and speech signals. In: International Joint Conference on Neural Networks (IJCNN). Killarney, Ireland (2015)
6. Comminiello, D., Scarpiniti, M., Azpicueta-Ruiz, L.A., Arenas-García, J., Uncini, A.: Functional link adaptive filters for nonlinear acoustic echo cancellation. IEEE Trans. Audio Speech Lang. Process. **21**(7), 1502–1512 (2013)
7. Comminiello, D., Scarpiniti, M., Parisi, R., Uncini, A.: Intelligent acoustic interfaces for immersive audio. In: 134th Audio Engineering Society Convention. Rome, Italy, 4-7 May 2013
8. Gil-Cacho, J.M., Signoretto, M., Van Waterschoot, T., Moonen, M., Jensen, S.H.: Nonlinear acoustic echo cancellation based on a sliding-window leaky kernel affine projection algorithm. IEEE Trans. Audio Speech Lang. Process. **21**(9), 1867–1878 (2013)
9. Guerin, A., Faucon, G., Le Bouquin-Jeannes, R.: Nonlinear acoustic echo cancellation based on Volterra filters. IEEE Trans. Speech Audio Process. **11**(6), 672–683 (2003)
10. Hofmann, C., Guenther, M., Huemmer, C., Kellermann, W.: Efficient nonlinear acoustic echo cancellation by partitioned-block significance-aware Hammerstein group models. In: 24th European Signal Processing Conference (EUSIPCO), pp. 1783–1787. Budapest, Hungary (2016)
11. Pao, Y.H.: Adaptive Pattern Recognition and Neural Networks. Addison-Wesley, Reading, MA (1989)
12. Patel, V., George, N.V.: Design of dynamic linear-in-the-parameters nonlinear filters for active noise control. In: 24th European Signal Processing Conference (EUSIPCO), pp. 16–20. Budapest, Hungary (2016)
13. Patra, J.C., Kot, A.C.: Nonlinear dynamic system identification using Chebyshev functional link artificial neural networks. IEEE Trans. Syst. Man Cybern. B Cybern. **32**(4), 505–511 (2002)

14. Rudin, W.: Principles of Mathematical Analysis. McGraw-Hill, New York, NY (1976)
15. Scardapane, S., Comminiello, D., Scarpiniti, M., Uncini, A.: Benchmarking functional link expansions for audio classification tasks. In: Bassis, S., Esposito, A., Morabito, F.C., Pasero, E. (eds.) Advances in Neural Networks: Computational Intelligence for ICT, Smart Innovation, Systems and Technologies, vol. 54, pp. 133–141. Springer (2016)
16. Sicuranza, G.L., Carini, A.: A generalized FLANN filter for nonlinear active noise control. IEEE Trans. Audio Speech Lang. Process. 19(8), 2412–2417 (2011)
17. Uncini, A.: Fundamentals of adaptive signal processing. In: Signal and Communication Technology. Springer International Publishing AG, Cham, Switzerland (2015). ISBN 978-3-319-02806-4
18. Van Vaerenbergh, S., Azpicueta-Ruiz, L.A., Comminiello, D.: A split kernel adaptive filtering architecture for nonlinear acoustic echo cancellation. In: 24th European Signal Processing Conference (EUSIPCO), pp. 1768–1772. Budapest, Hungary (2016)
19. Zhao, H., Zhang, J.: Functional link neural network cascaded with Chebyshev orthogonal polynomial for nonlinear channel equalization. Sig. Process. 88(8), 1946–1957 (2008). Aug

Chapter 11
Hierarchical Temporal Representation in Linear Reservoir Computing

Claudio Gallicchio, Alessio Micheli and Luca Pedrelli

Abstract Recently, studies on deep Reservoir Computing (RC) highlighted the role of layering in deep recurrent neural networks (RNNs). In this paper, the use of linear recurrent units allows us to bring more evidence on the intrinsic hierarchical temporal representation in deep RNNs through frequency analysis applied to the state signals. The potentiality of our approach is assessed on the class of Multiple Superimposed Oscillator tasks. Furthermore, our investigation provides useful insights to open a discussion on the main aspects that characterize the deep learning framework in the temporal domain.

Keywords Reservoir computing · Deep learning · Deep echo state network Multiple time-scales processing

11.1 Introduction

In the last years, the extension of deep neural network architectures towards recurrent processing of temporal data has opened the way to novel approaches to effectively learn hierarchical representations of time-series featured by multiple time-scales dynamics [1, 9, 10, 18, 19]. Recently, within the umbrella of randomized neural network approaches [4], Reservoir Computing (RC) [15, 21] has proved to be a useful tool for analyzing the intrinsic properties of stacked architectures in recurrent neural networks (RNNs), allowing at the same time to exploit the extreme efficiency of RC training algorithms in the design of novel deep RNN models. Stemming from the Echo State Network (ESN) approach [12] the study of the dynamics of multi-layered

C. Gallicchio (✉) · A. Micheli · L. Pedrelli
Department of Computer Science, University of Pisa, Largo B. Pontecorvo 3, Pisa, Italy
e-mail: gallicch@di.unipi.it

A. Micheli
e-mail: micheli@di.unipi.it

L. Pedrelli
e-mail: luca.pedrelli@di.unipi.it

© Springer International Publishing AG, part of Springer Nature 2019
A. Esposito et al. (eds.), *Neural Advances in Processing Nonlinear Dynamic Signals*, Smart Innovation, Systems and Technologies 102,
https://doi.org/10.1007/978-3-319-95098-3_11

recurrent reservoir architectures has been introduced with the deepESN model in [5, 7]. In particular, the outcomes of the experimental analysis in [5, 7] as well as theoretical results in the field of dynamical systems [6, 8], highlighted the role of layering in the inherent development of progressively more abstract temporal representations in the higher layers of deep recurrent models.

In this paper, we take a step forward in the study of the structure of the temporal features naturally emerging in layered RNNs. To this aim, we resort to classical tools in the area of signal processing to analyze the differentiation among the state representations developed by the different levels of a deepESN in a task involving signals in a controlled scenario. In particular, we simplify the deepESN design by implementing recurrent units with linear activation function, i.e. we adopt linear deepESN (L-deepESN). In the analysis of the frequency spectrum of network's states, this approach brings the major advantage of avoiding the effects of harmonic distortion due to non-linear activation functions. To provide a quantitative support to our analysis, we experimentally assess the L-deepESN model on a variety of progressively more involving versions of the Multiple Superimposed Oscillator (MSO) task [22, 23]. Note that the class of MSO tasks is of particular interest for the aims of this paper, especially in light of previous literature results that pointed out the relevant need for multiple time-scales processing ability [13, 20, 23] as well as the potentiality of linear models in achieving excellent predictive results in base settings of the problem [2]. Another example of application of linear RNNs is in [17].

As a further contribution, our investigation would offer interesting insights on the nature of compositionality in deep learning architectures. Typically, deep neural networks consist in a hierarchy of many non-linear hidden layers that enable a distributed information representation (through learning) where higher layers specialize to progressively more abstract concepts. Removing the characteristic of non-linearity, and focusing on the ability to develop a hierarchical diversification of temporal features (prior to learning), our analysis sheds new light into the true essence of layering in deep RNN even with *linear* recurrent units.

The rest of this paper is organized as follows. In Sect. 11.2 we introduce the L-deepESN model. In Sect. 11.3 we analyze the hierarchical nature of temporal representations in L-deepESN, presenting the experimental results on the MSO tasks and the outcomes of the signal processing analysis of the developed system dynamics. Finally, in Sect. 11.4 we draw the conclusions.

11.2 Linear Deep Echo State Networks

A deepESN architecture [7] is composed by a stack of N_L recurrent reservoir layers, where at each time step t the first layer receives the external input $\mathbf{u}(t) \in \mathbb{R}^{N_U}$, while successive layers are fed by the output of the previous layer in the hierarchy. We denote the state of layer i at time t by $\mathbf{x}^{(i)}(t) \in \mathbb{R}^{N_R}$, where we assume the same state dimension N_R for every layer for the sake of simplicity. A schematic representation of the reservoir architecture in a deepESN is provided in Fig. 11.1.

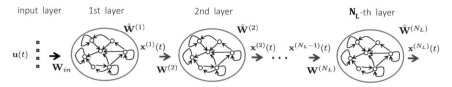

Fig. 11.1 Layered reservoir architecture in a deepESN

By referring to the case of leaky integrator reservoir units [13], and omitting the bias terms for the ease of notation, the state transition function of the first layer is given by the following equation:

$$\mathbf{x}^{(1)}(t) = (1 - a^{(1)})\mathbf{x}^{(1)}(t - 1) + a^{(1)}\mathbf{f}(\mathbf{W}_{in}\mathbf{u}(t) + \hat{\mathbf{W}}^{(1)}\mathbf{x}^{(1)}(t - 1)), \qquad (11.1)$$

whereas the state transition of layer $i > 1$ is ruled by the equation:

$$\mathbf{x}^{(i)}(t) = (1 - a^{(i)})\mathbf{x}^{(i)}(t - 1) + a^{(i)}\mathbf{f}(\mathbf{W}^{(i)}\mathbf{x}^{(i-1)}(t) + \hat{\mathbf{W}}^{(i)}\mathbf{x}^{(i)}(t - 1)), \qquad (11.2)$$

where $a^{(i)} \in [0, 1]$ is the leaking rate parameter at layer i, $\mathbf{W}_{in} \in \mathbb{R}^{N_R \times N_U}$ is the input weight matrix, $\mathbf{W}^{(i)} \in \mathbb{R}^{N_R \times N_R}$ is the weight matrix of the inter-layer connections from layer $i - 1$ to layer i, $\hat{\mathbf{W}}^{(i)} \in \mathbb{R}^{N_R \times N_R}$ is the matrix of recurrent weights of layer i, and $\mathbf{f}$ denotes the element-wise application of the activation function of the recurrent units. A null initial state is considered for the reservoirs in all the layers, i.e. $\mathbf{x}^{(i)}(0) = \mathbf{0}$ for all $i = 1, \ldots, N_L$.

The case of L-deepESN is obtained from Eqs. 11.1 and 11.2 when a linear activation function is used for each recurrent unit, i.e. $\mathbf{f} = \mathbf{id}$. As in standard RC, all the reservoirs parameters, i.e. all the weight matrices in Eqs. 11.1 and 11.2, are left untrained after initialization subject to stability constraints. According to the necessary condition for the Echo State Property of deep RC networks [6], stability can be accomplished by constraining the maximum among the spectral radii of matrices $\left((1 - a^{(i)})\mathbf{I} + a^{(i)}\hat{\mathbf{W}}^{(i)} \right)$, individually denoted by $\rho^{(i)}$, to be not above unity. Thereby, a simple initialization condition for L-deepESNs consists in randomly selecting the weight values in matrices $\mathbf{W}_{in}$ and $\{\mathbf{W}^{(i)}\}_{i=2}^{N_L}$ from a uniform distribution in $[-scale_{in}, scale_{in}]$, whereas the weights in recurrent matrices $\{\hat{\mathbf{W}}^{(i)}\}_{i=1}^{N_L}$ are initialized in a similar way and are then re-scaled to meet the condition on max $\rho^{(i)}$.

In this context it also interesting to observe that the use of linearities allows us to express the evolution of the whole system by means of an algebraic expression that describes the dynamics of an equivalent single-layer recurrent system with the same total number of recurrent units. Specifically, denoting by $\mathbf{x}(t) = (\mathbf{x}^{(1)}(t), \mathbf{x}^{(2)}(t), \ldots, \mathbf{x}^{(N_L)}(t)) \in \mathbb{R}^{N_L N_R}$ the global state of the network, the dependence of $\mathbf{x}(t)$ from $\mathbf{x}(t - 1)$ can be expressed as $\mathbf{x}(t) = \mathbf{V}\mathbf{x}(t - 1) + \mathbf{V}_{in}\mathbf{u}(t)$, where both $\mathbf{V} \in \mathbb{R}^{N_L N_R \times N_L N_R}$ and $\mathbf{V}_{in} \in \mathbb{R}^{N_L N_R \times N_U}$ can be viewed as block matrices, with block elements denoted respectively by $\mathbf{V}_{i,j} \in \mathbb{R}^{N_R \times N_R}$ and $\mathbf{V}_{in,i} \in \mathbb{R}^{N_R \times N_U}$, i.e.:

$$\mathbf{x}(t) = \begin{bmatrix} \mathbf{V}_{1,1} & \cdots & \mathbf{V}_{1,N_L} \\ \vdots & \ddots & \vdots \\ \mathbf{V}_{N_L,1} & \cdots & \mathbf{V}_{N_L,N_L} \end{bmatrix} \mathbf{x}(t-1) + \begin{bmatrix} \mathbf{V}_{in,1} \\ \vdots \\ \mathbf{V}_{in,N_L} \end{bmatrix} \mathbf{u}(t). \qquad (11.3)$$

Noticeably, the layered organization imposes a lower triangular block matrix structure to $\mathbf{V}$ such that in the linear case its blocks can be computed as:

$$\mathbf{V}_{i,j} = \begin{cases} \mathbf{0} & \text{if } i < j \\ (1 - a^{(i)})\mathbf{I} + a^{(i)}\hat{\mathbf{W}}^{(i)} & \text{if } i = j \\ \left(\prod_{k=j+1}^{i} a^{(k)}\mathbf{W}^{(k)}\right)\left((1 - a^{(j)})\mathbf{I} + a^{(j)}\hat{\mathbf{W}}^{(j)}\right) & \text{if } i > j. \end{cases} \qquad (11.4)$$

Moreover, as concerns the input matrix, we have:

$$\mathbf{V}_{in,i} = \begin{cases} a^{(1)}\mathbf{W}_{in} & \text{if } i = 1 \\ \left(\prod_{k=2}^{i} a^{(k)}\mathbf{W}^{(k)}\right) a^{(1)}\mathbf{W}_{in} & \text{if } i > 1. \end{cases} \qquad (11.5)$$

The mathematical description provided here for the L-deepESN case is particularly helpful in order to highlight the characterization resulting from the layered composition of recurrent units. Indeed, from an architectural perspective, a deep RNN can be seen as obtained by imposing a set of constraints to the architecture of a single-layer fully connected RNN with the same total number of recurrent units. Specifically, a deep RNN can be obtained from the architecture of a fully connected (shallow) RNN by removing the recurrent connections corresponding in the deep version to the connections from higher layers to lower layers and from each layer to higher layers different from the successive one, removing as well the input connections to the levels higher than 1. In this respect, the use of linear activation functions has the effect of enhancing the emergence of such constrained characterization and making it visible through the peculiar algebraic organization of the state update as described by Eqs. 11.3, 11.4 and 11.5. Indeed, the constrained structure given by the layering factor is reflected in the (lower triangular block) structure of the matrix $\mathbf{V}$ that rules the recurrence of the whole network dynamics in Eq. 11.3. In particular, the last line of Eq. 11.4 highlights the progressive filtering effect on the state information propagated towards the higher levels in the network, modulated by the leaking rates and through the magnitude of the inter-layer weights values. Similarly, the last line of Eq. 11.5 shows the analogous progressive filtering effect operated on the external input information for increasing level's depth.

Thereby, although from the system dynamics viewpoint it is possible to find a shallow recurrent network that is equivalent to an L-deepESN, the resulting form of the matrices that rules the state evolution, i.e. $\mathbf{V}$ and $\mathbf{V}_{in}$, has a distinct characterization that is due to the layered construction. Moreover, note that the probability of obtaining such matrices $\mathbf{V}$ and $\mathbf{V}_{in}$ by means of standard random reservoir initialization is negligible. Noteworthy, the aforementioned architectural constraints imposed by the hierarchical construction are reflected also in the ordered structure of the temporal features represented in higher levels of the recurrent architecture, as investigated for linear reservoirs in Sect. 11.3, and as observed, under a different perspective and using different mathematical tools, in the non-linear case in [7].

As regards network training, as in standard RC, the only learned parameters of the L-deepESN are those pertaining to the readout layer. This is used for output computation by means of a linear combination of the reservoir units activations in all the levels, allowing the linear learner to weight differently the contributions of the multiple dynamics developed in the network state. In formulas, at each time step t the output $\mathbf{y}(t) \in \mathbb{R}^{N_Y}$ is computed as $\mathbf{y}(t) = \mathbf{W}_{out}\mathbf{x}(t)$, where $\mathbf{W}_{out} \in \mathbb{R}^{N_Y \times N_L N_R}$ is the output weight matrix whose values are learned from a training set. Typically, as in the standard RC framework, the values in $\mathbf{W}_{out}$ are found in closed form by using direct methods such as pseudo-inversion or ridge regression [15].

11.3 Experimental Assessment

In this section we present the results of the experimental assessment of L-deepESN on the class of MSO tasks.

An MSO task consists in a next-step prediction on a 1-dimensional time-series, i.e. for each time step t the target output is given by $y_{target}(t) = u(t+1)$, where $N_U = N_Y = 1$. The considered time-series is given by a sum of sinusoidal functions:

$$u(t) = \sum_{i=1}^{n} sin(\varphi_i t) \tag{11.6}$$

where n denotes the number of sinusoidal functions, φ_i determines the frequency of the i-th sinusoidal function and t is the index of the time step. In the following, we use the notation MSOn to specify the number n of sinusoidal functions that are accounted in the task definition. The φ_i coefficients in Eq. 11.6 are set as in [14, 16], i.e. $\varphi_1 = 0.2$, $\varphi_2 = 0.331$, $\varphi_3 = 0.42$, $\varphi_4 = 0.51$, $\varphi_5 = 0.63$, $\varphi_6 = 0.74$, $\varphi_7 = 0.85$, $\varphi_8 = 0.97$, $\varphi_9 = 1.08$, $\varphi_{10} = 1.19$, $\varphi_{11} = 1.27$, $\varphi_{12} = 1.32$. In particular, in our experiments we focus on versions of the MSO task with a number of sine waves n ranging from 5 to 12. This allows us to exercise the ability of the RC models to develop a hierarchy of temporal representations in challenging cases where the input signal is enriched by the presence of many different time-scales dynamics. Besides, note that summing an increasing number of sine waves with frequencies that are not integer multiples of each other makes the prediction task harder due to the increasing signal period. An example of the input signal for the MSO12 task is given in Fig. 11.2. For all the considered settings of the MSO task, the first 400 steps are used for training (with a washout of length 100), time steps from 401 to 700 are used for validation and the remaining steps from 701 to 1000 are used for test.

In our experiments, we used L-deepESN with N_L levels, each consisting in a fully connected reservoir with N_R units. We assumed that $\mathbf{W}_{in}$ and $\{\mathbf{W}^{(i)}\}_{i=2}^{N_L}$ are initialized with the same scaling parameter $scale_{in}$, and we used the same value of the spectral

Fig. 11.2 A 400 time step long excerpt of the input sequence for the MSO12 task

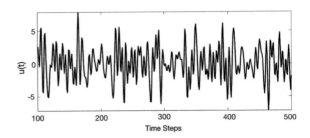

radius and of the leaking rate in every level, i.e. $\rho^{(i)} = \rho$ and $a^{(i)} = a$ for every i. For readout training we used ridge regression. Table 11.1 reports the range of values considered for every hyper-parameter considered in our experiments.

In order to evaluate the predictive performance on the MSO tasks, we used the normalized root mean square error (NRMSE), calculated as follows:

$$NRMSE = \sqrt{(\sum_{t=1}^{T}(y_{target}(t) - y(t))^2)/(T\sigma^2_{y_{target}(t)})}, \qquad (11.7)$$

where T denotes the sequence length, $y_{target}(t)$ and $y(t)$ are the target and the network's output at time t, and $\sigma^2_{y_{target}(t)}$ is the variance of y_{target}. For each reservoir hyper-parametrization, we independently generated 10 reservoir guesses, the predictive performance in the different cases has been averaged over such guesses and then the model's hyper-parameterization has been selected on the validation set.

In the following Sects. 11.3.1 and 11.3.2 we respectively evaluate our approach from a quantitative point of view, comparing the predictive performance of L-deepESN with related literature models, and from a qualitative perspective, by analyzing the frequencies of the state activations developed in the different reservoir levels.

Table 11.1 Hyper-parameters values considered for model selection on the MSO tasks

Hyper-parameter	Values considered for model selection
Number of levels N_L	10
Reservoir size N_R	100
Input scaling $scale_{in}$	0.01, 0.1, 1
Leaking rate a	0.1, 0.3, 0.5, 0.7, 0.9, 1.0
Spectral radius ρ	0.1, 0.3, 0.5, 0.7, 0.9, 1.0
Ridge regression regularization λ_r	$10^{-11}, 10^{-10}, ..., 10^0$

11.3.1 Predictive Performance

In this section we compare the quantitative results of the proposed approach with the performance reported in literature on recent (more complex and richer) variants of the MSO task, with a number of sine waves n varying from 5 to 12. Table 11.2 provides a comparison among the NRMSE achieved on the test set by L-deepESN and the available results in literature obtained by neuro-evolution [16], balanced ESN [14], ESN with infinite impulse response units (IIR ESN) [11] and Evolino [20] on the considered MSO tasks. Furthermore, in the same table, we report the performance achieved by linear ESN built with a single fully connected reservoir (L-ESN), considering the same range of hyper-parameters and total number of recurrent units as in the L-deepESN case.

Noteworthy, the proposed L-deepESN model outperformed the best literature results of about 3 or 4 orders of magnitude on all the MSO settings. Furthermore, test errors obtained by L-ESN are always within one order of magnitude of difference with respect to the best state-of-the-art results. These aspects confirms the effectiveness of the linear activation function on this task, as also testified by our preliminary results that showed poorer performance for RC networks with *tanh* units, unless forcing the operation of the activation function in the linear region. Moreover, L-deepESN always performed better then L-ESN. On the basis of the known characterization of the MSO task, our results confirm the quality of the hierarchical structure of recurrent reservoirs in representing multiple time-scales dynamics with respect to its shallow counterpart.

The literature approaches in Table 11.2 have different architectures with different configurations used in their model selections (for simplicity, we avoid listing them, see relative papers for details). For the sake of completeness, we performed further comparisons (out of table) considering L-deepESNs with the same total number of recurrent units used in model selections of ESN-based approaches, i.e. balanced ESN and IIR ESN, respectively described in [11, 14]. In particular, balanced ESN used

Table 11.2 Test NRMSE obtained by L-deepESN, L-ESN, neuro-evolution (n. evolution), balanced ESN, IIR ESN and Evolino on the MSO5-12 tasks

Task	L-deepESN	L-ESN	n.-evolution [16]	balanced ESN [14]	IIR ESN [11]	Evolino [20]
MSO5	6.75×10^{-13}	7.14×10^{-10}	4.16×10^{-10}	1.06×10^{-6}	8×10^{-5}	1.66×10^{-1}
MSO6	1.68×10^{-12}	5.40×10^{-9}	9.12×10^{-9}	8.43×10^{-5}	–	–
MSO7	5.90×10^{-12}	5.60×10^{-8}	2.39×10^{-8}	1.01×10^{-4}	–	–
MSO8	1.07×10^{-11}	2.08×10^{-7}	6.14×10^{-8}	2.73×10^{-4}	–	–
MSO9	5.34×10^{-11}	4.00×10^{-7}	1.11×10^{-7}	–	–	–
MSO10	8.22×10^{-11}	8.21×10^{-7}	1.12×10^{-7}	–	–	–
MSO11	4.45×10^{-10}	1.55×10^{-6}	1.22×10^{-7}	–	–	–
MSO12	5.40×10^{-10}	1.70×10^{-6}	1.73×10^{-7}	–	–	–

a maximum number of 250 units on the MSO5, MSO6, MSO7 and MSO8 tasks, while IIR ESN implemented 100 units on the MSO5 task. L-deepESN with $N_L = 10$ and $N_R = 25$ (i.e. a total of 250 recurrent units) performed better than balanced ESN, obtaining a test NRMSE of 1.20×10^{-11}, 8.73×10^{-11}, 2.42×10^{-10} and 9.06×10^{-10}, on the MSO5, MSO6, MSO7 and MSO8 tasks, respectively. Moreover, even L-deepESN with $N_L = 10$ and $N_R = 10$ (i.e. a total of 100 recurrent units) obtained a better performance than IIR ESN, achieving a test error of 7.41×10^{-11} on the MSO5 task.

11.3.2 Hierarchical Temporal Representation Analysis

In this section we investigate the temporal representation developed by the reservoirs levels in an L-deepESN, using as input signal the sequence considered for the MSO12 task, featured by rich dynamics with known multiple time-scales characterization (see Eq. 11.6). We used the same reservoir hyper-parameterization selected for the predictive experiments on the MSO12 task in Sect. 11.3.1, namely $N_R = 100$, $N_L = 10$, $scale_{in} = 1$, $a = 0.9$ and $\rho = 0.7$, averaging the results over 100 reservoir guesses. In our analysis, we first computed the states obtained by running the L-deepESN on the input sequence. Then, we performed the Fast Fourier Transform (FFT) [3] algorithm on the states of all the recurrent units over the time, normalizing the obtained values in order to enable a qualitative comparison. Finally, we averaged the FFT values on a layer-by-layer basis.

The FFT values obtained for progressively higher levels of L-deepESN are shown in Figs. 11.3a–d, which respectively focus on levels 1, 4, 7 and 10. These figures represent the state signal in the frequency domain, where it is possible to see 12 spikes corresponding to the 12 sine waves components of the input. Looking at the magnitude of the FFT components, i.e. at the height of the spikes in plots, we can have an indication of how the signals are elaborated by the individual recurrent levels. We can see that the state of the reservoir at level 1 shows FFT components all with approximately the same magnitude. The FFT components of reservoir states at levels 4, 7 and 10, instead, show diversified magnitudes. Specifically, we can see that in higher levels of the network higher frequency components are progressively filtered, and lower frequency components tend to have relative higher magnitudes. This confirms the insights on the progressive filtering effect discussed in Sect. 11.2 in terms of mathematical characterization of the system.

Results in Fig. 11.3 show that the hierarchical construction of recurrent models leads, even in the linear case, to a representation of the temporal signal that is sparsely distributed across the network, where different levels tend to focus on a different range of frequencies. Moreover, the higher is the level, the stronger is the focus on lower frequencies, hence the state signals emerging in deeper levels are naturally featured by coarser time-scales and slower dynamics. Thereby, the layered organization of the recurrent units determines a temporal representation that has an intrinsic hierarchical structure. According to this, the multiple time-scales in the network dynamics are

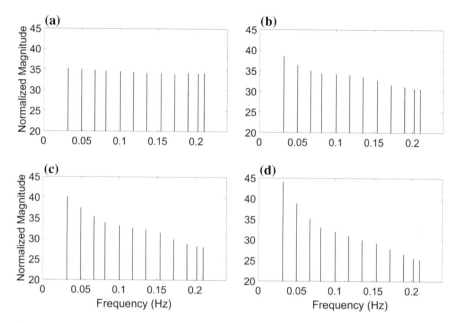

Fig. 11.3 FFT components of reservoir states in progressively higher levels of L-deepESN, **a** level 1, **b** level 4, **c** level 7, **d** level 10

ordered depending to the depth of reservoirs' levels. Such inherent characterization of the hierarchical distributed temporal representation can be exploited when training the readout, as testified by the excellent predictive performance of L-deepESN on the MSO tasks reported in Sect. 11.3.1.

11.4 Conclusions

In this paper, we have studied the inherent properties of hierarchical linear RNNs by analyzing the frequency of the states signals emerging in the different levels of the recurrent architecture. The FFT analysis revealed that the stacked composition of reservoirs in a L-deepESN tends to develop a structured representation of the temporal information. Exploiting an incremental filtering effect, states in higher levels of the hierarchy are biased towards slower components of the frequency spectrum, resulting in progressively slower temporal dynamics. In this sense, the emerging structure of L-deepESN states can be seen as an echo of the multiple time-scales present in the input signal, distributed across the layers of the network. The hierarchical representation of temporal features in L-deepESN has been exploited to address recent challenging versions of the MSO task. Experimental results showed that the proposed approach dramatically outperforms the state-of-the-art on the MSO tasks, emphasizing the relevance of the hierarchical temporal representation and also confirming the effectiveness of linear signal processing on the MSO problem.

Overall, we showed a concrete evidence that layering is an aspect of the network construction that is intrinsically able to provide a distributed and hierarchical feature representation of temporal data. Our analysis pointed out that this is possible even without (or prior to) learning of the recurrent connections, and releasing the requirement for non-linearity of the activation functions. We hope that the considerations delineated in this paper could contribute to open an intriguing research question regarding the merit of shifting the focus, from the concepts of learning and non-linearities, to the concepts of hierarchical organization and distribution of representation to define the salient aspects of the deep learning framework for recurrent architectures.

References

 1. Angelov, P., Sperduti, A.: Challenges in deep learning. In: Proceedings of the 24th European Symposium on Artificial Neural Networks (ESANN), pp. 489–495. i6doc.com (2016)
 2. Čerňanský, M., Tiňo, P.: Predictive modeling with echo state networks. Artif. Neural Netw ICANN **2008**, 778–787 (2008)
 3. Frigo, M., Johnson, S.G.: FFTW: An adaptive software architecture for the FFT. In: Proceedings of the 1998 IEEE International Conference on Acoustics, Speech and Signal Processing, vol. 3, pp. 1381–1384. IEEE (1998)
 4. Gallicchio, C., Martin-Guerrero, J., Micheli, A., Soria-Olivas, E.: Randomized machine learning approaches: Recent developments and challenges. In: Proceedings of the 25th European Symposium on Artificial Neural Networks (ESANN), pp. 77–86. i6doc.com (2017)
 5. Gallicchio, C., Micheli, A.: Deep reservoir computing: a critical analysis. In: Proceedings of the 24th European Symposium on Artificial Neural Networks (ESANN), pp. 497–502. i6doc.com (2016)
 6. Gallicchio, C., Micheli, A.: Echo state property of deep reservoir computing networks. Cogn. Comput. 337–350 (2017). https://doi.org/10.1007/s12559-017-9461-9
 7. Gallicchio, C., Micheli, A., Pedrelli, L.: Deep reservoir computing: a critical experimental analysis. Neurocomputing 87–99 (2017). https://doi.org/10.1016/j.neucom.2016.12.089
 8. Gallicchio, C., Micheli, A., Silvestri, L.: Local Lyapunov Exponents of Deep RNN. In: Proceedings of the 25th European Symposium on Artificial Neural Networks (ESANN), pp. 559–564. i6doc.com (2017)
 9. Hermans, M., Schrauwen, B.: Training and analysing deep recurrent neural networks. In: NIPS, pp. 190–198 (2013)
10. Hihi, S.E., Bengio, Y.: Hierarchical recurrent neural networks for long-term dependencies. In: NIPS, pp. 493–499 (1995)
11. Holzmann, G., Hauser, H.: Echo state networks with filter neurons and a delay& sum readout. Neural Netw. **23**(2), 244–256 (2010)
12. Jaeger, H., Haas, H.: Harnessing nonlinearity: predicting chaotic systems and saving energy in wireless communication. Science **304**(5667), 78–80 (2004)
13. Jaeger, H., Lukoševičius, M., Popovici, D., Siewert, U.: Optimization and applications of echo state networks with leaky-integrator neurons. Neural Netw. **20**(3), 335–352 (2007)
14. Koryakin, D., Lohmann, J., Butz, M.: Balanced echo state networks. Neural Netw. **36**, 35–45 (2012)
15. Lukoševičius, M., Jaeger, H.: Reservoir computing approaches to recurrent neural network training. Comput. Sci. Rev. **3**(3), 127–149 (2009)

16. Otte, S., Butz, M.V., Koryakin, D., Becker, F., Liwicki, M., Zell, A.: Optimizing recurrent reservoirs with neuro-evolution. Neurocomputing **192**, 128–138 (2016)
17. Pasa, L., Sperduti, A.: Pre-training of recurrent neural networks via linear autoencoders. In: Advances in Neural Information Processing Systems, pp. 3572–3580 (2014)
18. Pascanu, R., Gülçehre, Ç., Cho, K., Bengio, Y.: How to construct deep recurrent neural networks, pp. 1–13. arXiv preprint arXiv:1312.6026v5 (2014)
19. Schmidhuber, J.: Deep learning in neural networks: an overview. Neural Netw. **61**, 85–117 (2015)
20. Schmidhuber, J., Wierstra, D., Gagliolo, M., Gomez, F.: Training recurrent networks by evolino. Neural Comput. **19**(3), 757–779 (2007)
21. Verstraeten, D., Schrauwen, B., d'Haene, M., Stroobandt, D.: An experimental unification of reservoir computing methods. Neural Netw. **20**(3), 391–403 (2007)
22. Wierstra, D., Gomez, F.J., Schmidhuber, J.: Modeling systems with internal state using evolino. In: Proceedings of the 7th Annual Conference on Genetic and Evolutionary Computation, pp. 1795–1802. ACM (2005)
23. Xue, Y., Yang, L., Haykin, S.: Decoupled echo state networks with lateral inhibition. Neural Netw. **20**(3), 365–376 (2007)

Chapter 12
On 4-Dimensional Hypercomplex Algebras in Adaptive Signal Processing

Francesca Ortolani, Danilo Comminiello, Michele Scarpiniti
and Aurelio Uncini

Abstract The degree of diffusion of hypercomplex algebras in adaptive and non-adaptive filtering research topics is growing faster and faster. The debate today concerns the usefulness and the benefits of representing multidimensional systems by means of these complicated mathematical structures and the criterions of choice between one algebra or another. This paper proposes a simple comparison between two isodimensional algebras (quaternions and tessarines) and shows by simulations how different choices may determine the system performance. Some general information about both algebras is also supplied.

Keywords Adaptive filters · Quaternions · Tessarines · Hypercomplex
Widely linear · Least mean square

12.1 Introduction

One of the trends in the last two decades in digital signal processing has been the exploration of hypercomplex algebras for multidimensional signal processing with particular regard to adaptive filtering and intelligent systems. Since complex numbers were widely experimented and studied in both linear and nonlinear environments [9], the immediate step forward in hypercomplex algebras has considered quaternions [19] and octonions [6]. Scientists paid special attention to quaternion adaptive filtering [4, 5, 7, 17, 19] and the authors of this paper themselves investigated quaternion

F. Ortolani (✉) · D. Comminiello · M. Scarpiniti · A. Uncini
Dpt. of Information Engineering, Electronics and Telecommunications (DIET),
"Sapienza" University of Rome, Via Eudossiana 18, 00184 Rome, Italy
e-mail: francesca.ortolani@uniroma1.it

D. Comminiello
e-mail: danilo.comminiello@uniroma1.it

M. Scarpiniti
e-mail: michele.scarpiniti@uniroma1.it

A. Uncini
e-mail: aurelio.uncini@uniroma1.it

© Springer International Publishing AG, part of Springer Nature 2019
A. Esposito et al. (eds.), *Neural Advances in Processing Nonlinear
Dynamic Signals*, Smart Innovation, Systems and Technologies 102,
https://doi.org/10.1007/978-3-319-95098-3_12

131

algebra with an interest in frequency-domain adaptive filters [11]. Besides the traditional usage of quaternions in 3D graphics and navigation (it is known that quaternion rotations are not subject to deadlocks from space degeneration from 3D to 2D) [18], the new revival of hypercomplex processing consists in the experimentation of the peculiar algebraic properties in engineering multidimensional problems. Multidimensionality is somehow intrinsic to the nature of the data: it arises from the need for processing correlated data (not necessarily homogeneous, as in [19]).

The investigation topic in this paper considers two hypercomplex algebras having the same dimensions: quaternions and tessarines (the latter also known as *bicomplex numbers*). Both of them are a 4-dimensional algebra. So, which are the reasons why we should choose one algebra or another? This paper presents a couple of examples where different results are obtained from different mathematical representations of the systems. A line of reasoning is also suggested. With regard to adaptive signal processing, the first adaptive algorithm implemented in a hypercomplex algebra has been the Quaternion Least Mean Square (QLMS) algorithm [19]. Because of its straight comprehensibility and ease of implementation, this algorithm offered an instrument on hand for studying quaternion algebra combined with adaptive filtering. In this work, we adopted this algorithm to make a comparison of the two 4-dimensional algebras above mentioned. On this occasion, we derived and implemented a tessarine version of the LMS algorithm, namely TLMS. In a second step, we searched for a modification of both the QLMS and TLMS algorithms into a *widely linear* form (including full second order statistics) [8, 20, 21] and their behaviour was tested with *proper* and *improper* input signals. The highlight in this paper is the evidence that the choice of a specific algebra may condition a filter behaviour. We analysed this fact by introducing Ambisonic 3D audio signals into a 4-dimensional system. Ambisonics is a 3D audio recording and rendering technique developed in the '70s [2, 3, 16] and, in recent research, it was experimented that its so-called B-Format can be condensed and processed in a quaternion formalism [12, 13]. The main goal of the current work is to examine whether the good results obtained with quaternion algebra persist in other 4-dimensional algebras.

This paper is organized as follows: Sect. 12.2 introduces both quaternion and tessarine algebras underlining the fundamental differences between them; Sect. 12.3 presents a short summary of the 4-dimensional least mean square algorithms (QLMS and TLMS) and their *widely linear* modifications. Finally, Sect. 12.4 reports some interesting results from simulations with both widely linear and non-widely linear algorithms.

12.2 Introduction to 4-Dimensional Hypercomplex Algebras

As just introduced, both quaternions and tessarines are 4-dimensional algebras. Even though a quaternion q and a tessarine t look just alike ($q = q_0 + q_1\mathbf{i} + q_2\mathbf{j} + q_3\mathbf{k}$ with $q \in \mathbb{H}$ and $t = t_0 + t_1\mathbf{i} + t_2\mathbf{j} + t_3\mathbf{k}$ with $t \in \mathbb{T}$), their algebras have very little in common. The imaginary axes $(\mathbf{i}, \mathbf{j}, \mathbf{k})$ form a basis on which the two algebras are built. However, in the two cases, different fundamental algebraic properties are defined:

In quaternion algebra,

$$\mathbf{ij} = \mathbf{k}, \quad \mathbf{jk} = \mathbf{i}, \quad \mathbf{ki} = \mathbf{j}, \tag{12.1}$$

$$\mathbf{i}^2 = \mathbf{j}^2 = \mathbf{k}^2 = -1. \tag{12.2}$$

In tessarine algebra,

$$\mathbf{ij} = \mathbf{k}, \quad \mathbf{jk} = \mathbf{i}, \quad \mathbf{ki} = -\mathbf{j}, \tag{12.3}$$

$$\mathbf{i}^2 = \mathbf{k}^2 = -1, \quad \mathbf{j}^2 = +1. \tag{12.4}$$

Equations (12.1)–(12.4) embody the foremost difference between quaternion and tessarine algebras: the former is non-commutative, the latter is commutative. As a consequence of this, quaternion product and tessarine product are defined in different ways.

For quaternions $q_1, q_2 \in \mathbb{H}$, their product is calculated as

$$
\begin{aligned}
q_1 q_2 &= (a_0 + a_1\mathbf{i} + a_2\mathbf{j} + a_3\mathbf{k})(b_0 + b_1\mathbf{i} + b_2\mathbf{j} + b_3\mathbf{k}) \\
&= (a_0 b_0 - a_1 b_1 - a_2 b_2 - a_3 b_3) \\
&\quad + (a_0 b_1 + a_1 b_0 + a_2 b_3 - a_3 b_2)\,\mathbf{i} \\
&\quad + (a_0 b_2 - a_1 b_3 + a_2 b_0 + a_3 b_1)\,\mathbf{j} \\
&\quad + (a_0 b_3 + a_1 b_2 - a_2 b_1 + a_3 b_0)\,\mathbf{k}.
\end{aligned}
\tag{12.5}
$$

For tessarines $t_1, t_2 \in \mathbb{T}$, their product is calculated as

$$
\begin{aligned}
t_1 t_2 &= (a_0 + a_1\mathbf{i} + a_2\mathbf{j} + a_3\mathbf{k})(b_0 + b_1\mathbf{i} + b_2\mathbf{j} + b_3\mathbf{k}) \\
&= (a_0 b_0 - a_1 b_1 + a_2 b_2 - a_3 b_3) \\
&\quad + (a_0 b_1 + a_1 b_0 + a_2 b_3 + a_3 b_2)\,\mathbf{i} \\
&\quad + (a_0 b_2 - a_1 b_3 + a_2 b_0 - a_3 b_1)\,\mathbf{j} \\
&\quad + (a_0 b_3 + a_1 b_2 + a_2 b_1 + a_3 b_0)\,\mathbf{k}.
\end{aligned}
\tag{12.6}
$$

Moreover, (12.1) can be expressed by the *cross products* $\mathbf{i} \times \mathbf{j} = \mathbf{k}$, $\mathbf{j} \times \mathbf{k} = \mathbf{i}$, $\mathbf{k} \times \mathbf{i} = \mathbf{j}$. In fact, the cross product is non-commutative (anti-commutative). Regarding the sum, it is computed the same way with either quaternions or tessarines:

$$
\begin{aligned}
q \pm p &= (q_0 + q_1\mathbf{i} + q_2\mathbf{j} + q_3\mathbf{k}) \pm (p_0 + p_1\mathbf{i} + p_2\mathbf{j} + p_3\mathbf{k}) \\
&= (q_0 \pm p_0) + (q_1 \pm p_1)\,\mathbf{i} + (q_2 \pm p_2)\,\mathbf{j} + (q_3 \pm p_3)\,\mathbf{k}.
\end{aligned}
\tag{12.7}
$$

The *poly-conjugation*, with respect to the specific algebraic rules, conveniently defines both the conjugates of a quaternion and a tessarine:

$$p^* = p_0 + \sum_{\nu=1}^{3} p_\nu \mathbf{e}_\nu^3 \tag{12.8}$$

and typified in the two algebras becomes $q^* = q_0 - q_1\mathbf{i} - q_2\mathbf{j} - q_3\mathbf{k} \in \mathbb{H}$ and $t^* = t_0 - t_1\mathbf{i} + t_2\mathbf{j} - t_3\mathbf{k} \in \mathbb{T}$.

12.3 4D Least Mean Square Algorithms

In the 4-dimensional algebra of quaternions, the least mean square algorithm was formerly presented in [19] with a minor modification in [1]. Following an approach similar to [19], the TLMS algorithm has been derived by the authors of this paper. This section compares and comments QLMS and TLMS.

12.3.1 Algorithm Overview

The least mean square algorithm is an online error-correction-based adaptive algorithm: the cost function to be minimized during the adaptation is defined as the mean square error (MSE) $J_n(\mathbf{w}_n) = E\{e[n]e^*[n]\}$. The error $e[n]$ is the difference between a desired signal and the filter output ($e[n] = d[n] - y[n]$). The adaptive filter output can be defined by the scalar product $y[n] = \mathbf{w}_{n-1}^T \mathbf{x}_n$, thus resulting in

$$y[n] = \mathbf{w}_{n-1}^T \mathbf{x}_n = \begin{bmatrix} \mathbf{w}_a^T \mathbf{x}_a - \mathbf{w}_b^T \mathbf{x}_b - \mathbf{w}_c^T \mathbf{x}_c - \mathbf{w}_d^T \mathbf{x}_d \\ \mathbf{w}_a^T \mathbf{x}_b + \mathbf{w}_b^T \mathbf{x}_a + \mathbf{w}_c^T \mathbf{x}_d - \mathbf{w}_d^T \mathbf{x}_c \\ \mathbf{w}_a^T \mathbf{x}_c + \mathbf{w}_c^T \mathbf{x}_a + \mathbf{w}_d^T \mathbf{x}_b - \mathbf{w}_b^T \mathbf{x}_d \\ \mathbf{w}_a^T \mathbf{x}_d + \mathbf{w}_d^T \mathbf{x}_a + \mathbf{w}_b^T \mathbf{x}_c - \mathbf{w}_c^T \mathbf{x}_b \end{bmatrix} \in \mathbb{H} \tag{12.9}$$

for quaternions, and

$$y[n] = \mathbf{w}_{n-1}^T \mathbf{x}_n = \begin{bmatrix} \mathbf{w}_a^T \mathbf{x}_a - \mathbf{w}_b^T \mathbf{x}_b + \mathbf{w}_c^T \mathbf{x}_c - \mathbf{w}_d^T \mathbf{x}_d \\ \mathbf{w}_a^T \mathbf{x}_b + \mathbf{w}_b^T \mathbf{x}_a + \mathbf{w}_c^T \mathbf{x}_d + \mathbf{w}_d^T \mathbf{x}_c \\ \mathbf{w}_a^T \mathbf{x}_c + \mathbf{w}_c^T \mathbf{x}_a - \mathbf{w}_d^T \mathbf{x}_b - \mathbf{w}_b^T \mathbf{x}_d \\ \mathbf{w}_a^T \mathbf{x}_d + \mathbf{w}_d^T \mathbf{x}_a + \mathbf{w}_b^T \mathbf{x}_c + \mathbf{w}_c^T \mathbf{x}_b \end{bmatrix} \in \mathbb{T} \tag{12.10}$$

for tessarines, where $\mathbf{x}_n$ is the filter input vector at iteration n and $\mathbf{w}_{n-1}$ are the filter weights at iteration $n-1$: $\mathbf{x}_n = [x[n] \; x[n-1] \; \cdots \; x[n-M]]^T$, $\mathbf{w}_n = [w_0[n] \; w_1[n] \; \cdots \; w_M[n]]^T$, with M the filter length.

The computation of the gradient of the cost function $J_n(\mathbf{w}_n)$ (required for finding its minimum) leads to an adaptation equation which is the same for both QLMS and TLMS for outputs defined as in (12.9) and (12.10):

$$\mathbf{w}_n = \mathbf{w}_{n-1} + \mu e\,[n]\,\mathbf{x}_n^*. \tag{12.11}$$

where μ is the step size along the direction of the gradient.

12.3.2 Widely Linear Modification

Recent works about both complex and hypercomplex filtering showed a particular interest in widely linear algorithms [8, 10, 21, 22]. It has been observed that most real world signals are *improper* (or *noncircular*) in nature [14] and, in this case, a filter performance can be improved significantly if the full second order statistics of the signals is taken into account and included into the algorithm. If a random variable has a rotation-invariant probability distribution (with respect to all six pairs of rotation axes $(\mathbf{1}, \mathbf{i}),(\mathbf{1}, \mathbf{j}),(\mathbf{1}, \mathbf{k}), (\mathbf{i}, \mathbf{j}), (\mathbf{k}, \mathbf{j}), (\mathbf{k}, \mathbf{i}))$, it must be considered *proper*, or second-order circular. Signal properness in $\mathbb{H}$ can be checked by considering that for a proper quaternion random variable $q = q_a + q_b\mathbf{i} + q_c\mathbf{j} + q_d\mathbf{k}$ the following properties hold [21]:

1. $E\left\{q_m^2\right\} = \sigma^2$, $\forall m = a, b, c, d$ (all four components of q have equal power).
2. $E\left\{q_m q_n\right\} = 0$, $\forall m, n = a, b, c, d$ and $m \neq n$ (all four components of q are uncorrelated).
3. $E\left\{qq\right\} = -2E\left\{q_m^2\right\} = -2\sigma^2$, $\forall m = a, b, c, d$ (the pseudocovariance matrix does not vanish).
4. $E\left\{|q|^2\right\} = 4E\left\{q_m^2\right\} = 4\sigma^2$, $\forall m = a, b, c, d$ (the covariance of a quaternion variable is the sum of the covariances of all components).

Since for a quaternion proper signal the complementary covariance matrices, defined as $C_{\mathbf{q}}^{\mathbf{i}} = E\left\{\mathbf{q}\mathbf{q}^{\mathbf{i}H}\right\}, C_{\mathbf{q}}^{\mathbf{j}} = E\left\{\mathbf{q}\mathbf{q}^{\mathbf{j}H}\right\}, C_{\mathbf{q}}^{\mathbf{k}} = E\left\{\mathbf{q}\mathbf{q}^{\mathbf{k}H}\right\}$, vanish, widely linear algorithms incorporate and exploit this second order information on purpose. We need to define the quaternion *involutions* here:

$$
\begin{aligned}
q^{\mathbf{i}} &= -\mathbf{i}q\mathbf{i} = q_a + \mathbf{i}q_b - \mathbf{j}q_c - \mathbf{k}q_d \\
q^{\mathbf{j}} &= -\mathbf{j}q\mathbf{j} = q_a - \mathbf{i}q_b + \mathbf{j}q_c - \mathbf{k}q_d \\
q^{\mathbf{k}} &= -\mathbf{k}q\mathbf{k} = q_a - \mathbf{i}q_b - \mathbf{j}q_c + \mathbf{k}q_d.
\end{aligned} \tag{12.12}
$$

Involutions are functions $f(.)$ chosen in a way that, given $q, p \in \mathbb{H}$, we have the conditions: 1) $f(f(q)) = q$, 2) $f(q + p) = f(q) + f(p)$ and $f(\lambda q) = \lambda f(q)$, 3) $f(qp) = f(q)f(p)$.

The WL-QLMS algorithm updates four sets of filter weights: $\mathbf{w}, \mathbf{h}, \mathbf{u}, \mathbf{v} \in \mathbb{H}^{M \times 1}$, where M is the filter length. Accordingly, the filter output is computed by convoluting each weight vector with its corresponding input involution:

$$y_w\,[n] = \mathbf{w}_{n-1}^T\mathbf{x}_n, \quad y_h\,[n] = \mathbf{h}_{n-1}^T\mathbf{x}_n^{\mathbf{i}}, \quad y_u\,[n] = \mathbf{u}_{n-1}^T\mathbf{x}_n^{\mathbf{j}}, \quad y_v\,[n] = \mathbf{v}_{n-1}^T\mathbf{x}_n^{\mathbf{k}} \tag{12.13}$$

and summing all four contributions:

$$y\,[n] = y_w\,[n] + y_h\,[n] + y_u\,[n] + y_v\,[n]. \tag{12.14}$$

In conclusion, we have four adaptation equations:

$$\mathbf{w}_n = \mathbf{w}_{n-1} + \mu e\,[n]\,\mathbf{x}_n^*, \quad \mathbf{h}_n = \mathbf{h}_{n-1} + \mu e\,[n]\,\mathbf{x}_n^{\mathbf{i}*}$$
$$\mathbf{u}_n = \mathbf{u}_{n-1} + \mu e\,[n]\,\mathbf{x}_n^{\mathbf{j}*}, \quad \mathbf{v}_n = \mathbf{v}_{n-1} + \mu e\,[n]\,\mathbf{x}_n^{\mathbf{k}*}. \tag{12.15}$$

Is it possible to obtain a similar algorithm with tessarines? Well, if we apply the three conditions above in order to find tessarine involutions we obtain the following results ($t \in \mathbb{T}$):

$$t^{\mathbf{i}} = -\mathbf{i}t\mathbf{i} = t_a + \mathbf{i}t_b + \mathbf{j}t_c + \mathbf{k}t_d = t$$
$$t^{\mathbf{j}} = +\mathbf{j}t\mathbf{j} = t_a + \mathbf{i}t_b + \mathbf{j}t_c + \mathbf{k}t_d = t \tag{12.16}$$
$$t^{\mathbf{k}} = -\mathbf{k}t\mathbf{k} = t_a + \mathbf{i}t_b + \mathbf{j}t_c + \mathbf{k}t_d = t.$$

From (12.16), we see that tessarines are auto-involutive, so a widely linear model is possible to the extent that it is defined the same way as for complex numbers [15].

12.3.3 Computational Cost

Since both quaternions and tessarines are 4-dimensional algebras, the computational cost of QLMS and TLMS is the same. In fact, the computation of the filter output requires $4 \cdot 4 \cdot M$ multiplications per sample. The same effort is required in the weight update equation. Overall, the computational cost of QLMS and TLMS is $32 \cdot M \cdot n_{samples}$ multiplications. The situation changes when working with widely linear filters. In tessarine algebra, only the vector $\mathbf{x}_n$ and its conjugate are necessary in the algorithm definition, so whereas the WL-QLMS requires $4 \cdot 32 \cdot M \cdot n_{samples}$ multiplications, the computational cost in the WL-TLMS is reduced to $2 \cdot 32 \cdot M \cdot n_{samples}$.

12.4 Simulations

In this section, we propose two examples in order to make a performance comparison between quaternion and tessarine filtering according to the input signals. The simulation layout is represented in Fig. 12.1. In both simulations we have a system $\mathbf{w}_0$ to be identified, which is defined in the time domain by a set of random weights, uniformly distributed in the range $[-1, 1]$.

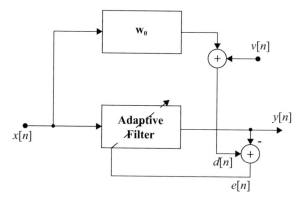

Fig. 12.1 Simulation layout

12.4.1   Generic Circular Input Signals

In this first example, we apply QLMS and TLMS in a context where the input signal $x[n]$ is considered as either a quaternion-valued or tessarine-valued colored noise with unit variance and it was obtained by filtering the white Gaussian noise $\eta[n]$ as $x[n] = bx[n-1] + \frac{\sqrt{1-b^2}}{\sqrt{4}}\eta[n]$, where b is a filtering parameter (here it was chosen as $b = 0.7$). The additive signal $v[n]$ is defined the same way as $x[n]$, but the parameter b is set to zero.

Signal $x[n]$ is circular, all its components are uncorrelated to each other, so, at first glance, it seems to be equivalent to consider it as a quaternion or a tessarine. In effect, our results are concordant with the expectations (Fig. 12.2): given the same filter parameters ($M = 12$, $\mu = 0.008$), the QLMS and TLMS exhibit the same MSE.

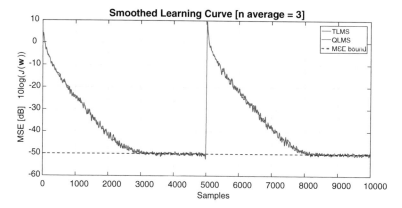

Fig. 12.2 MSE: QLMS versus TLMS with proper input signal

In this simulation, after 5000 samples, the weights $\mathbf{w}_0$ change abruptly. The two filters run after the variation with the same rate.

12.4.2 Ambisonic Improper Audio Input Signals

The second example we propose in this paper makes use of a well-structured 3D audio input signal. This signal has 4 components which were recorded by 4 microphones in accordance with the 3D audio technique called Ambisonics (B-Format). The first order Ambisonic B-Format technique mounts 4 coincident microphones, orthogonal to one another: one omnidirectional microphone (W) and three figure-of-eight microphones (X, Y, Z). Each microphone signal can be assigned to a 4-dimensional algebra component as

$$x[n] = x_W[n] + x_X[n]\mathbf{i} + x_Y[n]\mathbf{j} + x_Z[n]\mathbf{k}. \tag{12.17}$$

However, we want to prove that this assignment is not merely a matter of convenience. In fact, Fig. 12.3 shows how the choice of a different algebra, defining the mathematical space, determines the filter performance. Giving an interpretation of Fig. 12.3, we understand that a B-Format signal is rather inclined to be represented by quaternion algebra than by tessarines (the QLMS converges faster than TLMS on equal terms). In truth, in previous works [12, 13], the authors of this paper found a relation between the sound field as decomposed by Ambisonics and a quaternion-valued representation. Ambisonics decomposes the sound pressure field into a linear combination of *spherical harmonics* and the subgroup SO(3) of 3D Euclidean rotations does have a representation on the $(2m + 1)$-dimensional Hilbert space with spherical harmonics (span $\{Y_{mn}^\sigma(\theta, \phi), 0 \le n \le m, \sigma = \pm 1\}$, where $Y_{mn}^\sigma(\theta, \phi)$ are the spherical harmonics). It is known that the subspace of pure quaternions (those quaternions with null real component) is isomorphic to rotations. That said, the quaternion representation of Ambisonics does not simply consist in a compact formalism, but it has a physical and geometrical meaning.

In our simulation, the source is a monodimensional unit-variance white Gaussian noise in a computer-generated anechoic room. The source was placed at a distance of 20 cm from the B-Format array, 45° off-axis with the X microphone. Additive unit-variance white Gaussian noise $\nu[n] \in \mathbb{H}$, with $n = 0, 2..., P - 1$, was summed to the output signal of the system to be identified ($d[n] = \mathbf{w}_0^T\mathbf{x} + \nu[n]$). The filter parameters were chosen as $M = 12, \mu = 0.3$.

In addition, in Sect. 12.3.2, we emphasized the possibility to build a widely linear algorithm (WL-QLMS, WL-TLMS). Since the Ambisonic B-Format is improper, we expect the WL-QLMS and WL-TLMS to outperform QLMS and TLMS, respectively. The results from our simulation meet the expectations.

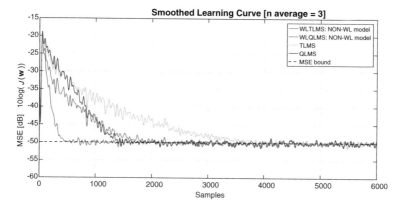

Fig. 12.3 MSE: (WL-)QLMS versus (WL-)TLMS with Ambisonic improper input signal. *NON-WL model* in the legend box refers to a system to be identified which only has $\mathbf{w}_0$ weights

12.5   Conclusion

Hypercomplex algebras are decisively making their own way in adaptive filtering applications. The question today is whether we really need such hypercomplex models to represent our systems. Besides that, what determines the rejection of one algebra in favor of another? In this paper, we proposed a simple comparison between two 4-dimensional hypercomplex algebras: quaternions and tessarines. We learned from simulations that some systems can be considered either quaternion-valued or tessarine-valued. In other cases, the choice of the algebraic representation determines the performance of the whole system. For instance, we introduced the Ambisonic B-Format signals into a 4-dimensional system and we saw that a quaternion adaptive algorithm converges much faster than its tessarine counterpart. We have found a relation between spherical harmonics/rotations and quaternions. The group of rotations and quaternions are both non-commutative. There is no equivalent in tessarine algebra, which is in fact commutative. However, further investigation may discover environments where tessarine processing is the most appropriate. In a next work, we are going to publish results from simulations in Ambisonic and Uniform Linear Array contexts, where in both cases the signals at the sensors are correlated. We saw that there are geometries in which a faster convergence is reached by means of tessarine algorithms.

References

1. Barthélemy, Q., Larue, A., Mars, J.I.: About qlms derivations. IEEE Trans. Signal Process. Letters **21**(2), 240–243 (2014)
2. Gerzon, M.A.: Ambisonics part two: studio techniques. Stud Sound 17, 24–26, 28–30 (1975)

3. Gerzon, M.A.: Ambisonics in multichannel broadcastingand video. J. Audio Eng. Soc. **33**(11), 859–871 (1985)
4. Jahanchahi, C., Took, C., Mandic, D.: A class of quaternion valued affine projection algorithms. Signal Process. **93**, 1712–1723 (2013)
5. Jahanehahi, C., Took, C.C., Mandic, D.P.: The widely linear quaternion recursive least squares filter. In: Proceedings of the 2nd International Workshop Cognitive Information Processing (CIP 2010)
6. Katunin, A.: Three-dimensional octonion wavelet transform. J. Appl. Math. Comput. Mech. **13**(1), 33–38 (2014)
7. Kraft, E.: A quaternion-based unscented kalman filter for orientation tracking. In: Proceedings of the 6th International Conference Information Fusion, pp. 47 – 54 (ISIF 2003)
8. Li, X., Adalı, T.: Complex-valued linear and widely linear filtering using mse and gaussian entropy. IEEE Trans. Sig. Proc. **60**(11), 5672–5684 (2012)
9. Mandic, D.P., Goh, V.S.L.: Complex valued nonlinear adaptive filters: noncircularity, widely linear and neural models. Wiley (2009)
10. Neeser, F.D., Massey, J.L.: Proper complex random processes with applications to information theory. IEEE Trans. Inf. Theory **39**(4), 1293–1302 (1993)
11. Ortolani, F., Comminiello, D., Scarpiniti, M., Uncini, A.: Frequency domain quaternion adaptive filters: algorithms and convergence performance. Sig. Process. **136**, 69–80 (2017)
12. Ortolani, F., Comminiello, D., Uncini, A.: The widely linear block quaternion least mean square algorithm for fast computation in 3d audio systems. In: Proceedings of the 26th International Workshop on Machine Learning for Signal Processing (Sep MLSP, 2016)
13. Ortolani, F., Uncini, A.: A new approach to acoustic beamforming from virtual microphones based on Ambisonics for adaptive noise cancelling. In: IEEE 36th International Conference on Electronics and Nanotechnology (ELNANO) (2016)
14. Picinbono, B.: On circularity. IEEE Trans. Sig. Proc. **42**(11), 3473–3482 (1994)
15. Picinbono, B., Chevalier, P.: Widely linear estimation with complex data. IEEE Trans. Sig. Proc. **43**(8), 2030–2033 (1995)
16. Rumsey, F.: Spatial Audio. Focal Press (2001)
17. Said, S., Bihan, N.L., Sangwine, S.J.: Fast complexified quaternion fourier transform. IEEE Trans. Signal Process. **56**(4), 1522–1531 (2008)
18. Shoemake, K.: Animating rotation with quaternion calculus. ACM SIGGRAPH Course Notes
19. Took, C.C., Mandic, D.P.: The quaternion lms algorithm for adaptive filtering of hypercomplex processes. IEEE Trans. Signal Process. **57**(4), 1316–1327 (2009)
20. Took, C., Mandic, D.P.: A quaternion widely linear adaptive filter. IEEE Trans. Signal Process. **58**(8), 4427–4431 (2010)
21. Took, C., Mandic, D.P.: Augmented second-order statistics of quaternion random signals. Signal Process. **91**(2), 214–224 (2011)
22. Vakhania, N.N.: Random vectors with values in quaternion Hilbert spaces. Theor. Probab. Appl. **43**(1), 18–40 (1998)

Chapter 13
Separation of Drum and Bass from Monaural Tracks

Michele Scarpiniti, Simone Scardapane, Danilo Comminiello, Raffaele Parisi and Aurelio Uncini

Abstract In this paper, we propose a deep recurrent neural network (DRNN), based on the Long Short-Term Memory (LSTM) unit, for the separation of drum and bass sources from a monaural audio track. In particular, a single DRNN with a total of six hidden layers (three feedforward and three recurrent) is used for each original source to be separated. In this work, we limit our attention to the case of only two, challenging sources: drum and bass. Some experimental results show the effectiveness of the proposed approach with respect to another state-of-the-art method. Results are expressed in terms of well-known metrics in the field of source separation.

Keywords Deep recurrent neural networks · Long short-term memory · Monaural audio source separation · Non-negative matrix factorization

13.1 Introduction

The separation of multiple sound sources from a single recording, or Single Channel Source Separation (SCSS), is a challenging problem that is receiving a great attention from researchers in the last decade [14]. However, while the standard Blind Source

M. Scarpiniti (✉) · S. Scardapane · D. Comminiello · R. Parisi · A. Uncini
Department of Information Engineering, Electronics and Telecommunications (DIET),
"Sapienza" University of Rome, Rome, Italy
e-mail: michele.scarpiniti@uniroma1.it

S. Scardapane
e-mail: simone.scardapane@uniroma1.it

D. Comminiello
e-mail: danilo.comminiello@uniroma1.it

R. Parisi
e-mail: raffaele.parisi@uniroma1.it

A. Uncini
e-mail: aurelio.uncini@uniroma1.it

© Springer International Publishing AG, part of Springer Nature 2019
A. Esposito et al. (eds.), *Neural Advances in Processing Nonlinear
Dynamic Signals*, Smart Innovation, Systems and Technologies 102,
https://doi.org/10.1007/978-3-319-95098-3_13

Separation (BSS) is a well-known and well-studied field in the adaptive signal processing and machine learning communities [4, 5], the problem of recovering original sources from only one recorded signal provides no very good solution yet.

SCSS is a challenging problem because, without any additional constraints, an infinite number of solutions can be found since the problem is generally ill-posed [10, 14, 16, 18]. In order to solve this ill-posed problem, several approaches have been derived in literature [14].

Specifically, [6] proposed the use of Empirical Mode Decomposition (EMD), [13] tries an approach based on subspaces in an Hilbert space, while [12] operates in the Wavelet domain by using a Bark scale. Moreover, the work [1] is based on a sparsification transformation in order to enhance the source separation. However, results obtained with such approaches are not satisfactory.

A class of approaches that has raised a great attention in literature is composed by a maximum likelihood method in a Gaussian framework [2, 10], in which some basis functions are learned and then used to reconstruct the single sources. Another powerful method for SCSS is the non-negative matrix factorization (NMF) [16], used to learn the non-negative reconstruction bases and weights of different sources and use them to factorize time-frequency spectral representations, followed by a masking operation [15]. This latter approach provides good results [20].

However, since audio and music are sequential by nature, recurrent neural networks (RNNs) respecting temporal dynamics have emerged as a powerful tool for this class of signals. In particular, RNNs have been used for automatic speech recognition (ASR) and speech enhancement [21].

Due to the difficulty of the proposed task, some authors have recently introduced deep neural networks (DNNs) for the SCSS [7], obtaining very interesting results. DNNs are neural networks with many (two or more) hidden layers. To this purpose, [9] has approached SCSS by a deep recurrent neural network (DRNN). However, the main issue in using RNNs is that their learning could be very difficult due to the vanishing or exploding gradient problem. In order to overcome these issues, Hochreiter and Schmidhuber have introduced in [8] some RNNs that use a new block called Long-Short Term Memory (LSTM) unit instead of the classical neuron.

In this paper, we propose a DRNN based on the LSTM unit to separate two audio sources from a single channel recording. In particular, we focus our attention on the separation of drum and bass sources. The choice of these two kinds of musical instruments is due to the fact that the bass and drum are the backbone of any music track, since they form the rhythm section. Moreover, the separation of these sources is very challenging because they are strongly correlated at both the spectral and the temporal level, and, in addition, the drum is also made up of different synchronous components (bass drum, snare, toms, hi-hat, cymbals, etc.).

The rest of the paper is organized as follows. In Sect. 13.2 we briefly introduce the model of SCSS, while Sect. 13.3 describes the LSTM unit. The proposed approach is provided in Sect. 13.4. Finally, we validate our approach in Sect. 13.5 and we conclude with some final remarks in Sect. 13.6.

Fig. 13.1 Data organization
in a tensor form

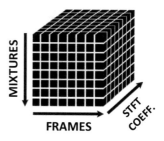

13.2 The Model of Single Channel Source Separation

Let us consider a set of P unknown and statistically independent sources denoted as $\mathbf{s}_n = [s_1[n], \ldots, s_P[n]]^T$, such that the components $s_i[n]$ are zero-mean and mutually independent [10]. The recorded signal $\mathbf{x}_n$ can be seen as

$$\mathbf{x}_n = \lambda_1 s_1[n] + \lambda_2 s_2[n] + \ldots + \lambda_P s_P[n] = \boldsymbol{\lambda}^T \mathbf{s}_n, \qquad (13.1)$$

where $\boldsymbol{\lambda} = [\lambda_1, \lambda_2, \ldots, \lambda_P]^T$ collects the gain λ_i of the i-th source, which is constant over time.

The mixture signal $\mathbf{x}_n$ is divided in frames $x[n]$ by a window function $w[n]$ of length N and then a Short-Time Fourier Transform (STFT) is evaluated

$$X_n(\omega) = \sum_{l=-\infty}^{\infty} x[l] w[l - nN] e^{-j\omega l}. \qquad (13.2)$$

In this work we used a Hanning window, an overlap of 50% and $N = 8192$ samples. Moreover, in order to provide a compact and powerful data representation, we organize data in a tensor form, arranging them by frames, frequency bins and (eventually) different mixtures, as shown in Fig. 13.1.

One of the first attempt to recover original sources $s_i[n]$ from the single channel mixture $\mathbf{x}_n$ in (13.1) is based on the modeling of sources $s_i[n]$ with suitable basis function that are learned by an ICA algorithm based on the maximum likelihood approach [10].

In order to overcome the poor solution obtained by [10], the author in [20] proposed an approach based on the NMF. The idea is to apply the NMF to factorize the magnitude spectra obtained by (13.2) to obtain spectra of original sources and a matrix of the related mixing weights.

13.3 The Long Short-Term Memory Cells

Recurrent neural networks (RNNs) are particular neural networks with feedback loops, meaning that outputs of a hidden layer are fed again to its inputs. The learning of such a network is generally performed by time unfolding of the whole layers and then updating the weight vector by an extension of the backpropagation algorithm, called backpropagation through time (BPTT). The main issue in this approach, particularly emphasized in deep RNNs (DRNNs), is the vanishing gradient problem that obstacles the network performance. In order to solve this issue, a variation of RNNs with the so-called Long Short-Term Memory (LSTMs) units, was proposed by Hochreiter and Schmidhuber [8]. LSTMs are gated cells and they preserve the error by maintaining it quite constant over many time steps.

The LSTM units are capable of learning long-term dependencies reducing the problems on the gradient. An LSTM introduces a new structure called *memory cell*, that is composed of four main elements: an input gate, a neuron with a self-recurrent connection (a connection to itself), a forget gate and an output gate, as shown in Fig. 13.2 (compared with a standard recurrent neuron). The basic idea is to adopt three gates that can avoid the gradient to diverge. Let us denote with $\mathbf{x}_n$ and $\mathbf{h}_n$ the input to the LSTM unit and the hidden state at time-step n. Then, denoting with $\mathbf{W}_i$, $\mathbf{W}_f$, $\mathbf{W}_c$, $\mathbf{W}_o$, $\mathbf{U}_i$, $\mathbf{U}_f$, $\mathbf{U}_c$, $\mathbf{U}_o$ and $\mathbf{V}_o$ some weight matrices and with $\mathbf{b}_i$, $\mathbf{b}_f$, $\mathbf{b}_c$ and $\mathbf{b}_o$ suitable bias vectors, we first compute the values for $\mathbf{i}_n$, the input gate, and $\mathbf{z}_n$, the candidate value for the states of the memory cells at time n:

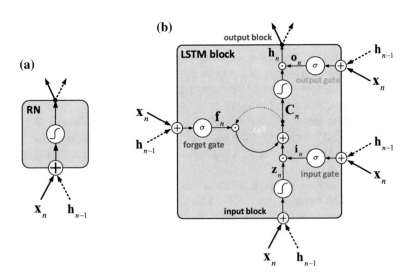

Fig. 13.2 Details of a simple recurrent neuron (RN) (**a**) and a LSTM unit (**b**), used in hidden layers

$$\mathbf{i}_n = \sigma \left(\mathbf{W}_i \mathbf{x}_n + \mathbf{U}_i \mathbf{h}_{n-1} + \mathbf{b}_i \right), \tag{13.3}$$

$$\mathbf{z}_n = \tanh \left(\mathbf{W}_c \mathbf{x}_n + \mathbf{U}_c \mathbf{h}_{n-1} + \mathbf{b}_c \right), \tag{13.4}$$

where $\sigma(\cdot)$ represents the sigmoid function. Second, we compute the value for $\mathbf{f}_n$, the activation of the memory cells' forget gates at time n:

$$\mathbf{f}_n = \sigma \left(\mathbf{W}_f \mathbf{x}_n + \mathbf{U}_f \mathbf{h}_{n-1} + \mathbf{b}_f \right), \tag{13.5}$$

Then, given the values of the input gate activation $\mathbf{i}_n$, the forget gate activation $\mathbf{f}_n$ and the candidate state value $\mathbf{z}_n$, we can compute $\mathbf{C}_n$, the memory cells' new state at time n:

$$\mathbf{C}_n = \mathbf{i}_n \mathbf{z}_n + \mathbf{f}_n \mathbf{C}_{n-1}, \tag{13.6}$$

Finally, with the new state of the memory cells $\mathbf{C}_n$, we can compute the value of their output gates $\mathbf{o}_n$ and, subsequently, their outputs $\mathbf{h}_n$:

$$\mathbf{o}_n = \sigma \left(\mathbf{W}_o \mathbf{x}_n + \mathbf{U}_o \mathbf{h}_{n-1} + \mathbf{V}_o \mathbf{C}_n + \mathbf{b}_o \right), \tag{13.7}$$

$$\mathbf{h}_n = \mathbf{o}_n \tanh \left(\mathbf{C}_n \right). \tag{13.8}$$

13.4 Proposed Approach

The proposed network for the solution of the SCSS problem is a DRNN containing six hidden layers: the first three layers are feed-forward layers, here called *dense* layers, while the latter ones are recurrent layers with LSTM units, as schematically shown in Fig. 13.3 (for simplicity only two layers of each type are shown).

The whole architecture is trained by the RMSprop algorithm. RMSprop is an unpublished and adaptive learning rate method proposed by Geoff Hinton in [17]. The idea of RMSprop is that the learning rate of a gradient adaptation is recursively rescaled as a decaying average of all past squared gradients. Let us denote with $\boldsymbol{\theta}_n$ and $\mathbf{g}_n$ the parameters vector and gradient vector at time n, respectively. Then the RMSprop algorithm assumes the following formulation

$$\boldsymbol{\theta}_{n+1} = \boldsymbol{\theta}_n - \frac{\eta}{\sqrt{E\left\{\mathbf{g}_n^2\right\} + \varepsilon}} \mathbf{g}_n, \tag{13.9}$$

$$\sqrt{E\left\{\mathbf{g}_n^2\right\}} = \gamma \sqrt{E\left\{\mathbf{g}_{n-1}^2\right\}} + (1-\gamma)\,\mathbf{g}_n^2, \tag{13.10}$$

where γ is a forgetting factor similar to the Momentum term and ε is a regularization parameter. Hinton suggests to set $\gamma = 0.9$ and the learning rate to $\eta = 0.001$.

The proposed network works in the frequency domain by considering the magnitude spectra of source signals. In our case, with reference to the work in [9], the cost function J_{MSE} to be optimized is the squared error

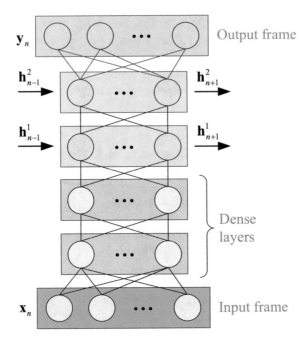

Fig. 13.3 Scheme of the DRNN used for SCSS. The hidden layers are obtained by stacking dense layers and recurrent layers

$$J_{MSE} = \frac{1}{2} \sum_{n=1}^{L} \left(\|\mathbf{y}_{1n} - \mathbf{s}_{1n}\|_2^2 + \|\mathbf{y}_{2n} - \mathbf{s}_{2n}\|_2^2 \right), \tag{13.11}$$

where $\mathbf{y}_{1n}$ and $\mathbf{y}_{2n}$ are the estimates of the original sources $\mathbf{s}_{1n}$ and $\mathbf{s}_{2n}$, respectively, and $n = 1, 2, \ldots, L$ with L being the length of the input sequence.

In the following we restrict our analysis to the case of a mixture composed by only two sources. However, a generalization to a greater number of sources is straightforward.

The training phase is shown in Fig. 13.4 and consist in the evaluation of the magnitude of the STFT after the division of the input signal in overlapped frames and a scaling procedure to ensure that every sample is inside the interval $[-1, \ 1]$ in order to avoid numeric problems during the learning procedure. The overall frames are arranged in the tensor representation shown in Fig. 13.1. Hence, the obtained signal is used to train two separate DRNNs, one for each original source.

In the testing phase, the two learned DRNNs are used to separate the two original sources from an unknown mixture and to verify the effectiveness of the proposed network. The scheme for the testing phase is shown in Fig. 13.5.

From the unknown mixture, the system evaluates the magnitude spectrum after the division in frames. The phase of the spectrum is extracted and then utilized in the reconstruction procedure. After the usual scaling and tensor data organization, the

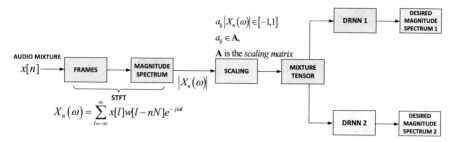

Fig. 13.4 Scheme of the training phase

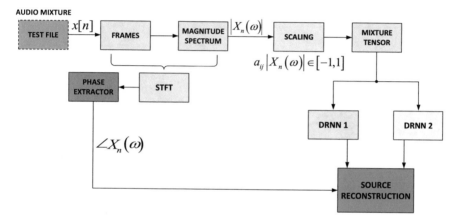

Fig. 13.5 Scheme of the testing phase

two DRNNs provide an estimate of magnitude spectra of the original sources. These estimates, along with the mixture phase, are sent to a reconstruction block, shown in Fig. 13.6 for a single source.

The reconstruction of the estimated sources is done by unscaling the network output and by a post-processing consisting in a time-frequency mask $\mathbf{m}_{kn}$ applied to the k-th solution. This mask is defined as

$$\mathbf{m}_{kn} = \frac{|Z_{1n}(\omega)|}{|Z_{1n}(\omega)| + |Z_{2n}(\omega)|} \tag{13.12}$$

where the $Z_{kn}(\omega) = b_{ij} Y_{kn}(\omega)$ is the unscaled version of the network output. This mask is then used to filter the magnitude spectrum of the mixture, obtaining the reconstructed signal after the addition of the mixture phase. The final signal $z_k[n]$ in the time domain is obtained by applying the inverse STFT. An eventual block to evaluated the effectiveness of the obtained solution can also be used.

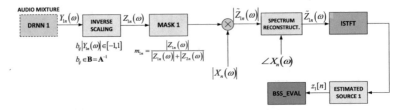

Fig. 13.6 Scheme for the reconstruction of the separated sources

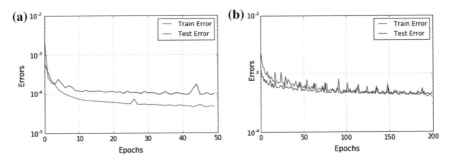

Fig. 13.7 Training and test errors for: bass (**a**) and drum (**b**) sources

13.5 Experimental Results

Results of the proposed approach are evaluated in terms of three metrics: Source to Interference Ratio (SIR), Source to Artifacts Ratio (SAR), and Source to Distortion Ratio (SDR). The definition of these metrics is available in [19] and evaluated according to the related BSS-EVAL toolbox. These metrics are evaluated by splitting the estimated source as in

$$z[n] = s[n] + e_{\text{interf}}[n] + e_{\text{noise}}[n] + e_{\text{artif}}[n], \tag{13.13}$$

where $e_{\text{interf}}[n]$, $e_{\text{noise}}[n]$ and $e_{\text{artif}}[n]$ are terms that quantify the interference, additive noise and artifacts over the target signal $s[n]$. Comparisons are performed with respect the separation method using the NMF approach in [11, 16].

The training has been performed on the MedleyDB dataset,[1] a 40GB dataset of annotated, royalty-free multitrack recordings for academic research and collected from the New York University (NYU) [3]. From this dataset, we collected all tracks of drum and bass, and then separately mixed them according to (13.1) in order to construct the polyphonic track to train and test our architecture. A total of 100 different tracks are used for the training and 20 tracks for testing. Each hidden layer of the DRNN used for bass (resp. drum) separation has 100 (resp. 200) neurons. The training and test errors for the proposed network are shown in Fig. 13.7.

---

[1] Available at: http://medleydb.weebly.com/.

Table 13.1 Results in terms of SIR, SAR and SDR of some test tracks for the proposed DRNN

	Bass			Drum		
	SIR	SAR	SDR	SIR	SAR	SDR
Track 1	29.87	19.38	17.13	16.27	13.28	10.68
Track 2	17.56	17.14	14.21	14.82	15.92	11.74
Track 3	23.76	22.32	18.89	16.77	15.22	12.16
Track 4	22.44	20.87	18.43	17.03	16.51	13.32

Table 13.2 Results in terms of SIR, SAR and SDR of some test tracks for the NMF approach

	Bass			Drum		
	SIR	SAR	SDR	SIR	SAR	SDR
Track 1	25.31	18.72	16.85	15.86	11.74	9.28
Track 2	13.74	9.76	9.26	10.69	15.45	7.05
Track 3	20.75	19.71	17.65	15.05	12.03	11.47
Track 4	20.51	19.63	16.83	16.07	14.01	11.29

Results in terms of the metrics SIR, SAR and SDR for four different test tracks, randomly selected from the dataset, are summarized in Table 13.1, while the related comparisons of NMF approach are shown in Table 13.2. From these tables we can

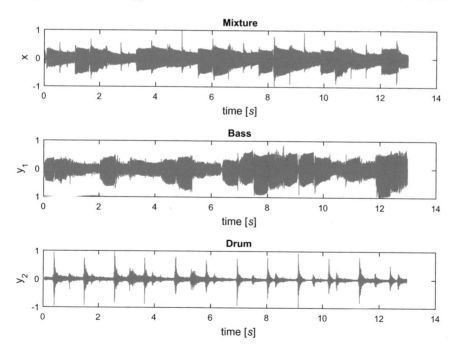

Fig. 13.8 Waveform of the mixture track (first row), estimated bass source (second row) and estimated drum source (third row)

see that the proposed approach outperforms the NMF one in each metric. This fact is also validated by listening to the separated sources and by the inspection of Fig. 13.8 that show the mixture of a random track in the first row and the estimated bass and drum sources in the second and third row respectively.

13.6 Conclusions

In this paper, we have proposed a deep recurrent neural network (DRNN) that uses the LSTM units for solving the challenging problem of the source separation from a single channel audio mixture. In particular, we focus our attention on the separation of bass and drum instruments from monophonic music. However, the approach is simply extendible to a greater number of sources. Some experimental results, implemented by using a DRNN with six hidden layers (three feed-forward and three fully recurrent) and evaluated in terms of the well known metrics SIR, SAR and SDR, have shown the effectiveness of the proposed approach.

References

1. Asari, H., Olsson, R.K., Pearlmutter, B.A.: Sparsification for monaural source separation. In: Makino, S., Lee, T.W., Sawada, H. (eds.) Blind Speech Separation, Chap. 14, pp. 387–410. Springer (2007)
2. Beierholm, T., Dam Pedersen, B., Winthert, O.: Low complexity bayesian single-channel source separation. In: Proceedings of IEEE International Conference on Acoustics, Speech, and Signal Processing (ICASSP 2004) (2004)
3. Bittner, R., Salamon, J., Tierney, M., Mauch, M., Cannam, C., Bello, J.P.: MedleyDB: a multitrack dataset for annotation-intensive MIR research. In: 15th International Society for Music Information Retrieval Conference, pp. 1–6. Taipei, Taiwan (2014)
4. Cichocki, A., Amari, S.: Adaptive Blind Signal and Image Processing. Wiley (2002)
5. Comon, P., Jutten, C. (eds.): Handbook of Blind Source Separation. Springer (2010)
6. Gao, B., Woo, W.L., Dlay, S.S.: Single-channel source separation using EMD-subband variable regularized sparse features. IEEE Trans. Audio Speech Lang. Process. **19**(4), 961–976 (2011)
7. Grais, E.M., Sen, M.U., Erdogan, H.: Deep neural networks for single channel source separation. In: 2014 IEEE International Conference on Acoustic, Speech and Signal Processing (ICASSP 2014), pp. 1–5. Florence, Italy, 4–9 May 2014
8. Hochreiter, S., Schmidhuber, J.: Long short-term memory. Neural Comput. **9**(8), 1735–1780 (1997)
9. Huang, P.S., Kim, M., Hasegawa-Johnson, M., Smaragdis, P.: Joint optimization of masks and deep recurrent neural networks for monaural source separation. IEEE/ACM Trans. Audio, Speech Lang. Process. **23**(12), 1–12 (2015)
10. Jang, G.J., Lee, T.W.: A maximum likelihood approach to single-channel source separation. J. Mach. Learn. Res. **4**(12), 1365–1392 (2003)
11. Lee, D.D., Seung, H.S.: Learning the parts of objects by nonnegative matrix factorization. Nature **401**(6755), 788–791 (1999)
12. Litvin, Y., Cohen, I.: Source separation using Bark-scale wavelet packet decompostion. In: Proceedings of the IEEE International Workshop on Machine Learning for Signal Processing (MLSP2009), pp. 1–4. Grenoble, France, 1–4 Sept 2009

13. Molla, K., Hirose, K.: Single-mixture audio source separation by subspace decomposition of Hilbert spectrum. IEEE Trans. Audio Speech Lang. Process. **15**(3), 893–900 (2004)
14. Patki, K.: Review of single channel source separation techniques. In: Proceedings of the 14th International Society for Music Information Retrieval Conference (ISMIR 2013), pp. 1–5. Curitiba, Brasil, 4–8 Nov 2013
15. Reddy, A.M., Raj, B.: Soft mask methods for single-channel speaker separation. IEEE Trans. Audio Speech Lang. Process. **15**(6), 1766–1776 (2007)
16. Smaragdis, P., Brown, J.C.: Non-negative matrix factorization for polyphonic music transcription. In: Proceedings of the IEEE Workshop on Applications of Signal Processing to Audio and Acoustics, pp. 177–180, 19–22 Oct 2003
17. Tieleman, T., Hinton, G.: Lecture 6.5—RMSProp. Tech. rep., COURSERA: Neural Networks for Machine Learning (2012)
18. Uncini, A.: Fundamentals of adaptive signal processing. In: Signals and Communication Technology. Springer International Publishing, Switzerland (2015)
19. Vincent, E., Gribonval, R., Fevotte, C.: Performance measurement in blind audio source separation. IEEE Trans. Audio Speech Lang. Process. **14**(4), 1462–1469 (2006)
20. Virtanen, T.: Monaural sound source separation by non-negative matrix factorization with temporal continuity and sparseness criteria. IEEE Trans. Audio Speech Lang. Process. **15**(3), 1066–1074 (2007)
21. Weninger, F., Eyben, F., Schuller, B.: Single-channel speech separation with memory-enhanced recurrent neural networks. In: Proceedings of the 2014 IEEE International Conference on Acoustics, Speech and Signal Processing (ICASSP 2014), pp. 3709–3713. Florence, Italy, 4–9 May 2014

Chapter 14
Intelligent Quality Assessment of Geometrical Features for 3D Face Recognition

G. Cirrincione, F. Marcolin, S. Spada and E. Vezzetti

Abstract This paper proposes a methodology to assess the discriminative capabilities of geometrical descriptors referring to the public Bosphorus 3D facial database as testing dataset. The investigated descriptors include histogram versions of Shape Index and Curvedness, Euclidean and geodesic distances between facial soft-tissue landmarks. The discriminability of these features is evaluated through the analysis of single block of features and their meanings with different techniques. Multilayer perceptron neural network methodology is adopted to evaluate the relevance of the features, examined in different test combinations. Principle Component Analysis (PCA) is applied for dimensionality reduction.

Keywords 3D face recognition · Geometrical descriptors
Dimensionality reduction · Principal component analysis · Neural network

14.1 Introduction

3D face recognition has been deeply investigated in the last decades due to the large number of applications in both security and safety domains, even in real-time scenarios. The third dimension improves accuracy and avoids problems like lighting and make-up variations. In addition, it allows the adoption of geometrical features to study and describe the facial surface.

In this work, the second principal curvature, indicated by k_2, the shape index (S) and the curvedness (C) are used. We rely on the formulations given by Do

F. Marcolin · S. Spada (✉) · E. Vezzetti
DIGEP, Politecnico Di Torino, Turin, Italy
e-mail: s219665@studenti.polito.it

G. Cirrincione
Laboratory LTI, Université de Picardie Jules Verne, Amiens, France
e-mail: exin@u-picardie.fr

G. Cirrincione
University of South Pacific, Suva, Fiji

© Springer International Publishing AG, part of Springer Nature 2019
A. Esposito et al. (eds.), *Neural Advances in Processing Nonlinear Dynamic Signals*, Smart Innovation, Systems and Technologies 102,
https://doi.org/10.1007/978-3-319-95098-3_14

Carmo [1] for the principal curvatures and by Koenderink and van Doorn for the shape index and curvedness [2]. The shape index describes the shape of the surface. Koenderink and van Doorn proposed a partition of the range $[-1, 1]$ in 9 categories, which correspond to 9 different surfaces, ranging from cup to dome/cap, but other representations exist [3, 4]. The partition taken into consideration in this work relies on 7 categories, ranging from cup ($S \in [-1; -0.625 [$) to dome ($S \in] 0.625; 1]$); values of S approximately equal to 0 ($S \in [-0.125; +0.125]$) correspond to saddle-type surfaces. The curvedness is a measure of how highly or gently curved a point is and is defined as the distance from the origin in the (k_1, k_2)-plane. Point-by-point maps of these descriptors of faces we refer to previous work of Vezzetti and Marcolin [4, 5].

The shape index was recently adopted by Quan et al. [6] to build a 3D shape representation scheme for FR relying on the combination of statistical shape modelling and non-rigid deformation matching. The method was tested and compared on the BU-3DFE and GavabDB databases and obtained competitive results even in presence of various expressions. The same descriptor was introduced by Ming [7] for searching nose borders as saddle rut-like shapes in a 3D facial regional segmentation preceding FR relying on a regional and global regression mapping (RGRM) technique. Experiments run on FRGC v2, CASIA, and BU-3DFE databases involving variously "emotioned" faces proved the satisfactory accuracy of the methodology. Ganguly et al. proposed a Range Face image Recognition System (RaFaReS) where the principal curvatures and the curvedness are used as descriptors of the facial surface. The usability of the system was validated on the Frav3D database and results ranged from 89.21 to 94.09% RR depending on the features selected and on the classifier (k-NN and three-layer MLP backpropagation neural network) [8].

In the present work, 7-bins histogram versions of the descriptors are adopted. The choice of 7 bins is given by the definition of the shape index give above. For manageability reasons, the same number of bins is kept for the other descriptors. Besides these descriptors, other features are taken into consideration: the nose volume, Euclidean distances and geodesic distances between typical facial fiducial points, which lie on the skin, called landmarks. The landmarks used here to evaluate these distances are: OE-outer eyebrow, IE-inner eyebrow, EX-exocanthion, EN-endocanthion, N-nasion, AL-alare, PRN-pronasal, SN-subnasal. Except for the eyebrow points, which are not considered real soft-tissue landmarks, their morphometric definitions are provided by Swennen et al. [9].

This paper proposes a methodology to analyze the geometric descriptors and assess their discriminative capabilities. These descriptors are divided in four classes: Euclidean, curvature, shape index and geodesic. Section 14.2 classifies the whole database by using all the descriptors. Section 14.3 introduces the methodology and analyzes the descriptors.

Fig. 14.1 The confusion matrix from MLP as a colored heat map

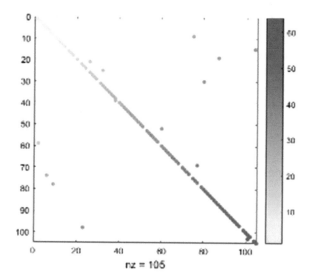

$$rz = 105$$

14.2   The 3D Face Database and Its Classification

The dataset is composed of 211 geometrical descriptors (grouped in 4 classes) of 741 faces from 105 subjects. The associated matrix is 741×211. It is statistically normalized by columns (z-score) for having the same range for each descriptor.

The 3D face recognition (classification) is performed by using a Multilayer Perceptron (MLP) [10]. It has a unique hyperbolic tangent hidden layer and a softmax output activation function. The error cost is given by the cross-entropy error. Consequently, the networks outputs the conditional probabilities of class membership.

Training is performed by SCG, which makes use of the error derivatives estimated by the backpropagation algorithm. The network is validated by dividing the dataset in 636 samples for the training set and the remaining ones (one face per person) for validation. All descriptors are taken in account. Hence MLP has 191 inputs. The number of hidden neurons for the best bias variance trade-off is 300 for the 90.04% classification rate in validation. Figure 14.1 represents the confusion matrix as a heat map, just showing that is very sparse. However, the small size of the training set may imply the hidden neurons work as templates for the faces. Therefore, this result is not enough satisfactory.

The fact of using all features, with the goal of exploiting their redundancy, yields a maximum achievable accuracy for the classification. The analysis which follows does not try to improve on it. Instead, by deleting groups of descriptors, uses the classification rate of success in order to detect their relative importance. In this sense, the neural approach is here considered as a probe of the sensitivity of the classification to the choice of features.

14.3 Analysis of Geometrical Descriptors

The methodology for the analysis of the geometric descriptors has a two-fold aspect. It does not only consider the advantages and limits for each descriptor in itself and with regard to the other ones, but also makes use of statistical and neural techniques in an unconventional way, i.e. as probes driven by data, for assessing the conclusions.

The statistical tool used here are the following:

- The Principal Component Analysis (PCA), here estimated by solving the SVD problem. Basically, the first principal component (PC) describes the average features of a face, while the subsequent PCs are related to the variance, and so are more important for face recognition.
- The k-means algorithm [11] for clustering.
- The biplots [12] for high dimensional data visualization. They are a generalization of scatter plots. A biplot allows information on both samples and variables of a data matrix to be displayed graphically. Samples are displayed as points while variables are displayed as vectors.

The neural techniques are given by:

- The curvilinear component analysis (CCA [13]). It is a self-organizing neural network for dimensionality reduction whose neurons are equipped with two weights, the first for the input vector quantization and the second for the projection, which is performed by preserving as many distances as possible in the input space (only distances below a certain threshold have to be invariant in the projection). In particular, the associated dy-dx diagram is important for the following study. This plot represents distances between pairs of points in the input space (the dx value) and in the reduced (latent) space (the dy value) as a pair (dy, dx). If a distance is preserved in the projection, the corresponding pair is on the bisector. If the pair is under the bisector, it represents the projection as an unfolding of the input data manifold. If the manifold is linear, all points tend to lie on or around (because of noise) of the bisector.
- The MLP, configured as seen before, for outputting the conditional probabilities of class membership [10].

More in detail, the steps followed in this work are:

- Analysis of the data (face) manifold: by means of CCA and PCA, the intrinsic dimensionality and the nonlinearity are evaluated, with the aim of defining the smallest set of independent descriptors, in the sense of minimum valuable information.
- Estimation of the descriptor class importance for classification: MLP is used in order to evaluate the quality of one class in classification, both in using only the class of descriptors (single case) and in using the whole matrix except that one (subtractive case); the results are only qualitative and cannot be exactly complementary.

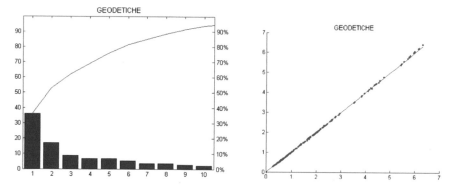

Fig. 14.2 Explained variance by PCA (left) and dy-dx plot by CCA (right)

- Assessment of the ability both in discriminating subjects and in clustering the faces as belonging to the same person: it is achieved by the k-means algorithm and checked in a supervised way by counting the number of correct (recognized) labels for each cluster, under the assumption all faces of the same subject (same labels) belong to the same cluster. However, this is a weak hypothesis, because of the possibly different scenarios (e.g. illumination), noise in data and so on. Indeed, using all the descriptors yields a success rate of only 58.44%. Considering that a good clustering cannot be appreciated only in the single case, only the subtractive case is taken in account.
- Study of the statistical properties (covariance) of the descriptor classes and mutual interaction between faces and descriptors; classes are analyzed by using the whole database and in relation to the face information in a double PCA analysis: one in the face space, the other in the descriptor space by interpreting the corresponding biplot.

14.3.1   Geodesic Distances

These distances are more accurate than the Euclidean distances Indeed, in 3D, the geodesic distance between two points on a surface is computed as the length of the shortest path connecting the two points. As a consequence, true distances between facial soft-tissue landmarks are estimated. Instead, the Euclidean distances do not capture this information because they do not follow the shape of the face.

By using PCA on this group of 22 descriptors, it follows that the percentage of explained variance of the first 10 PCs is 91.16% (see Fig. 14.2 left). This fact suggests an intrinsic dimension of 10. At this purpose, CCA is performed for a dimensionality reduction from 22 to 10 ($\lambda = 20, 80$ epochs). The corresponding dy-dx diagram is plotted in Fig. 14.2 right. It shows that the manifold is linear and confirms the validity of the intrinsic dimensionality estimation.

Fig. 14.3 3D biplot: faces
(red points), blue vectors
(geodesic class), green
vectors (all the other classes)

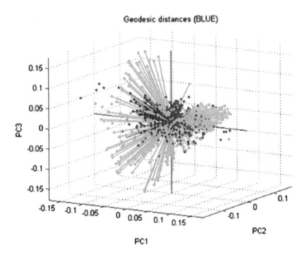

It can be deduced that all faces represented by these descriptors lie on a 10-dimensional hyperplane. This results in a redundancy of the descriptors. Only 10 geodesic distances (or linear combinations of these ones) are needed. Also, faces outside this hyperplane cannot be well represented.

The analysis by MLP yields the following results:

1. Single case: test rate of about 27% with 150 hidden neurons.
2. Subtractive case: test rate of about 89% with 300 hidden neurons.

Considering that the subtractive result is similar to the complete one in Sect. 14.2, it follows that these descriptors are not important in itself, but only in conjunction with other geometric descriptors.

The k-means approach yields a success rate of 56.01% for the subtractive case. This is only slightly lower than in the global case. It means that this class is not necessary for the clustering ability (in the sense that all other descriptors are enough for achieving nearly the same accuracy).

The statistical analysis uses the 3D biplot (see Fig. 14.3). The blue vectors, which represent the geodesic distances, are nearly orthogonal to the plane PC1-PC2, which means they are insensitive to the average values of the faces, but depend on their variance. However, they are clustered, which implies the intrinsic dimension is low. The fact their moduli are large implies their variance is high (they well represent a portion of the descriptor space). With regard to the position of faces (red points), the orthogonality implies the PCA scores are very low, thus confirming the little importance of this class in the classification.

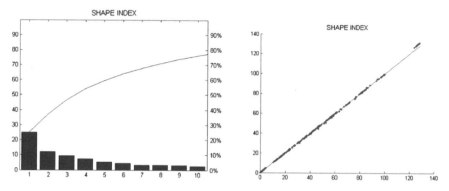

Fig. 14.4 Explained variance by PCA (left) and dy-dx plot by CCA (right)

14.3.2   Shape Index

The class of shape indices contains 42 descriptors. By using PCA on this group, it follows that the percentage of explained variance of the first 30 PCs is 96.74%. (see Fig. 14.4 left). This fact suggests the importance of most of these features. CCA is per-formed for a dimensionality reduction from 42 to 30 (λ = 50, 120 epochs). The corresponding dy-dx diagram is plotted in Fig. 14.4 right. It shows that the manifold is linear and suggests the idea of a 30-dimensional set of hyperplanes. Indeed, the clusters in the CCA diagram show large distances: they represent intercluster distances, which can be justified either as distant clusters or, more probably, as (face) outliers.

The analysis by MLP yields the following results:

1. Single case: best test rate of about 78.09% with 200 hidden neurons.
2. Subtractive case: best test rate of about 58.02% with 150 hidden neurons.

It can be deduced these indices have a strong impact in face recognition.

The k-means approach yields a success rate of 51.69% for the subtractive case. This is lower than for the geodetic subtractive case and is a confirmation of the better validity of the shape index class.

The statistical analysis uses the 3D biplot (see Fig. 14.5). The blue vectors, which represent the shape indices, are nearly orthogonal to the plane PC1-PC2, which means they are insensitive to the average values of the faces as for the geodetic distances. On the contrary, they spread the PC2-PC3 plane, which implies a good detection of variance of data and confirms the high intrinsic dimensionality. The length of the vectors suggests a high variance for each descriptor (better description of the face).

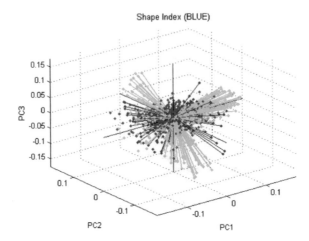

Fig. 14.5 3D biplot: faces (red points), blue vectors (geodesic class), green vectors (all the other classes)

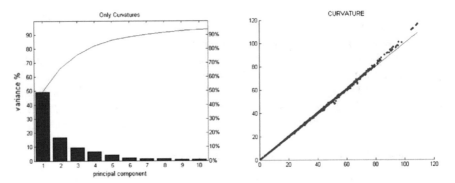

Fig. 14.6 Explained variance by PCA (left) and dy-dx plot by CCA (right)

14.3.3 Indices of Curvature

The curvature class is composed of 84 indices from histograms of the coefficient (42) and the Curvedness (42). The percentage of explained variance (see Fig. 14.6 left) of the first 30 PC components is 98.44% (the first PC is about 50%). It means that a lot of indices are meaningful. This intrinsic dimension is confirmed by CCA performed for a dimensionality reduction from 84 to 50 ($\lambda = 50$, 120 epochs), whose dy-dx diagram in Fig. 14.6 right shows that the manifold is linear and suggests the idea of a 50-dimensional f hyperplane with some outliers.

MLP yields the following results:

1. Single case: best test rate of about 56.19% with 100 hidden neurons.
2. Subtractive case: best test rate of about 60% with 170 hidden neurons.

Fig. 14.7 3D biplot: faces
(red points), blue vectors
(curvature class), green
vectors (all the other classes)

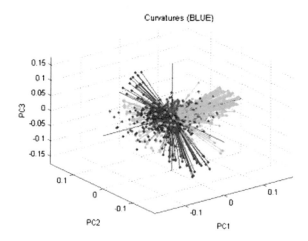

Fig. 14.8 Right facial points
used to compute distances.
The same are obtained in the
left part of the face

It can be deduced these indices have a strong impact as for the shape index, but
are slightly less important.

The k-means approach yields a success rate of 58.84% for the subtractive case.
This is of the same order of the geodesic subtractive case and proves worse discrimination w.r.t. the shape index.

The statistical analysis by the 3D biplot (see Fig. 14.7) shows the curvature vectors
partially spread the PC2-PC3 plane, in a negative correlated way, which implics a
good detection of variance, also because of the high score of their projection on data.
The length of the vectors also suggests a high variance for each descriptor (better
description of the face). However, some average statistics of faces is captured.

14.3.4 Euclidean Distances

The Euclidean class is composed of 62 distances derived from facial soft-tissue
landmarks located around the nose and the eyes (see Fig. 14.8).

The percentage of explained variance (see Fig. 14.9 left) of the first 10 PC components is 93.7%%. The intrinsic dimensionality is very low. This is confirmed by CCA

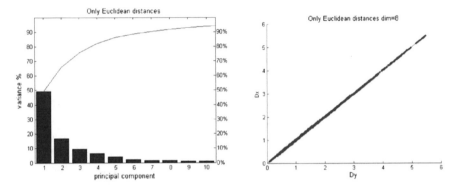

Fig. 14.9 Explained variance by PCA (left) and dy-dx plot by CCA (right)

for a dimensionality reduction from 62 to 8 ($\lambda = 70$, 120 epochs), whose *dy-dx* diagram in Fig. 14.9 right. It proves the Euclidean manifold is at least an 8-dimensional hyperplane.

However, after a correlation analysis which has selected 8 independent features, it has been proved that 5 of them have a bimodal statistics and are not able to discriminate 100 faces (they always yield the same value). These distances are Pronasal (PN) with right Outer eyebrow (OEdx), right Ala of nose (ALA) with right Outer eyebrow, right Exocantion with right Outer eyebrow, left Exocantion with left Outer eyebrow and right Outer eyebrow with left Outer eyebrow. This observation is probably justified by the fact that these facial soft-tissue landmarks are located in positions in which the face movement is difficult. It follows that these distances do not vary for people with different facial expressions. After a detailed examination of these points, the conclusion is that the point 'right Outer eyebrow' (also, the same point in the left position) is a couple of coordinates repeated in all the five distances. This means that: OEdx tracks the change of movement of points PN or ALA when a person modifies his expression. The point Exocantion is located in the inner corner of the eye where there is no possibility of movement. The last distance is between right Outer eyebrow and the analogous in the left position. This distance does not vary for the reason that it is parallel to the frontal anatomical plane and the maximum movement permitted for these points is to shift among the same plane. Resuming, 5 features are certainly unable to classify. There remain only 3 features. MLP confirms the uselessness of this class:

1. Single case: best test rate of about 45.4% with 200 hidden neurons.
2. Subtractive case: best test rate of about 80.95% with 200 hidden neurons.

The k-means algorithm yields a success rate of 70.05% for the subtractive case. It means that the discrimination is better without the Euclidean distances than with the whole database. They worsen the classification.

The 3D biplot (see Fig. 14.10) shows that: the Euclidean distances are clustered around the first component, which represents the average behavior of the faces.

Fig. 14.10 3D biplot: faces (red points), blue vectors (Euclidean class), green vectors (all the other classes)

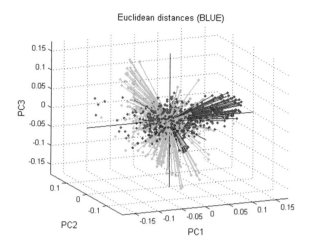

Hence, they are not able to well distinguish the images. This clustering also proves the fact they do not discriminate well.

14.4 Conclusion

This work analyzes four classes of geometric descriptors for 3D face recognition, by using the public Bosphorus 3D facial database as dataset. At this aim, statistical and neural techniques are employed in an original way. Classes can be ranked according to their behavior in classification. Certainly, the worst features are the Euclidean distances. They do not represent at all the variance in faces. They worsen the face recognition. They are easy to estimate, but not meaningful with regard to the true distances in the face, unlike the geodesics, which require a higher computational cost. However, only 10 geodesics are needed. Both curvatures and geodesic descriptors work far better, but they are not enough for capturing the peculiarity of the face.

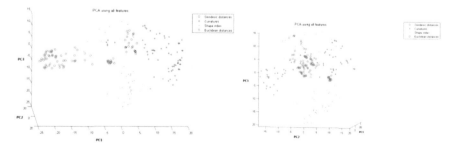

Fig. 14.11 Two views of the scatter plot of the first 3 PCs for all descriptors

At least 50 curvature descriptors are needed for a good discrimination. The best descriptors belong to the shape index class, which requires 30 descriptors. However all these classes but the Euclidean distances have to work in a coordinated way for achieving the best possible result. Figure 14.11 shows the different regions of competence with regard to the first 3 PCs. However, consider that this is only a reduced representation of the feature space (except for the Euclidean class).

Future work will deal with the use of the parallel coordinate plots and other neural networks.

References

1. Do Carmo, M.: Differential Geometry of Curves and Surfaces. Prentice-Hall Inc., Englewood Cliffs (1976)
2. Koenderink, J.J., van Doorn, A.J.: Surface shape and curvature scales. Image Vis. Comput. **10**(8), 557–564 (1992)
3. Dorai, C., Jain, A.K.: COSMOS-A representation scheme for 3D free-form objects. IEEE Trans. Pattern Anal. Mach. Intell. **19**(10), 1115–1130 (1997)
4. Vezzetti, E., Marcolin, F.: Geometrical descriptors for human face morphological analysis and recognition. Robot. Auton. Syst. **60**(6), 928–939 (2012)
5. Vezzetti, E., Marcolin, F.: Novel descriptors for geometrical 3D face analysis. Multimed. Tools Appl. **X**, 1–30, (2016)
6. Quan, W., Matuszewski, B.J., Shark, L.K.: Statistical shape modelling for expression-invariant face analysis and recognition. Pattern Anal. Appl. 1–17 (2015)
7. Ming, Y.: Robust regional bounding spherical descriptor for 3D face recognition and emotion analysis. Image Vis. Comput. **35**, 14–22 (2015)
8. Ganguly, S., Bhattacharjee, D., Nasipuri, M.: RaFaReS: range image-based 3D face recognition system by surface descriptor. In: First International Conference on Next Generation Computing Technologies (NGCT), pp. 903–908, September 2015
9. Swennen, G.R., Schutyser, F.A., Hausamen, J.E.: Three-Dimensional Cephalometric: A Color Atlas and Manual. Springer Science & Business Media (2005)
10. Bishop, C.: Pattern Recognition and Machine Learning. Springer Science Business Media, LLC
11. MacQueen, J.B.: Some methods for classification and analysis of multivariate observations, In: Proceedings of Fifth Berkeley Symposium on Mathematical Statistics and Probability, Berkeley, vol. 1, pp. 281–297. University of California Press (1967)
12. Gower, J.C., Hand, D.J.: Biplots. Chapman & Hall, London (1996)
13. Demartines, P., Hérault, J.: Curvilinear component analysis: a self-organizing neural network for nonlinear mapping of data sets. IEEE Trans. Neural Netw. **8**(1), 148–154 (1997)

Part III
Computational Intelligence and Related Techniques in Industrial and ICT Engineering

Chapter 15
Convolutional Neural Networks for the Identification of Filaments from Fast Visual Imaging Cameras in Tokamak Reactors

Barbara Cannas, Sara Carcangiu, Alessandra Fanni, Ivan Lupelli, Fulvio Militello, Augusto Montisci, Fabio Pisano, Giuliana Sias and Nick Walkden

Abstract The paper proposes a region-based deep learning convolutional neural network to detect objects within images able to identify the filamentary plasma structures that arise in the boundary region of the plasma in toroidal nuclear fusion reactors. The images required to train and test the neural model have been synthetically generated from statistical distributions, which reproduce the statistical properties in terms of position and intensity of experimental filaments. The recently proposed Faster Region-based Convolutional Network algorithm has been customized to the

B. Cannas · S. Carcangiu (✉) · A. Fanni · A. Montisci · F. Pisano · G. Sias
Department of Electrical and Electronic Engineering, University of Cagliari, Via Marengo 1, 09123 Cagliari, Italy
e-mail: s.carcangiu@diee.unica.it

B. Cannas
e-mail: cannas@diee.unica.it

A. Fanni
e-mail: fanni@diee.unica.it

A. Montisci
e-mail: amontisci@diee.unica.it

F. Pisano
e-mail: fabio.pisano@diee.unica.it

G. Sias
e-mail: giuliana.sias@diee.unica.it

I. Lupelli · F. Militello · N. Walkden
Culham Centre for Fusion Energy, Abingdom, UK
e-mail: Ivan.Lupelli@ukaea.uk

F. Militello
e-mail: Fulvio.Militello@ukaea.uk

N. Walkden
e-mail: Nick.Walkden@ukaea.uk

© Springer International Publishing AG, part of Springer Nature 2019
A. Esposito et al. (eds.), *Neural Advances in Processing Nonlinear Dynamic Signals*, Smart Innovation, Systems and Technologies 102, https://doi.org/10.1007/978-3-319-95098-3_15

167

problem of identifying the filaments both in location and size with the associated
score. The results demonstrate the suitability of the deep learning approach for the
filaments detection.

Keywords Deep learning · Convolutional neural network · Object detector
Nuclear fusion reactors

15.1 Introduction

One of the main issues in the operation of a nuclear fusion tokamak reactor [1] is to
understand the dynamics and phenomenology of the plasma edge, in order to avoid
the erosion of the reactor walls and the energy loss, due to the occurrence of plasma
turbulence. In fact, it has been recognized that some turbulent intermittent events are
the cause of plasma transport at edge in a radial direction beyond the last closed flux
surface (LCFS), with various degrees of penetration into the scrap-off layer (SOL).
These events directly influence the location and strength of the heat and particle
flux to plasma facing components [2]. Hence, a main topic is the study of particular
turbulences, called blobs or filaments, which is crucial for the design of the SOL
profiles.

Various filaments detection methods have been proposed in literature. In [3], the
plasma blobs are determined by statistically setting some thresholds on the local
plasma density signal. This method requires the threshold optimization in every
experiment, due to the intrinsic variability and complexity of the filamentary structures. In [4–6], different blob identification algorithms have been presented based on
the gas puff imaging (GPI) diagnostic. These methods are not suitable for real-time
application. The aim of this paper is automatically identifying the filamentary plasma
structures, that arise in the boundary region of the plasma in the Mega-Amp Spherical Tokamak (MAST) located at the Culham Science Center of Abingdon (UK). The
identification of the filaments has to be done starting from the 2D images, acquired
with a fast camera installed in MAST. The camera records the light emission associated with the filaments during a plasma discharge, at time scale relevant for their
dynamics.

The imaging data, obtained by the fast camera, has to be analyzed in order to
extract filaments information, such as their position, size, shape, intensity value and
the dynamics with which it moves along the magnetic field lines. Hence an algorithm,
able to identify the filaments and to evaluate their properties, has been developed. For
our investigation only the mid plane images have been used, and the aim is to provide
a fair filaments 3D reconstruction from 2D images, so that all the filaments features
could be known and the control of plasma edge could be better developed. Knowing
that filaments are generally aligned to magnetic field lines [7], the reconstruction of
the field lines at the equilibrium can be used as an artifact to describe the filaments
position and size inside images. In particular, in [8] an image processing tool (Elzar
code) has been presented in which the light emission is integrated along the field

lines associated with the position of the filaments. The images resulting from this process are mapped into a specific plane whose coordinates are the major radius and the toroidal angle in the MAST cylindrical coordinates. In fact, each field line can be uniquely defined at any 3D spatial point and it is identified by its position at the mid-plane of the tokamak (Z = 0 in the MAST cylindrical coordinate system). These 2-D maps have been used in this paper to extract information on the location and dimension of the filaments.

To this purpose, Convolutional Neural Networks (CNNs) [9] have been used. CNNs are a family of multi-layer neural networks particularly designed for use on two-dimensional data, such as images and videos. In this paper, the potentiality of CNN has been investigated in order to implement the filaments identification system. In particular, a recently proposed neural deep learning algorithm, called Faster Region-based Convolutional Neural Network (Faster-RCNN) [10] has been used to detect the filaments from the 2D-maps, allowing to extract the information on their position and size in the original images.

15.2 Filaments Simulations

In principle, the emission of a filament can be described in a 2D plane as a shape parameterized by the major radius R and the toroidal angle φ. Figure 15.1 shows on the left the image taken from the mid-plane fast visible camera, and on the right its projection in the R, φ plane done by using the Elzar code [8], a tool written in Python which integrates the light emission of each filament along the associated magnetic field lines and plots the image processing result into the 2D grid. The bright light emissions of the filaments visible in the camera image can be identified as elliptical objects in the R, φ plane.

In the simplest assumption, each filament can be modelled as a two-dimensional Gaussian function of R and φ:

$$f(R, \varphi) = A \cdot exp\left[-\frac{(R - R_0)^2}{\sigma_R^2} - \frac{(\varphi - \varphi_0)^2}{\upsilon_\varphi^2} \right] \qquad (15.1)$$

where A is the filament amplitude, R_0, φ_0 is the position of the its center, σ_R, σ_φ are the R and φ spreads of the blob. Figure 15.2 shows an example of synthetic Gaussian filament in the R, φ plane, in which some noise has been added to the amplitude in order to simulate experimental conditions.

The values of A, R_0, φ_0, σ_R, σ_φ are generated randomly from statistical distributions, in order to reproduce the statistical properties of experimental filaments. As it happens experimentally, not all the entire set of generated filaments is visible by the camera, whose field of view doesn't cover all the volume of the tokamak. Thus, only a subset of generated filaments is visible in the R, φ plane.

The synthetic image I is thus created as a summation of several Gaussian functions, each of which represents one filament:

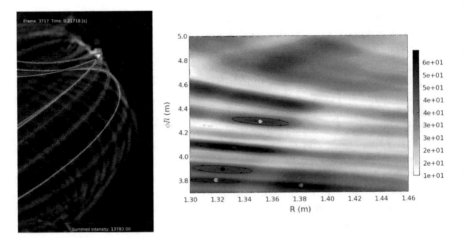

Fig. 15.1 Image taken from the mid-plane fast visible camera (left) and projection of experimental filaments with Elzar code (right)

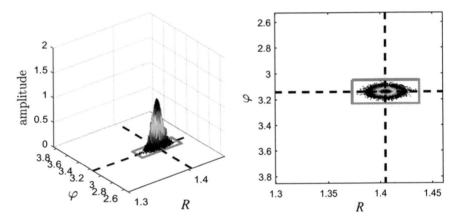

Fig. 15.2 View of a generated synthetic filament

$$I(R, \varphi) = \sum_i f_i(R, \varphi) = \sum_i A_i \cdot exp\left[-\frac{(R - R_{i,0})^2}{\sigma_{R,i}^2} - \frac{(\varphi - \varphi_{i,0})^2}{\sigma_{\varphi,i}^2} \right] \quad (15.2)$$

and then normalized in amplitude. The synthetic dataset consists of 10 000 images of size 227×227, with radial and angular intervals $1.3\,m \leq R \leq 1.46\,m$ and $2.53\,rad \leq \varphi \leq 3.85\,rad$ respectively.

These synthetic images have been used to build the filaments detection deep learning model (training images) and to test its performance (testing images).

15.3 Convolutional Neural Networks (CNNs)

A Convolutional Neural Network is a deep learning neural network particularly suited to image processing [9, 11]. The architecture of a CNN is inspired by the structure of the animals' visual system, and it consists of several blocks of different layers each one performing a prescribed task:

- the convolutional layers perform a convolution filtering of the previous layer. Small regions of input neurons, called Local Receptive Fields (LRF), are connected to neurons in the first convolutional hidden layer. These input networks correspond to pixels nearby the image. Sliding the LFR across the input image without changing the weights, which act as a filter, a feature map is created from the input layer to the hidden layer neurons. This task is efficiently performed by convolution. Being the weights the same for all the hidden neurons, all the hidden neurons detect the same feature from the different regions of the original image;
- in order to increase the non-linear capability of the decision functions, an activation layer is used, named ReLU (Rectified Linear Unit) layer. It applies a transformation to the output of the convolution layer by means of an elementwise activation function without changing the dimension of the output. Different activation functions can be used [12]; among them, one of the most common is the non-saturating activation function $f(x) = \max(0, x)$;
- the pooling layers subsample small rectangular blocks taken from the convolutional layer to produce a single output from that block.

This cascade of Convolution-ReLU-Pooling blocks extracts the more significant features from the original image, which constitute the input of the final layer. Just as in the conventional neural networks, this is a fully connected layer that performs the association among these features and the desired output labels corresponding to the classification of the objects into the image. Figure 15.3 reports a schematic representation of a CNN.

The network weights have to be optimized by means of a training process, which can be performed from scratch or using CNN trained to learn similar problems. The CNN can be used just to extract the relevant features from the image. These features are used as input to a machine-learning model, which performs the desired classification. This last approach is less accurate but requires a limited amount of data and computation resources.

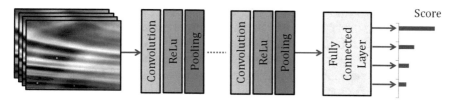

Fig. 15.3 Schema of a convolutional neural network

Recently, several CNN algorithms have been proposed to accurately detect objects within images, such as the Region-based Convolutional Networks (RCNN) [13], and their extensions, Fast RCNN [14] and Faster RCNN [10], which differ in the way they select the regions of the image (or proposals) to be selected and how to select them. In particular, an RCNN is an object detection algorithm that combines region proposals with features computed by a convolutional neural network. However, for each object proposal, it performs a training step, without sharing computation. The Fast RCNN algorithm performs a single-stage training that jointly learns to classify object proposals and refine their spatial locations [14]. Even if Fast RCNN overcomes RCNN in terms of required computation resources, the time spent to compute the object proposals is still a limit, at least implementing it on a CPU. The Faster RCNN [10] computes proposals with a deep convolutional neural network realizing a system entirely based on the paradigm of the deep-learning and with reduced calculation times. To this purpose, a Region Proposal Network (RPN) is used, which shares convolutional layers with the detection network. An RPN has as input an image and produced as outputs a set of rectangular object proposals, each associated to its classification score referred to a set of object classes versus the background.

The global deep neural network is trained by alternating a fine tuning of the RPN module with a fine tuning of the object detection Fast RCNN, while keeping the proposals fixed. This scheme demonstrated to fast converge to a unified system where the convolutional features are shared between the region proposal and the object detection tasks.

15.4   Faster-RCNN for Filaments Detection

Based on the Faster RCNN described in [10] a filament detector has been developed. The input data set is composed of 10 000 images. Each image contains several instances of filaments (from 1 to 19) with their associated bounding boxes specifying the object locations within the image. Each bounding box is defined by a four tuple (x, y, h, w) that specifies its top-left corner (x, y) and its height and width (h, w). The data set is split into a training set (6 000 images) for training the detector, and a test set (4 000 images) for evaluating the detector.

In order to construct the detector, firstly, a CNN is built. Because of the detector needs to analyze smaller sections of the image, the input size of the CNN must be similar in size to the smallest object in the data set. In this data set all the objects are larger than $[3 \times 3]$, so an input size of $[9 \times 9]$ has been selected. The middle layers of the CNN are then defined. The middle layers are made up of repeated blocks of convolutional, ReLU, and pooling layers.

As previously mentioned, the Faster RCNN consists of two networks that share several convolutional layers; these layers produce convolutional feature maps. The first network is the RPN that, from the convolutional feature maps, generates region proposals which are used by the second network, the Fast RCNN, for the detection

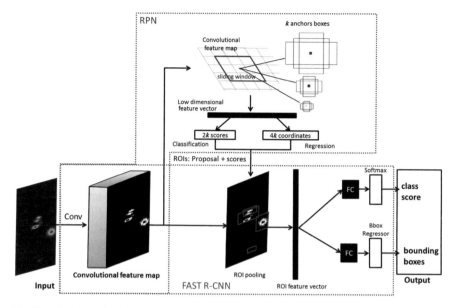

Fig. 15.4 Schema of the faster RCNN

(Fig. 15.4). Hence, starting from the basic structure of the previously defined CNN, its different layers are assigned to the RPN and to the Fast RCNN.

As shown in Fig. 15.4, in order to obtain region proposals, a sliding window is run spatially on the feature maps and, for each sliding window, a set of k anchors (i.e. reference boxes) are generated all having the same center of the sliding window but different aspect ratios and scales. Furthermore, for each of these anchors, using the Intersection-over-Union (IoU) metric, defined as the ratio between the area of overlap and the area of union, a value p^* is computed which indicates how much the anchors overlap with the ground-truth bounding boxes. The value p^* has been computed considering the following rules:

$$p^* = \begin{cases} 1 & \text{if } IoU > 0.3 \\ -1 & \text{if } IoU < 0.3 \\ 0 & \text{otherwise} \end{cases}$$

If $p^* = 1$, the anchor is labeled positive, if $p^* = -1$ it is labeled non-positive whereas, if $p^* = 0$, the anchor does not contribute to the training objective.

The output of each convoluted sliding window is then mapped to a low dimensional feature vector, followed by a ReLU nonlinearity. Finally, these low-dimensional features are fed into two fully connected layers that identify the proposals: a bounding box regression layer with $4k$ outputs encoding the 4 coordinates of k reference boxes and a box-classication layer with $2k$ outputs (2 scores for each of the k boxes indicating how likely each of the k regions contains an object or the background).

For each image, the top-ranked 2000 proposals generated by the RPN have been selected to construct the training set for the Fast RCNN. For each proposal, a region of interest (RoI) pooling layer extracts a fixed length feature vector from the feature maps.

Each feature vector is then fed into a sequence of fully connected (FC) layers that finally branch into two sibling output layers: one that produces *softmax* probability estimates over filament class, plus a catch-all "background" class and another layer that outputs 4 real-valued numbers that encode refined bounding-box positions for each filament.

In order to optimize the detector performance, the weights of the network obtained by the previous steps are used to retrain the RPN. In particular, the weights of convolutional layers shared with the Fast RCNN are kept fixed, whereas the other layers of the RPN are fine-tuned. Finally, the process is iterated by fine tuning the layers of the Fast RCNN while keeping fixed the convolutional layers shared with the RPN.

Both RPN and Fast RCNN are trained using Stochastic Gradient Descent with Momentum, for 10 epochs with initial learning rate of 10^{-6}.

15.5   Results

Figure 15.5 shows some examples of results for the filaments detection on the test set. For each example, the target boxes are plotted with dashed gray lines, whereas the boxes provided by the detector are reported as black lines. Figure 15.5a refers to a test example where five filaments were present. As can be seen, the filament detector has worked very well, detecting all the filament structures, which are perfectly framed by the black rectangles. In Fig. 15.5b the filament detector produced a false alarm, detecting more filaments than those actually existing; in particular, two very close filaments (see bottom of Fig. 15.5b) have been identified by the detector as three different structures. Both in Fig. 15.5c and d, even if the large majority of filamentary structures have been precisely detected, the filament detector produced missed alarms, missing some filaments than those actually existing; in particular, the lowest filament magnitude in Fig. 15.5c and the boundary filament in the top-right of Fig. 15.5d have not been detected.

A statistical analysis on the entire test set showed that the leading causes of missed alarms have been given by filaments having low magnitude and/or lying in the boundary region of the image.

In order to evaluate the performance of the filament detector, the following criteria have been used:

- a filament is considered as correctly detected if its center lies inside some of the boxes created by the filament detector. In this case the filament detector produces a True Positive (TP) response;
- a filament is considered missed if its center lies outside any of the boxes created by the filament detector, and a False Negative (FN) response is produced;

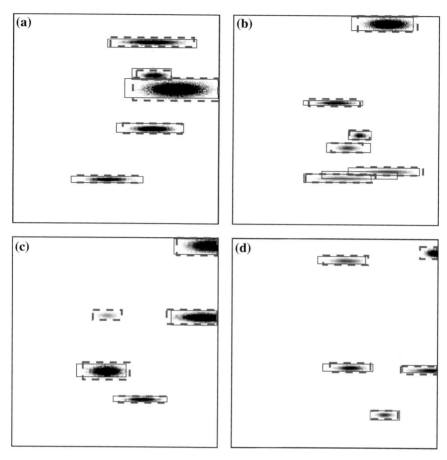

Fig. 15.5 Detection examples on the test set with indication of target boxes (dashed gray lines) and boxes provided by the detector (black lines)

Table 15.1 Detection performance in terms of true positive (TP), false negative (FN), and false positive (FP)

	♯ Actual filaments	TP	FN	FP
Training set	50085	33808	16277	866
Test set	33962	22911	11051	556

- if a box created by the filament detector does not contain the center of any actual filament, it is considered as a False Positive (FP).

Table 15.1 summarizes the obtained results, in terms of the number of actual filaments detected (TP) or not detected (FN), and the number of detected filaments that were not existent (FP).

Table 15.2 Detection performance in terms of precision and recall

	Pr	Rc
Training set	0.976	0.675
Test set	0.975	0.675

In order to evaluate the performance of the filament detector, some global indices have been considered:

- Precision (*Pr*), also called as positive predictive value, defined as the ratio between the number of true positives (*TP*) and the sum of the true positives and false positives (*FP*):

$$Pr = \frac{TP}{TP + FP}$$

- Recall (*Rc*), also known as sensitivity, defined as the ratio between the number of true positives and the sum of the true positives and false negatives (*FN*):

$$Rc = \frac{TP}{TP + FN}$$

Table 15.2 report these indices for both the training and the test set.

As it can be noticed, the results in terms of performances on the training and test set are quite equivalent, showing that the network has been able to learn the training set properties without overfitting problems. The high level of precision tells us that the most of detected objects are actually filaments. On the other side, a recall of 0.675 gives evidence that there is a good part of filaments which the network is still not able to detect.

15.6   Conclusions

This paper proposed an object detector, which is able to identify particular kind of structures, called filaments, appearing in the plasma edge, and recorded with a fast camera installed on MAST. To this purpose a deep learning convolutional neural network has been trained with synthetic filaments images which reproduce the statistical properties of experimental filaments in terms of position and intensity. This choice allowed assessing the algorithm performances through the evaluation of commonly used indices. In particular, the precision of the method was very high, overcoming 97.5% both in the training and test sets. But, the values of the recall index was considerably lower, achieving 67.5% in both the sets. This is due to the difficulty of the detector to identify filaments with low pixels intensity or located on the edges of the frames. Further investigations will be performed in the future to decrease these missed detections.

The long-term goal is to get good performance of the algorithm even on experimental data, so that to enable an objective automation of the identification process in the view of analyzing the filaments dynamics. The obtained results indicate that Faster-RCNN can be a powerful tool for image analysis in the nuclear fusion research.

References

1. Wesson, J., Campbell, D.J.: Tokamaks, vol. 149. Oxford university Press (2011)
2. Ayed, N.B., Kirk, A., Dudson, B., Tallents, S., Vann, R.G.L., Wilson, H.: Inter-ELM filaments and turbulent transport in the Mega-Amp Spherical Tokamak. Plasma Phys Controll. Fusion **51**(3), 035016 (2009)
3. D'Ippolito, D.A., Myra, J.R., Zweben, S.J.: Convective transport by intermittent blob-filaments: Comparison of theory and experiment. Phys. Plasmas **18**(6), 060501 (2011)
4. Love, N.S., Kamath, C.: Image analysis for the identification of coherent structures in plasma. Applications of Digital Image Processing. In: Tescher, AG. (eds.) Proceedings of the SPIE, vol. 6696 (2007)
5. Davis, W.M., Ko, M.K., Maqueda, R.J., Roquemore, A.L., Scotti, F., Zweben, S.J.: Fast 2-D camera control, data acquisition, and database techniques for edge studies on NSTX. Fus. Eng. Des. **89**(5), 717–720 (2014)
6. Myra, J.R., Davis, W.M., D'Ippolito, D.A., LaBombard, B., Russell, D.A., Terry, J.L., Zweben, S.J.: Edge sheared flows and the dynamics of blob-filaments. Nucl. Fus. **53**(7), 073013 (2013)
7. Dudson, B.D., Ayed, N.B., Kirk, A., Wilson, H.R., Counsell, G., Xu, X., et al.: Experiments and simulation of edge turbulence and filaments in MAST. Plasma Phys. Controll. Fus. **50**(12), 124012 (2008)
8. Militello, F., et al.: Multi-code analysis of scrape-off layer filament dynamics in MAST. Plasma Phys. Controll. Fus. **58**(10), 105002 (2016)
9. Jarrett, K., Kavukcuoglu, K., LeCun, Y.: What is the best multi-stage architecture for object recognition? In: IEEE 12th International Conference on Computer Vision, pp. 2146–2153 (2009)
10. Ren, S., He, K., Girshick, R., Sun, J.: Faster r-cnn: Towards real-time object detection with region proposal networks. In: Advances in Neural Information Processing Systems, pp. 91–99 (2015)
11. Bengio, Y.: Learning deep architectures for AI. Foundations and trends®. Mach. Learn. **2**(1), 1–127 (2009)
12. Laudani, A., Lozito, G.M., Fulginei, F.R., Salvini, A.: On training efficiency and computational costs of a feed forward neural network: a review. Comput. Intell. Neurosci. **83**(818243) (2015)
13. Girshick, R., Donahue, J., Darrell, T., Malik, J.: Rich feature hierarchies for accurate object detection and semantic segmentation. In: Proceedings of the IEEE Conference on Computer Vision and Pattern Recognition, pp. 580–587 (2014)
14. Girshick, R.: Fast r-cnn. In: Proceedings of the IEEE International Conference on Computer Vision, pp. 1440–1448 (2015)

Chapter 16
Applying Network Analysis for Extracting Knowledge About Environment Changes from Heterogeneous Sensor Data Streams

Francesco Cauteruccio, Paolo Lo Giudice, Giorgio Terracina
and Domenico Ursino

Abstract Sensor network analysis has become a challenging task. The detection of sensor anomalies is one of the most prominent topics in this research area. In the past, researchers mainly focused on the detection and analysis of single-sensor anomalies. In this paper, we shift the focus from a local approach, aimed to detect anomalies on single sensors, to a global one, aiming at detecting and investigating the consequences, on the whole sensor network and/or its subnetworks, of anomalies present in one or more (heterogeneous) sensors.

Keywords Wireless sensor networks · Anomaly detection · Network analysis
Dashboard

F. Cauteruccio · G. Terracina
DEMACS, University of Calabria, Rende, Italy
e-mail: cauteruccio@mat.unical.it

G. Terracina
e-mail: terracina@mat.unical.it

P. Lo Giudice
DIIES, University "Mediterranea" of Reggio Calabria, Reggio Calabria, Italy
e-mail: paolo.lo.giudice@unirc.it

D. Ursino (✉)
DICEAM, University "Mediterranea" of Reggio Calabria, Reggio Calabria, Italy
e-mail: ursino@unirc.it

© Springer International Publishing AG, part of Springer Nature 2019
A. Esposito et al. (eds.), *Neural Advances in Processing Nonlinear Dynamic Signals*, Smart Innovation, Systems and Technologies 102,
https://doi.org/10.1007/978-3-319-95098-3_16

179

16.1   Introduction

In the last few years, research on Wireless Sensor Networks (WSNs) has been ignited by important advances in various technological areas, such as wireless communications, digital electronics and micro-electro-mechanical systems. These improvements allowed for an easy development of low-power and low-cost multi-functional sensors and networks thereof. Sensor networks usually include a large number of nodes, each of which may sense several measures. Cooperation among nodes is usually sought for in such networks. Sensor nodes are usually positioned either inside or very close to observed events, and the main objective is to provide users with a better understanding of the environment in which sensors are deployed, thus giving the opportunity to acquire new information and intelligence. While the management of sensor networks and the development of robust data acquisition layers received much attention in the literature, one big open challenge in this research area is anomaly detection [1, 2]. Anomalies can be generated by either malfunctioning sensors or changes in the monitored environment. In most cases, being able to distinguish between the two scenarios is a challenging task. Most of the past approaches for anomaly detection focused on the analysis of data produced by each single device [3]. The most notable approaches in this setting can be grouped in four categories, namely: (i) rule-based detection [4], (ii) statistical techniques [5], (iii) graph-based techniques [6], and (iv) data mining and computational intelligence-based techniques [7]. Instead, network-based approaches for anomaly detection in WSNs received less attention [8–11]. In fact, in spite of a strict complementarity and correlation between network analysis and WSNs, only in the latest years, researchers have begun to apply network analysis-based techniques to WSNs. However, they have only proposed the application of classical network analysis parameters to this context. Indeed, most of the proposed approaches employ centrality measures [12], which allow the detection of anomalies of only one node at a time.

   In this paper, we aim at introducing new solutions for the analysis of heterogeneous sensors organized as a network. In particular, our techniques will be based on the evaluation of the connectivity of the whole WSN and its subnetworks (instead of on node centrality), and are mainly focused on potential anomalies involving more sensors located therein. They adopt a metric capable of uniformly handling measures provided by heterogeneous sensors, as well as a dashboard of network analysis parameters. This way, they allow the detection of anomalies involving more (heterogeneous) sensors, and the evaluation of the impact of these anomalies on the whole sensor network and its subnetworks. The plan of this paper is as follows. In Sect. 16.2, we introduce our model used to represent WSNs and our anomaly detection approach. In Sect. 16.3, we present some preliminary results on tests carried out on a sensor network, along with some discussions. Conclusions and future work are illustrated in Sect. 16.4.

16.2 Methods

16.2.1 Network Construction

Let $\mathcal{W}$ be a WSN. Without loss of generality, assume that the corresponding sensors can be partitioned along two orthogonal dimensions.[1] In the scenario considered here, these dimensions are location and physical quantities to evaluate (in particular, we consider $p = 3$ physical quantities, i.e., temperature, lightness and humidity). Assume that the WSN covers l locations (in particular, we consider $l = 3$ locations, named A, B and C in the following) and that one location contains n devices, each measuring p physical quantities. As a consequence, the overall number of sensors is $s = pln$.

A network $\mathcal{N} = \langle V, E \rangle$ can be associated with $\mathcal{W}$. Here, V is the set of the nodes of $\mathcal{N}$. Each node $v_i \in V$ corresponds to a sensor and has associated a label $\langle l_i, p_i \rangle$, where l_i represents its location and p_i denotes the physical quantity it measures. E is the set of the edges of $\mathcal{N}$. Each edge e_{ij} connects the nodes v_i and v_j. It can be represented as $e_{ij} = (v_i, v_j, w_{ij})$. Here, w_{ij} is a measure of "distance" between v_i and v_j. It is an indicator of the non-correlation level of the sensors associated with v_i and v_j. Actually, each parameter representing this feature could be adopted in our model. In the experiments presented in this paper we adopted Multi-Parameterized Edit Distance (MPED) [13] for its capability of measuring the non-correlation level of sensors regarding heterogeneous physical quantities, characterized by different units of measure and possible data shifts.

$\mathcal{N}$ can be partitioned along one or both dimensions. We indicate by $\mathcal{N}_p = \langle V_p, E_p \rangle$ the subnets obtained by taking only the nodes that correspond to the sensors measuring the physical quantity p. Here, $p \in \{l, t, h\}$ can denote lightness, temperature and humidity, respectively. Analogously, we indicate by $\mathcal{N}_q = \langle V_q, E_q \rangle$ the subnets obtained by taking only the nodes that correspond to the sensors operating at the location q. Here, $q \in \{A, B, C\}$. Finally, we denote by $\mathcal{N}_{pq} = \langle V_{pq}, E_{pq} \rangle$ the subnet obtained by considering only the nodes corresponding to the sensors that measure the physical quantity p and operate in the location q, along with the edges linking them.

16.2.2 Network Parameters

As pointed out in the Introduction, we use several parameters to construct our dashboard supporting the extraction of knowledge about environment changes. The first four parameters are derived from classical network theory; the fifth is derived from a particular centrality measure proposed in [14]; the last is introduced by us. In

---

[1]Actually, the number of dimensions could be greater than two, without requiring any change of the approach.

this section, we present an overview of these parameters. In the following, we define all of them on a reference network $\mathcal{N} = \langle V, E \rangle$. The first parameter is the *Characteristic Path Length*, also known as the *Average Shortest Path Length*. It is defined as the average length of the shortest paths connecting all possible pairs of network nodes. More formally, let $l(v_i, v_j)$ be the length of the shortest path between v_i and v_j. The Characteristic Path Length $\mathcal{L}_\mathcal{N}$ of $\mathcal{N}$ is defined as: $\mathcal{L}_\mathcal{N} = \frac{1}{|V|(|V|-1)} \sum_{v_i \in V} \sum_{v_j \in V, v_j \neq v_i} l(v_i, v_j)$. The second parameter is the *Average Node Connectivity*. Given two nodes v_i and v_j, their connectivity $c(v_i, v_j)$ represents the minimum number of edges that need to be removed to disconnect them. The Average Node Connectivity $\mathcal{C}_\mathcal{N}$ is defined as: $\mathcal{C}_\mathcal{N} = \frac{1}{|V|(|V|-1)} \sum_{v_i \in V} \sum_{v_j \in V, v_j \neq v_i} c(v_i, v_j)$. The third parameter is the *Average Number of Simple Paths*. Given two nodes v_i and v_j, we indicate by $p(v_i, v_j)$ the number of simple paths (i.e., paths with no repeated nodes) between them. Then, we define the Average Number of Simple Paths $\mathcal{P}_\mathcal{N}$ as: $\mathcal{P}_\mathcal{N} = \frac{1}{|V|(|V|-1)} \sum_{v_i \in V} \sum_{v_j \in V, v_j \neq v_i} p(v_i, v_j)$. The fourth parameter is the *Average Clustering Coefficient*. In order to define it, we must preliminarily introduce the neighborhood $nbh(v_i)$ of a node v_i as follows: $nbh(v_i) = \{v_j | e_{ij} \in E\}$. Then, we define the Clustering Coefficient of a node v_i as: $s(v_i) = \frac{2 \cdot |\{e_{jk} | v_j, v_k \in nbh(v_i), e_{jk} \in E\}|}{|nbh(v_i)| \cdot (|nbh(v_i)|-1)}$. Finally, we define the Average Clustering Coefficient as: $\mathcal{S}_\mathcal{N} = \frac{1}{|V|} \sum_{v_i \in V} s(v_i)$. The fifth parameter is the *Average Closeness Vitality*. Given a node v_i, the closeness vitality $t(v_i)$ represents the increase in the sum of distances between all the pairs of nodes of $\mathcal{N}$, when v_i is excluded from $\mathcal{N}$ [14]. The Average Closeness Vitality $\mathcal{T}_\mathcal{N}$ is defined as: $\mathcal{T}_\mathcal{N} = \frac{1}{|V|} \sum_{v_i \in V} t(v_i)$. The sixth parameter (i.e., the one introduced by us) is the *Connection Coefficient*. It starts from the observation that, in network analysis, one of the most powerful tools for investigating the connection level of a network is the concept of clique. As a consequence, it is reasonable to adopt this concept to evaluate the cohesion of a network. This coefficient takes the following considerations into account: (i) both the dimension and the number of cliques are important as connectivity indicators; (ii) the concept of clique is intrinsically exponential; in other words, a clique of dimension $n + 1$ is exponentially more complex than a clique of dimension n.

In order to define the Connection Coefficient it is necessary to introduce a support network $\mathcal{N}^\pi = \langle V, E^\pi \rangle$, obtained by removing from $\mathcal{N}$ the edges with an "excessive" weight; observe that the nodes of $\mathcal{N}^\pi$ are the same as the nodes of $\mathcal{N}$. To formally define E^π, we employ the distribution of the weights of the edges of $\mathcal{N}$. Specifically, let max_E (resp., min_E) be the maximum (resp., minimum) weight of an edge of E. It is possible to define a parameter $step_E = \frac{max_E - min_E}{10}$, which represents the length of a "step" of the interval between min_E and max_E. We can define $d^k(E)$, $0 \leq k \leq 9$, as the number of the edges of E whose weights belong to the interval between $min_E + k \cdot step_E$ and $min_E + (k + 1) \cdot step_E$. All these intervals are closed on the left and open on the right, except for the last one that is closed both on the left and on the right. E^π can be defined as: $E^\pi = \{e_{ij} \in E | e_{ij} \in \bigcup_{k \leq th_{max}} d^k(E)\}$. We have experimentally set $th_{max} = 6$. We are now able to define the Connection Coefficient $\mathcal{Q}_\mathcal{N}$ of $\mathcal{N}$. In particular, let C be the set of the cliques of $\mathcal{N}^\pi$; let C_k be the set of cliques of dimension k of $\mathcal{N}^\pi$; finally, let $|C_k|$ be the cardinality (i.e., the number of cliques) of C_k. Then, $\mathcal{Q}_\mathcal{N}$ is defined as: $\mathcal{Q}_\mathcal{N} = \sum_{k=1}^{|V|} |C_k| \cdot 2^k$.

16.2.3 Approach to Knowledge Extraction

The idea underlying our approach is that, if some changes occur on sensor data streams, then some variations can be observed in some or all the dashboard parameters, when measured on the whole network, and/or on some of its subnetworks, depending on the number, the kind and the location of involved sensors. Our approach consists of a training phase and a testing phase. To carry out them, we employed available data (see Sect. 16.3.1) and, according to the holdout technique, we partitioned these data in such a way as to use 2/3 of them for the training phase and 1/3 of them for the testing phase. As for the training phase, we considered the following situations: (1) all sensors behaved correctly; (2) two sensors in location A and two sensors in location B were perturbed, in such a way as to decrease humidity; (3) two sensors in location B and two sensors in location C were perturbed, in such a way as to decrease lightness; (4) two sensors in location A and two sensors in location C were perturbed, in such a way as to increase lightness. Obtained results, along with the corresponding discussion, are presented in Sect. 16.3. After the training phase, we started the testing phase. In this case, we considered the following situations: (1) all sensors behaved correctly; (2) two sensors in location B and two sensors in location C were perturbed, in such a way as to decrease humidity; (3) two sensors in location A and two sensors in location C were perturbed, in such a way as to decrease lightness; (4) two sensors in location A and two sensors in location B were perturbed, in such a way as to increase lightness. Obtained results, along with the corresponding discussion, are presented in Sect. 16.3. Here, we simply point out that our approach behaved very well and was capable of correctly identifying all perturbations.

Finally, we applied our approach to the following situations: (1) one sensor in the location A and one sensor in the location B were perturbed, in such a way as to decrease humidity; (2) one sensor in the locations A and C was perturbed, in such a way as to increase lightness, and one sensor in the locations B and C was perturbed, in such a way as to decrease the same physical quantity; (3) three sensors in the location A and one sensor in the location B were perturbed, in such a way as to decrease humidity; (4) one sensor in the location A was perturbed, in such a way as to increase humidity; (5) one sensor in the location B was perturbed, in such a way as to increase lightness. Obtained results, along with the corresponding discussion, are presented in Sect. 16.3. Here, we anticipate that our approach showed its suitability to detect almost all perturbations.

16.3 Results

16.3.1 Testbed

To collect data for the experiments introduced in Sect. 16.2.3, we built a WSN by following specific guidelines. In particular, we organized devices in a multi-hop Wireless Sensor Area Network (WSAN) and managed them through the Building Management Framework (BMF) [15]. This is a framework for domain-specific networks, which offers an efficient and flexible management of WSANs deployed in indoor areas by allowing users to take advantage of sensing/actuation intelligent techniques and fast prototyping of WSAN applications. BMF enabled the use of heterogeneous WSANs through a base station, which acted both as data collector and network configurator. Communication between base station and devices was carried out by means of the BMF Communication Protocol, an application level protocol built on top of multi-hop network protocols [16, 17]. We composed the WSAN using MICAz sensor devices, providing 128 kB for program storage, 512 kB for data storage, and 4 kB of RAM. Devices were powered mainly by means of external power. They were configured to communicate with the base station, sending data every minute. To test our approach, we synthetically injected several anomalies at pre-determined time slots. In particular, to increase lightness, we employed artificial sources of lightness with controlled intensity, whereas to reduce lightness, we applied artificial lightness filters. Finally, humidity was controlled by chemicals. Our network consisted of 9 devices labeled by increasing numbers. Each device included 3 sensors, which retrieved values for humidity, lightness and temperature. Devices 1, 2 and 3 have been positioned in location A, devices 4, 5 and 6 operated in location B, devices 7, 8 and 9 were situated in location C. A, B and C were three different rooms on the same floor of a building. Finally, we collected data for 24 days without perturbations and other 36 days with several perturbations, as described in Sect. 16.2.3.

16.3.2 Obtained Results and Discussion

In this section, we report the results obtained by performing all the experiments mentioned in Sect. 16.2.3. Preliminarily, we observe that the definition of the six coefficients forming our dashboard suggests that a decrease of the connection level of a network or a subnetwork leads to: (i) an increase of $\mathcal{L_N}$ and $\mathcal{T_N}$; (ii) a decrease of $\mathcal{C_N}$, $\mathcal{P_N}$, $\mathcal{S_N}$ and $\mathcal{Q_N}$. The purpose of the training phase was to find the optimal values of some thresholds underlying our approach (for instance, the value of th_{max} in the definition of Connection Coefficient - see Sect. 16.2.2) and to have a first idea of its behavior. In Table 16.1, we report all the results regarding the training phase after the optimal values of thresholds were set. In particular, this table consists of four sub-tables, each corresponding to one of the four situations mentioned in Sect. 16.2.3.

Table 16.1 Results obtained by our approach during the training phase

Network	$\mathcal{L}_\mathcal{N}$	$\mathcal{C}_\mathcal{N}$	$\mathcal{P}_\mathcal{N}$	$\mathcal{T}_\mathcal{N}$	$\mathcal{Q}_\mathcal{N}$	$\mathcal{S}_\mathcal{N}$
Overall	1.1054	22.4387	6508290	64.2548	1163264	0.8944
$\mathcal{N}_t$	1.0322	7.1056	14232	15.1429	592	0.8413
$\mathcal{N}_l$	1.0451	7.1111	13200	16.6667	592	0.8595
$\mathcal{N}_h$	1.0278	7.5833	16758	16.9143	512	0.9722
$\mathcal{N}_A$	1.1944	5.6944	8012	23.7241	224	0.8339
$\mathcal{N}_B$	1.1667	5.9444	9274	22.4000	256	0.8582
$\mathcal{N}_C$	1.1944	6.0556	7896	23.7241	288	0.7794
Overall	1.1795	20.0684	4652472	74.7500	227328	0.8239
$\mathcal{N}_t$	1.1189	6.4444	10376	21.1613	384	0.8212
$\mathcal{N}_l$	1.1011	6.5833	11816	20.0000	320	0.7905
$\mathcal{N}_h$	1.4167	3.9444	2268	38.0952	96	0.5270
$\mathcal{N}_A$	1.3611	4.5000	3208	34.0870	120	0.5582
$\mathcal{N}_B$	1.3456	4.7778	4572	32.0800	144	0.5858
$\mathcal{N}_C$	1.1833	6.0444	7828	26.9091	248	0.7832
Overall	1.2194	19.1937	3790486	81.2263	99840	0.7796
$\mathcal{N}_t$	1.2556	5.8778	9924	20.8824	412	0.7392
$\mathcal{N}_l$	1.5000	4.1111	6102	26.3704	192	0.6000
$\mathcal{N}_h$	1.0556	7.2778	14924	17.8824	512	0.9392
$\mathcal{N}_A$	1.2111	5.4000	7990	23.0000	200	0.8571
$\mathcal{N}_B$	1.3222	4.5278	5990	29.1429	108	0.5630
$\mathcal{N}_C$	1.3333	4.7778	3824	32.0000	120	0.5407
Overall	1.2394	18.1937	3480632	80.2263	97650	0.7823
$\mathcal{N}_t$	1.2356	5.6648	9633	21.2435	408	0.7491
$\mathcal{N}_l$	1.5200	3.9345	6260	27.3221	192	0.5800
$\mathcal{N}_h$	1.0776	6.9318	13924	17.7623	512	0.9154
$\mathcal{N}_A$	1.3782	4.4987	5843	28.2322	108	0.661
$\mathcal{N}_B$	1.1911	5.1000	7232	23.0000	206	0.8200
$\mathcal{N}_C$	1.3433	4.6578	3126	31.6850	120	0.5207

For each situation, we report the values of the six parameters of the dashboard for the overall network and the subnetworks $\mathcal{N}_t, \mathcal{N}_l, \mathcal{N}_h, \mathcal{N}_A, \mathcal{N}_B$ and $\mathcal{N}_C$ (see Sect. 16.2.1). In this table, Situation 1 represents the correct one. In Situation 2, we observe: (i) a very high increase of $\mathcal{L}_\mathcal{N}$ and $\mathcal{T}_\mathcal{N}$, along with a very high decrease of $\mathcal{C}_\mathcal{N}, \mathcal{P}_\mathcal{N}, \mathcal{S}_\mathcal{N}$ and $\mathcal{Q}_\mathcal{N}$ for the network $\mathcal{N}_h$; (ii) a high increase of $\mathcal{L}_\mathcal{N}$ and $\mathcal{T}_\mathcal{N}$, along with a high decrease of $\mathcal{C}_\mathcal{N}, \mathcal{P}_\mathcal{N}, \mathcal{S}_\mathcal{N}$ and $\mathcal{Q}_\mathcal{N}$ for the networks $\mathcal{N}_A$ and $\mathcal{N}_B$; (iii) a moderate increase of $\mathcal{L}_\mathcal{N}$ and $\mathcal{T}_\mathcal{N}$, along with a moderate decrease of $\mathcal{C}_\mathcal{N}, \mathcal{P}_\mathcal{N}, \mathcal{S}_\mathcal{N}$ and $\mathcal{Q}_\mathcal{N}$ for the overall network. In Situation 3 (resp., 4), we observe: (i) a very high increase of $\mathcal{L}_\mathcal{N}$ and $\mathcal{T}_\mathcal{N}$, along with a very high decrease of $\mathcal{C}_\mathcal{N}, \mathcal{P}_\mathcal{N}, \mathcal{S}_\mathcal{N}$ and $\mathcal{Q}_\mathcal{N}$ for the network $\mathcal{N}_l$; (ii) a high increase of $\mathcal{L}_\mathcal{N}$ and $\mathcal{T}_\mathcal{N}$, along with a high decrease of $\mathcal{C}_\mathcal{N}$,

$\mathcal{P}_\mathcal{N}$, $\mathcal{S}_\mathcal{N}$ and $\mathcal{Q}_\mathcal{N}$ for the networks $\mathcal{N}_B$ and $\mathcal{N}_C$ (resp., $\mathcal{N}_A$ and $\mathcal{N}_C$); (iii) a moderate increase of $\mathcal{L}_\mathcal{N}$ and $\mathcal{T}_\mathcal{N}$, along with a moderate decrease of $\mathcal{C}_\mathcal{N}$, $\mathcal{P}_\mathcal{N}$, $\mathcal{S}_\mathcal{N}$ and $\mathcal{Q}_\mathcal{N}$ for the overall network. These results confirm that our approach is really capable of capturing the perturbations in wireless sensor networks or subnetworks caused by sensor anomalies (and, indirectly, it is able to evaluate the network and subnetwork resilience to sensor anomalies). The only weakness revealed by this first test is that, in its current version, our approach is not able to tell us if these perturbations are caused by an increase or a decrease of the corresponding physical quantity.

The purpose of the testing phase was to verify both the setting of the threshold values and the corresponding results detected during the training phase. In Table 16.2, we report all the results regarding this phase. Observe that the situations considered during this phase are the same as the ones examined during the training phase; however, we modified the subnetworks (among A, B and C) involved in each perturbation in such a way as to prevent overfitting. Obtained results confirm that the selection of the threshold values performed during the training phase was correct. They also confirm all the observations about the features of our approach, which we drew at the end of the training phase.

After the testing phase confirmed the suitability of our approach, we applied it to new situations not considered during the previous phases. These situations are described in detail in Sect. 16.2.3. In Table 16.3, we report the corresponding results. From their analysis we can draw very interesting observations. In particular, in Situation 1, we obtain the same trend as the one seen in Situation 2 of the training phase. However, the perturbation degree is more reduced. This is correct because, for locations A and B, we perturbed one sensor, instead of two. In Situation 2, we observe: (i) a very high increase of $\mathcal{L}_\mathcal{N}$ and $\mathcal{T}_\mathcal{N}$, along with a very high decrease of $\mathcal{C}_\mathcal{N}$, $\mathcal{P}_\mathcal{N}$, $\mathcal{S}_\mathcal{N}$ and $\mathcal{Q}_\mathcal{N}$ for the network $\mathcal{N}_l$; these increases and decreases are comparable with the ones observed in Situation 3 of the training phase; (ii) a moderate (resp., high, very high) increase of $\mathcal{L}_\mathcal{N}$ and $\mathcal{T}_\mathcal{N}$, along with a moderate (resp., high) decrease of $\mathcal{C}_\mathcal{N}$, $\mathcal{P}_\mathcal{N}$, $\mathcal{S}_\mathcal{N}$ and $\mathcal{Q}_\mathcal{N}$, for the networks $\mathcal{N}_A$ and $\mathcal{N}_B$ (resp., $\mathcal{N}_C$, $\mathcal{N}_l$); (iii) a moderate increase of $\mathcal{L}_\mathcal{N}$ and $\mathcal{T}_\mathcal{N}$, along with a moderate decrease of $\mathcal{C}_\mathcal{N}$, $\mathcal{P}_\mathcal{N}$, $\mathcal{S}_\mathcal{N}$ and $\mathcal{Q}_\mathcal{N}$, for the overall network. Observe that, since our approach considers perturbations, but it currently does not distinguish between increases and decreases, even if, in the network $\mathcal{N}_l$, there are opposite perturbations in two lightness sensors, their consequences are not nullified by our approach, but, on the contrary, are "combined" by it. In our opinion, this is a correct behavior of our approach.

In Situation 3, we observe: (i) an increase (resp., decrease) of $\mathcal{L}_\mathcal{N}$ and $\mathcal{T}_\mathcal{N}$ (resp., $\mathcal{C}_\mathcal{N}$, $\mathcal{P}_\mathcal{N}$, $\mathcal{S}_\mathcal{N}$ and $\mathcal{Q}_\mathcal{N}$), comparable with the one of Situation 2 of the training phase for both the overall network and the network $\mathcal{N}_h$; (ii) a significant (resp., moderate) increase of $\mathcal{L}_\mathcal{N}$ and $\mathcal{T}_\mathcal{N}$, along with a significant (resp., moderate) decrease of $\mathcal{C}_\mathcal{N}$, $\mathcal{P}_\mathcal{N}$, $\mathcal{S}_\mathcal{N}$ and $\mathcal{Q}_\mathcal{N}$ for the network $\mathcal{N}_A$ (resp., $\mathcal{N}_B$). In Situation 4 (resp., 5), we observe: (i) a very moderate increase of $\mathcal{L}_\mathcal{N}$ and $\mathcal{T}_\mathcal{N}$, along with a very moderate decrease of $\mathcal{C}_\mathcal{N}$, $\mathcal{P}_\mathcal{N}$, $\mathcal{S}_\mathcal{N}$ and $\mathcal{Q}_\mathcal{N}$ for the overall network and for the networks $\mathcal{N}_h$ and $\mathcal{N}_A$ (resp., $\mathcal{N}_l$ and $\mathcal{N}_B$). This reveals a second weakness of our approach, which shows a difficulty to find a single anomaly. Indeed, in this case, it found a slight change in the dashboard parameters for both the whole network and the involved

Table 16.2 Results obtained by our approach during the testing phase

Network	$\mathcal{L}_\mathcal{N}$	$\mathcal{C}_\mathcal{N}$	$\mathcal{P}_\mathcal{N}$	$\mathcal{T}_\mathcal{N}$	$\mathcal{Q}_\mathcal{N}$	$\mathcal{S}_\mathcal{N}$
Overall	1.1135	20.4387	7120293	65.3746	1163264	0.9144
$\mathcal{N}_t$	1.0411	6.5306	13939	15.1529	592	0.8712
$\mathcal{N}_l$	1.0361	6.2480	13737	17.1227	592	0.8891
$\mathcal{N}_h$	1.0235	7.3311	16123	16.8242	512	0.8920
$\mathcal{N}_A$	1.1826	5.4129	7910	22.7241	228	0.8451
$\mathcal{N}_B$	1.1700	5.8331	8992	21.4000	256	0.8112
$\mathcal{N}_C$	1.1929	6.2410	7786	23.7241	288	0.8042
Overall	1.1896	20.1224	4993459	72.63	294629	0.8484
$\mathcal{N}_t$	1.1289	6.2468	11001	22.1982	320	0.8391
$\mathcal{N}_l$	1.2133	6.6631	10829	21.0782	384	0.8081
$\mathcal{N}_h$	1.5177	3.8104	3124	37.1719	112	0.5328
$\mathcal{N}_A$	1.1922	6.2324	7128	27.8801	208	0.7312
$\mathcal{N}_B$	1.3232	4.9188	4492	31.9500	128	0.5558
$\mathcal{N}_C$	1.3511	4.4780	3198	33.0870	118	0.5182
Overall	1.2766	20.2308	4290486	81.3094	97744	0.7824
$\mathcal{N}_t$	1.3111	5.5833	9850	20.0000	258	0.7825
$\mathcal{N}_l$	1.4389	4.0833	3438	25.9750	96	0.6412
$\mathcal{N}_h$	1.0242	7.3611	13978	18.4421	384	0.9825
$\mathcal{N}_A$	1.3056	4.5278	4762	30.1515	108	0.5713
$\mathcal{N}_B$	1.1896	5.5278	7288	22.1429	216	0.8462
$\mathcal{N}_C$	1.2825	4.9444	3594	32.9143	96	0.5356
Overall	1.2251	17.9876	3990563	82.2263	97650	0.7769
$\mathcal{N}_t$	1.2944	5.8326	9112	22.7241	408	0.7839
$\mathcal{N}_l$	1.4678	4.6161	6383	26.3352	112	0.5455
$\mathcal{N}_h$	1.1111	6.5833	13816	17.6686	384	0.9005
$\mathcal{N}_A$	1.4001	4.7144	6152	27.8652	96	0.6148
$\mathcal{N}_B$	1.3675	4.3056	3886	30.9850	88	0.5198
$\mathcal{N}_C$	1.1887	6.2421	7341	22.7692	256	0.8825

subnetworks. This is mainly due to the purpose of our approach, which does not aim at performing anomaly detection in one sensor (actually, a long list of approaches carrying out this task—e.g., [4–7]—already exists) but, instead, it aims at detecting the consequences, on the whole network and its subnetworks, of anomalies involving more (heterogeneous) sensors installed in different locations. In fact, in this case, the interaction of these anomalies in the network could be extremely variegate and could depend on the number, the kind and the location of perturbed sensors, so that their detection, along with the detection of their effects, becomes extremely difficult and

Table 16.3 Results obtained by our approach during the examination of some situations of interest

Network	$\mathcal{L}_\mathcal{N}$	$\mathcal{C}_\mathcal{N}$	$\mathcal{P}_\mathcal{N}$	$\mathcal{T}_\mathcal{N}$	$\mathcal{Q}_\mathcal{N}$	$\mathcal{S}_\mathcal{N}$
Overall	1.1435	21.5534	5580928	70.0000	114264	0.8534
$\mathcal{N}_t$	1.0712	6.3159	11432	18.2221	384	0.8613
$\mathcal{N}_l$	1.0572	6.4354	11202	18.6667	384	0.8564
$\mathcal{N}_h$	1.2578	4.5673	4564	22.2124	144	0.8123
$\mathcal{N}_A$	1.2235	5.1843	6006	28.3673	200	0.7034
$\mathcal{N}_B$	1.2351	5.4992	8842	27.4332	224	0.6982
$\mathcal{N}_C$	1.1833	5.3556	7828	24.7347	248	0.7792
Overall	1.2199	19.3747	3948573	80.3252	97650	0.7856
$\mathcal{N}_t$	1.2456	5.6658	8562	21.9383	388	0.7467
$\mathcal{N}_l$	1.6100	3.5039	5987	28.2392	192	0.5971
$\mathcal{N}_h$	1.0877	6.4837	12527	17.3877	512	0.8672
$\mathcal{N}_A$	1.2292	4.5948	7873	27.223	228	0.6823
$\mathcal{N}_B$	1.2334	5.1229	7367	26.2391	228	0.6891
$\mathcal{N}_C$	1.2921	4.6578	3834	32.2320	120	0.5012
Overall	1.1235	21.9987	3977283	74.5673	231872	0.8223
$\mathcal{N}_t$	1.1312	6.2989	12345	21.3939	512	0.8323
$\mathcal{N}_l$	1.1433	6.5643	12234	20.3332	512	0.8340
$\mathcal{N}_h$	1.4872	3.9440	3542	38.9412	120	0.7795
$\mathcal{N}_A$	1.8342	2.2338	1987	35.1843	96	0.4032
$\mathcal{N}_B$	1.2151	4.4738	6932	25.6230	224	0.5820
$\mathcal{N}_C$	1.1933	6.0872	8239	23.3235	284	0.7780
Overall	1.1228	21.3789	6184736	67.3233	131872	0.8534
$\mathcal{N}_t$	1.0613	6.4599	12341	17.3939	592	0.8613
$\mathcal{N}_l$	1.0732	6.8865	12854	16.3452	592	0.8564
$\mathcal{N}_h$	1.1640	5.6534	9532	20.9482	288	0.8123
$\mathcal{N}_A$	1.2132	5.1928	6987	26.1212	288	0.7034
$\mathcal{N}_B$	1.1951	5.4738	9928	24.7210	320	0.6982
$\mathcal{N}_C$	1.19445	5.5872	8239	23.3235	320	0.7792
Overall	1.1289	21.8729	6857326	67.3252	131662	0.8556
$\mathcal{N}_t$	1.0782	6.7654	12662	17.2352	592	0.8467
$\mathcal{N}_l$	1.1728	5.9987	5987	20.4568	288	0.8023
$\mathcal{N}_h$	1.0654	6.2356	12277	16.4555	592	0.8553
$\mathcal{N}_A$	1.1892	5.6457	9854	25.3356	320	0.7061
$\mathcal{N}_B$	1.2234	5.0101	5346	26.4564	288	0.7072
$\mathcal{N}_C$	1.1921	5.5482	8899	23.2845	284	0.7843

justifies the employment of quite time-expensive approaches like ours. As for this issue, the results described in this section allow us to conclude that our approach reaches the objectives for which it was designed.

16.4 Conclusion

In this paper, we have presented a new approach to analyzing WSNs, which considers network organization as a whole; this shifts the focus of the analysis from single sensors to the whole network and its subnetworks. Our approach is based on network connectivity measures that, overall, contribute to a rich dashboard, which allows the effective detection of perturbations in WSNs. Our model also allows the network to be sliced in different subnetworks, supporting the investigation of this phenomenon under different perspectives, as well as a better characterization of perceived perturbations. Our experimental campaign confirms the effectiveness of our approach. In the future, we plan to remove the current weaknesses of our approach, as evidenced by our experiments. First, we aim at allowing our approach to distinguish perturbations caused by an increase or a decrease of a physical quantity. Then, we plan to integrate our approach with the ones detecting anomalies in single sensors. The ultimate goal is to construct an effective framework, which can detect anomalies on single sensors and can investigate their consequences on the whole network and its subnetworks, along with their resilience to sensor malfunctions.

References

1. Bosman, H., Liotta, A., Iacca, G., Wrtche, H.: Anomaly detection in sensor systems using lightweight machine learning. In: 2013 IEEE International Conference on Systems, Man, and Cybernetics, Manchester, pp. 7–13 (2013)
2. Bosman, H., Liotta, A., Iacca, G., Wrtche, H.: Online extreme learning on fixed- point sensor networks. In: IEEE 13th International Conference on Data Mining Workshops (ICDMW), pp. 319–326 (2013)
3. Zhang, Y., Jiang, J.: Bibliographical review on reconfigurable fault-tolerant control systems. Ann. Rev. Control 32(2), 229–252 (2008)
4. Ho, J.-W., et al.: Distributed detection of replica node attacks with group deployment knowledge in wireless sensor networks. Ad. Hoc Netw., 147688 (2009)
5. Li, G. et al.: Group-based intrusion detection system in wireless sensor networks. Comput. Commun., 4324–32 (2008)
6. Ngai, E.C.H., Liu, J., Lyu, M.R.: An efficient intruder detection algorithm against sinkhole attacks in wireless sensor networks. Comput. Commun. 30(1112), 2353–2364 (2007)
7. Yu, Z., Tsai, J.J.P.: A framework of machine learning based intrusion detection for wireless sensor networks. In: IEEE International Conference on Sensor Networks, Ubiquitous, and Trustworthy Computing (SUTC 2008), Taichung, pp. 272–279 (2008)
8. Curia, V., Tropea, M., Fazio, P., Marano, S.: Complex networks: Study and performance evaluation with hybrid model for Wireless Sensor Networks. In: Electrical and Computer Engineering (CCECE), 2014 IEEE 27th Canadian Conference, pp. 1–5. IEEE
9. Cuzzocrea, A., Papadimitriou, A., Katsaros, D., Manolopoulos, Y.: Edge betweenness centrality: a novel algorithm for QoS-based topology control over wireless sensor networks. J. Netw. Comput. Appl. 35(4), 1210–1217 (2012)
10. Jain, A.: Betweenness centrality based connectivity aware routing algorithm for prolonging network lifetime in wireless sensor networks. Wirel. Netw. 22(5), 1605–1624 (2016)
11. Xie, M., Han, S., Tian, B., Parvin, S.: Anomaly detection in wireless sensor networks: A survey. J. Netw. Comput. Appl. 34(4), 1302–1325 (2011)

12. Papadimitriou, A., Katsaros, D., Manolopoulos, Y.: Social network analysis and its applications in wireless sensor and vehicular networks. International Conference on e-Democracy, pp. 411–420. Springer, Berlin, Heidelberg
13. Cauteruccio, F., Stamile, C., Terracina, G., Ursino, D., Sappey-Marinier, D.: An automated string-based approach to extracting and characterizing White Matter fiber-bundles. Comput. Biol. Med. **77**(1), 64–75 (2016)
14. Koschützki, D., Lehmann, K.A, Peeters, L., Richter, S., Tenfelde-Podehl, D., Zlotowski, O.: Centrality indices. In: Network Analysis, pp. 16–61. Springer (2005)
15. Fortino, G., Guerrieri, A., OHare, G., Ruzzelli, A.: A flexible building management framework based on wireless sensor and actuator networks. J. Netw. Comput. Appl. **35**(2012), 1952 (1934)
16. Gnawali, O., Fonseca, R., Jamieson, K., Kazandjieva, M., Moss, D., Levis, P.: CTP: an efficient, robust, and reliable collection tree protocol for wireless sensor networks. ACM Trans. Sens. Netw. **10**(1) (2013) 16:116:49
17. Levis, P., Patel, N., Culler, D., Shenker, S.: Trickle: a self-regulating algorithm for code propagation and maintenance in wireless sensor networks. In: 1st Conference on Symposium on Networked Systems Design and Implementation - vol. 1. NSDI 04, Berkeley, CA, USA, USENIX Association (2004)

Chapter 17
Advanced Computational Intelligence Techniques to Detect and Reconstruct Damages in Reinforced Concrete for Industrial Applications

Salvatore Calcagno and Fabio La Foresta

Abstract In reinforced concrete, as known, the steel bar, damping totally the traction stress, they are mainly subject to breakage. Then, as required by current legislation, it is necessary a check protocol of the specimens characterizing any defects since the typology of defect, often, determines its intended use. From this, the choice to use non-invasive technique such as Non-Destructive Testing and Evaluation (NDT/NDE) based on Eddy Currents is necessary. Starting from a campaign of Eddy Currents measurements potentially affected by uncertainty and/or imprecision, in this work we propose a new fuzzy approach based on Computing with Words techniques where a *word* is considered a label of a fuzzy set of points shared by similarities coming to an adaptive bank of fuzzy rules structured by classes possibly updated by the Expert's knowledge. The numerical results obtained by means of the proposed approach are comparable with the results carried out by Fuzzy Similarities techniques already established in the literature.

17.1 Introduction to the Problem

The assessment of the integrity of a reinforced concrete specimen, both in a civil and industrial context, is imperative in evaluating the amount of risk of collapse of the structure [1]. Then, as in reinforced concrete structure the traction stresses are completely damped by the steel bars, and given that the concrete can only damp compression stresses, both detection and characterization of defectiveness in the steel bars is an indispensable task to assess the above risk without compromising the functionality of the structure. Under elastic-plastic conditions, (see [2]), the steel of

S. Calcagno (✉) · F. L. Foresta
DICEAM Department, Universitá "Mediterranea", Via Graziella Feo di Vito,
89122 Reggio Calabria, Italy
e-mail: calcagno@unirc.it

F. L. Foresta
e-mail: fabio.laforesta@unirc.it

© Springer International Publishing AG, part of Springer Nature 2019
A. Esposito et al. (eds.), *Neural Advances in Processing Nonlinear Dynamic Signals*, Smart Innovation, Systems and Technologies 102,
https://doi.org/10.1007/978-3-319-95098-3_17

191

Fig. 17.1 Reinforced
concrete specimen with a
steel bar exploited fro the
campaign of experimental
measurements

the bars has limits about its load capacity which, if passed, greatly adversely affect
the specimen's state of health as obvious fracture phenomena can propagate along
the bars. Moreover, the adherence between steel and concrete would be irrepara-
bly compromised. The Scientific Community is working hard to solve the problem
by developing reliable theoretical models for analyzing phenomena related to the
origin and diffusion of cracks in slim structural elements particularly sensitive to
these issues (see, e.g., [2–4]). On the other hand, although the load to which the
specimen is subjected is not sufficient to create cracks, however, yielding phenomena may arise in the bars creating dangerous mapping of the mechanical tension
maps, compromising seriously the quality of the specimen. But the theoretical models developed to achieve mechanical tension maps are extremely complex and require
high computational effort that are poorly combined with the objective requirement
of having low-cost computational investigation techniques particularly useful for
real-time analysis. So, the idea of using Non Destructive Testing and Evaluation
(NDT/NDE) based on Eddy Currents (ECs) [5] is matured because ECs build
maps of electrical voltages equivalent to the mechanical ones [6–8] but with an
acceptable computational load. In such a context, taking into account that measurements may be affected by uncertainties and/or inaccuracies, we propose a new fuzzy
approach to detecting defects based on the concept of Computing with Words (CWs)
in which *word* is the label of a fuzzy set of elements shared by similarity (the so-
called "granule") writing a bank of adaptive fuzzy rules potentially upgradeable and
improved by the expert's knowledge. In the following, the details of the approach
will be proposed, tested and validated through an experimental database achieved
at NDT/NDE Lab in "Mediterranea" University of Reggio Calabria. Numerical
results obtained were also compared to those obtained using Fuzzy Similarity (FS)
technique already established in the literature.

17.2 The Proposed Approach

17.2.1 ECs NDT & CWs: A Quick Look

Traditionally, computing means manipulating numbers and symbols. This contrasts with the way of thinking of a man who uses words to reason and calculate: any reasoning starts from a premise expressed in a natural language to reach conclusions also expressed in a natural language. *CWs*, born in 1996 with pioneering works [9–11, 13], imposes a fusion between natural language and computation with fuzzy variables. The starting point is the concept of *granule*, *g*, that is, a set of elements similar to each other. So, a word *w* is an instance of a granule *g*, as well as *g*, representing a fuzzy constraint on a variable, is the denotation of *w*. Often a word *w* can be simple or composed; for example, the extension of a defect in bar is an *atomic word*; instead *very extensive* is considered a *composite word*. The transition among granules is gradual and represents a fuzzy constraint on a given variable. In fact, in the proposition *the defect is extended*, *extended* is the label of the granule and the fuzzy set *extended* is the fuzzy constraint on the extension of the defect. Conceptually, *CWs* is the confluence of two streams: *fuzzy logic* and *test-score semantics* whose contact points are the collection of canonical forms obtained from the premises (expressed in a natural language) that explain the constraints. Starting from canonical forms, the propagation of constraints leads us to conclusions described with other fuzzy constraints. Finally, the induced constraints are translated into natural language using appropriate linguistic approximation. Formally, to construct a *CWs*-based system, you must start from a set of propositions expressed in a natural language, the Initial Data Set (*IDS*), from which you want to deduce a response to get to the Terminal Data Set (*TDS*) also expressed in a natural language. The transition from *IDS* to *TDS* occurs by explicit propagation of fuzzy constraints (and any changes) by means of a rule bank: the inference transforms the previous antecedents that are drawn in natural language by appropriate approximators (Fig. 17.2).

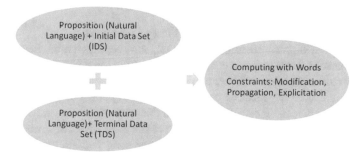

Fig. 17.2 An example of structure based on *CWs*: starting from the *IDS* we obtain the *TDS* by means of both propagation and manipulation of the bank of fuzzy constraints

Canonic Formulations: Fuzzy Constraints and Constraining Relations

Fuzzy Constraints formalization involves the writing of canonical forms such as
"*Z is Relation*" (also called canonical function) where *Z* is a fuzzy variable and
"Relation" is a fuzzy constraining relation. Usually, a canonical function can be disjunctive when all the terms are linked by the connective *O R*; it is conjunctive when all
the terms are linked by *AN D* and *Relation* puts out the fuzzy constraint contained in
a *proposition* by writing *proposition* → *Z is Relation* in which an Explanatory
Database (*ED*) explain the *proposition* by means of a set of relations that returns
the constrained variable *Z* acting on *ED* for extracting the constraining *Relation*.
In addition, if *Relation* is modified by any *manipulators* such as *not, very*, then it
is permissible to write *Z is manipulator*(*R*) → *Z is function*(*Relation*) where
function(*Relation*) is the modification of *Relation* due to the manipulator application. For example, if *manipulator* = *not, function*(*Relation*) = *not* − *Relation*
and its membership function μ takes the form $\mu_{not}(s) = 1 - \mu(s)$ in which *s* is an
element of *Relation*.

An Example of *ED* for *ECs NDT* Problems

If the proposition is é "*The Defect is not very extensive*" we extract the *ED* as
follows:

$$ED = Population_{Defects}\{Type\ of\ Defect;\ Extension\} + Extensive\{Extension;\ \mu()\}$$
$$(17.1)$$

where *Population*$_{defects}$ is the relation whose topics are the *Type of Defect* and
Extension while *Extensive* is the characterization of *Extension* and+ indicates
the conjunction. Then, *Z* é expressible as follows:

$$Z = Extension(Defect) =_{Extension} Population\{Type\ of\ Defect = Defect\}$$
$$(17.2)$$

providing the action acting on *ED* returning *Z*: *Type of Defect* is associated to
Defect whose results is projected on *Extension* producing the extension of the
defect. In addition, the constraining relation *not so extensive* is expressible as
Relation = *Extensive*{*Extension*; $\mu^2()$} while the negation is expressible by
means of *Relation* = *Extensive*{*Extension*; $1 - \mu()$}.

IDS Construction and Propagation of the Fuzzy Constraints

Starting from the simple structure *Z is Relation*, and keeping in mind that the goal is
to write a bank of fuzzy rules, it is advisable to rewrite the constructs in the form *Z
is variable copula Relation* in order to highlight how *Z* i constrained by the relation
assigning a value to the discrete variable *variable copula*. For example, *variable
copula= disjunctive, dis* for short, means that the constraint is equivalent to writing
Z is Relation thus expressing the proposition of the *IDS* terms of canonical forms.
For details of the value interpretations for the *variables copula*, see [9, 12]. After
having structure the *IDS*, it is necessary to propose an efficient procedure for the
propagation of the fuzzy constraints. In particular, if *Relation*$_1$ and *Relation*$_2$ are
two relations, it is easy to obtain rules of disjunction, projection and surjection as
show in Table 17.1.

Table 17.1 Examples of propagation of fuzzy constraints

Kind of rule	Structure
Disjunction	**IF** Z is $Relation_1$ **OR** Z is $Relation_2$ **THEN** Z is $(Relation_1 \cup Relation_2)$
Conjunction	**IF** Z is $Relation_1$ **AND** Z is $Relation_2$ **THEN** Z is $(Relation_1 \cap Relation_2)$
Projection	**IF** (Z_1, Z_2) is $Relation$ **THEN** Z_2 is $sup(Relation)$
Surjection	**IF** Z is $Relation$ **THEN** (Z_1, Z_2) is $(Relation \times M)$

Terminal Data Set

Generally, the construction of TDS starts from the generalized extension principle **IF** $f(Z_1, \ldots, Z_n)$ is Relation **THEN** is $query(Z_1, \ldots, Z_n) \, query(f^{-1}(Relation))$ where $Z_1, \ldots, Z_n$ are fuzzy variables while $f(Z_1, \ldots, Z_n)$ is Relation and $query(Z_1, \ldots, Z_n)$ are the constraint extracted by IDS and query constraining TDS respectively. It is known in the literature that starting from the principle of extension is possible, considering $constraint = q(s_1, s_2, \ldots, s_n)$, to formulate the TDS in terms of maximization problem as follows:

$$\mu_{query}(Z_1, \ldots, Z_n)(constraint) = sup_{s_1, s_2, \ldots, s_n} (\mu_{Relation}(f(s_1, s_2, \ldots, s_n)))$$

(17.3)

so the mean value of the extension of the defect, indicating by hpd_j, $j = 1, \ldots, n$ the entire population of defect extensions with membership functions $\mu_{extended}(hpd_j)$ is obtainable by:

$$\mu_{constraint} = sup_{hpd_1, \ldots, hpd_n} (\mu_{medium}(\frac{1}{n} \sum_j \mu_{extended}(hpd_j)))$$

(17.4)

17.3 Numerical Results

17.3.1 The Experimental Database

At the Laboratory of Electrotechnics and Non-Destructive Testing, DICEAM Department Mediterranea University of Reggio Calabria, a concrete block reinforced by a steel bar longitudinally located with an artificial cut has been investigated under gradually increasing axial loads (Fig. 17.1). Concerning the training step, the block has been subjected to loads more and more increasing starting from 180 kN up to 210 kN by increments of 10 kN. Microscopically, the structure of the material, under each load, changes locally so that the magnetic properties change together its degradation. The analysis has been carried out by means of magnetic changes induced in the bar after each deformation. The block, after each load application, has been subjected to exciting currents inducing an EC that, by means a FLUXSET probe [14] spaced

Table 17.2 Structure of the database in terms of the number of signals related to both training and testing sections

TRAINING DATABASE	Number of signals
No defects	60
CL_1 180 kN	60
CL_2 190 kN	60
CL_3 200 N	50
CL_4 210 kN	50
TESTING DATABASE	Number of signals
	50

10 cm from the specimen, a set of signals of the variation of the overall magnetic field, which takes into account the presence of the defect, have been carried out. Such magnetic variation is converted into an equivalent electric variation providing the pick-up voltage [mV] whose phase demodulation gives us a measurement proportional to the magnetic field parallel to the plane of the specimen. To take into account the skin effect, the analysis was performed with a sinusoidal signal of 10 Vpp with 1.022 Hz moving the sensor by a step-by-step scanning system over a square portion at the middle of the block containing the cut on the bar. And again, both an AC sinusoidal exciting field (1 kHz) and an electric current (120 mA RMS) have been applied while, concerning the driving signal, a triangular shape at the frequency of 100 kHz with 2 Vpp amplitude has been considered. After each load application, and concerning each point of the scanned area, four signals have been obtained: real and imaginary parts, module and phase of the pick-up voltage. Because mechanical loads produce almost overlapping deformations (and therefore magnetic behaviors close each other), the block has been subjected to increasing load sets by creating load classes with overlapping mechanical and magnetic properties. In particular, four load classes were created: CL_1 180 kN, CL_2 190 kN, CL_3 200N and CL_4 210 kN; remembering that in each class the signals produced by the nominal load (labeling the class) and the loads that are close to the latter are concerned. Furthermore, a campaign of measurements to extract a set of signals in the absence of mechanical loads was carried out on a block containing a faultless bar, thus a further class of signals in the absence of defects was built. Concerning the testing step, a high number of signals was produced under the action of different mechanical loads which, for our purposes, are supposedly unknown. As shown in in Table 17.2, the experimental database has been divided into two main sections (training and testing ones) where, for each class, the number of signals analyzed is highlighted. They were considered, as input data, the normalized impedance $\dot{Z}_{norm}$, the mean $Stat_{1_{norm}}^{12 \times 12}$, the variance $Stat_{2_{norm}}^{12 \times 12}$, the skewness $Stat_{3_{norm}}^{12 \times 12}$ and the Kurtosis $Stat_{4_{norm}}^{12 \times 12}$ of a portion 12 × 12 mm centered on the i^h pixel as specified in Table 17.3. Concerning the output of the procedure, they were considered the classes CL_i, $i = 1, 2, 3, 4$ and the class of signals in the absence of loads.

17.3.2 The Extracted Bank of Fuzzy Rules

The granulation procedure begins by dividing the signals between inputs and outputs as reported in Table 17.3 adding two additional outputs, *Near* and *Far* (coded with 1 and 2 respectively) indicating whether the defect is facing the probe or otherwise. Together the codification, Table 17.3 also shows the ranges of the possible values. In addition, we consider just two gaussian membership functions for $\dot{Z}_{norm}$, $(\dot{Z}_{norm})_{MF_1}$ and $(\dot{Z}_{norm})_{MF_2}$, because distributed around 0.1 and 0.4. Six other membership functions are sufficient for other inputs (Extra Small (ES), Very Small (VS), Small (S), Medium (M), Large (L), Very Large (VL)) to cover the distribution of the values appropriately. In order to achieve the bank of fuzzy rules, we construct a table where, for each training dataset signal, we associate the membership class carrying out a number of rules equal to the number of signals contained in the training dataset. For example, for the first signal, we perform the rule

Table 17.3 Inputs/outputs codification and ranges of possible values

Inputs	Details	Ranges
$\dot{Z}_{norm} = \dfrac{\dot{Z}_j}{\dot{Z}_{j0}}$	$\dot{Z}_j$ impedance of the ith point of the block with the defective bar; $\dot{Z}_j$ impedance of the ith point of the block with bar without defects;	0.07–0.46
$Stat_{1norm}^{12 \times 12} = \dfrac{Stat_{1j}^{12 \times 12}}{Stat_{1o}^{12 \times 12}}$	$Stat_{1j}^{12 \times 12}$ mean of the portion 12×12 centered on the jth pixel of the block with the defective bar; $Stat_{1o}^{12 \times 12}$ mean of the portion 12×12 centered on the jth pixel of the block with bar without defects;	0.24–0.73
$Stat_{2norm}^{12 \times 12} = \dfrac{Stat_{2j}^{12 \times 12}}{Stat_{2o}^{12 \times 12}}$	$Stat_{2j}^{12 \times 12}$ variance of the portion 12×12 centered on the jth pixel of the block with the defective bar; $Stat_{2o}^{12 \times 12}$ variance of the portion 12×12 centered on the jth pixel of the block without defects;	0.38–0.91
$Stat_{3norm}^{12 \times 12} = \dfrac{Stat_{3j}^{12 \times 12}}{Stat_{3o}^{12 \times 12}}$	$Stat_{3j}^{12 \times 12}$ skewness of the portion 12×12 centered on the jth pixel of the block with the defective bar; $Stat_{3o}^{12 \times 12}$ skewness of the portion 12×12 centered on the jth pixel of the block with bar without defects;	0.29–0.77
$Stat_{4norm}^{12 \times 12} = \dfrac{Stat_{4j}^{12 \times 12}}{Stat_{4o}^{12 \times 12}}$	$Stat_{4j}^{12 \times 12}$ kurtosis of the portion 12×12 centered on the jth pixel of the block with the defective bar; $Stat_{4o}^{12 \times 12}$ kurtosis of the portion 12×12 centered on the jth pixel of the block with bar without defects;	0.18–0.69
Outputs		
$CL_i, i = 1, 2, 3, 4$	Classes of signals in presence of certain loads	
No-Defects	Classes of signals without loads	

$$\textbf{IF } \dot{Z}_{norm} \text{ is } (\dot{Z}_{norm})_{MF_2} \text{ and } Stat_{1_{norm}}^{12 \times 12} \text{ is } ES \text{ and} \qquad (17.5)$$

$$Stat_{2_{norm}}^{12 \times 12} \text{ is } L \text{ and } Stat_{3_{norm}}^{12 \times 12} \text{ is } VL$$

$$\text{and } Stat_{4_{norm}}^{12 \times 12} \text{ is } S \textbf{ THEN } Defect \text{ is } CL_3 \text{ and } Position \text{ is } Far$$

that, after compaction and elimination of duplicate rules, leads to write a bank of 8 fuzzy rules on which the procedure explained in Sect. 17.2 is applied. For example, considering the first rule of the above-mentioned bank:

$$\textbf{IF } \dot{Z}_{norm} \text{ is } (\dot{Z}_{norm})_{MF_2} \text{ and } Stat_{1_{norm}}^{12 \times 12} \text{ is } VS \text{ and} \qquad (17.6)$$

$$Stat_{2_{norm}}^{12 \times 12} \text{ is } S \text{ and } Stat_{3_{norm}}^{12 \times 12} \text{ is } VL \text{ and}$$

$$Stat_{4_{norm}}^{12 \times 12} \text{ is } M \textbf{ THEN } Defect \text{ is } CL_1 \text{ and } Position \text{ } Near$$

the corresponding rule in terms of CWs becomes:

$$\textbf{IF } \dot{Z}_{norm}\{Parameter = \dot{Z}_{norm}\} \text{ is } (\dot{Z}_{norm})_{MF_2}\{0.38 - 0.5; \mu()\} \text{ and} \qquad (17.7)$$

$$Stat_{1_{norm}}^{12 \times 12}\{Parameter = Stat_{1_{norm}}^{12 \times 12}\} \text{ is } VS\{0.25 - 0.34; \mu()\} \text{ and}$$

$$Stat_{2_{norm}}^{12 \times 12}\{Parameter = Stat_{2_{norm}}^{12 \times 12}\} \text{ is } S\{0.56 - 0.68; \mu()\} \text{ and}$$

$$Stat_{3_{norm}}^{12 \times 12}\{Parameter = Stat_{3_{norm}}^{12 \times 12}\} \text{ is } VL\{0.61 - 0.78; \mu()\} \text{ and}$$

$$Stat_{4_{norm}}^{12 \times 12}\{Parameter = Stat_{4_{norm}}^{12 \times 12}\} \text{ is } M \{0.32 - 0.45; \mu()\} \textbf{ THEN}$$

$$_{Possible \; values}Characteristic_{Defect}\{Parameter = Class\} \text{ is } CL_1\{1; \mu()\} \text{ and}$$

$$_{Possible \; values}Characteristic_{Position}\{Parameter = Position\} \text{ is } Near\{1; \mu()\}$$

obtaining, therefore, a ban k formed by 8 rules from which the TDS is determined using the Eq. 17.3. In this regard, indicating with $typeDef_i$ the generic defect and with $\mu_{Defect}(typeDef_i)$ the corresponding membership function, the solution of Eq. 17.3 is carries out by means of the resolution of the following maximization problem:

$$\mu\left(N^{-1}\sum_i (typeDef_i)\right) = sup_{typeDef_i}\left\{\mu_{max}\left\{N^{-1}\sum_i \mu_{Defect}(typeDef_i)\right\}\right\}$$
$$(17.8)$$

Table 17.4 shows the results, in terms of both detection and classification, with the proposed procedure. In addition, by comparison, results were produce using both Fuzzy Inference System (FIS) managed by the bank of fuzzy rules (17.6) and Fuzzy Similarity (FS) concept among classes [15–17]. These results showed a good performance of the proposed approach. Specifically, in terms of detection, the proposed procedure showed the presence of defect in all the analyzed signals, while the classification procedure was correctly categorized by 99.6% of the cases. The remaining 0.4% of misclassification, although the defect has been detected, is not able to classify it. However, the problem can be solved by creating a new class of defects that the procedure detects and classifies.

Table 17.4 Comparison of the results (in terms of both detection and classification) achieved by CWs, Mamdani's FIS and FS

	CWs	FIS	FS
Detection (%)	100	100	100
Classification (%)	99.6	96.4	99.7

17.4 Conclusions

In this work, starting from a campaign of ECs uncertainties measurements, a fuzzy procedure based on CWs for the construction of a bank of fuzzy rules has been conceived, implemented and tested for detecting and classifying defects in steel bar of reinforced concrete specimens subjected to groups of axial traction loads gradually increasing. In particular, in advance, a Mamdani's multi-input/multi-output FIS was built where the inputs considered were electrical parameters and their particular statistical features while the outputs were their membership classes. From this FIS a new system based on CWs, but equivalent to it, was carried out able, on the one hand, to detect the presence of the defect with certainty and, on the other hand, classify it with a high percentage of success comparable with the results carried out by the established procedure based on FS. Finally, the proposed system is structured to consider misclassified cases as a further class in which the defect is however detected and classified as a case of doubt.

References

1. Mietz, J., Fischer, J.: Evaluation of NDT methods for detection of prestressing steel damage at post-tensioned concrete structures. Mater. Corros. **58**, 789–794 (2007)
2. Bazant, Z.P., Planas, J.: Fracture and Size Effect. CRC Press, New York (1998)
3. Alonso Rodriguez, A., Valli, A.: Eddy Current Approximation of Maxwell Equations, Theory, Algorithms and Applications. MS&A Series Vol. 4, Springer-Verlag Mailand (2010)
4. American Concrete Institute: Fracture Mechanics of Concrete. Technical Report (2013)
5. Chady, T., Enokizono, M., Takeuchi, K., Kinoshita, T., Sikora, R.: Eddy current testing of concrete structures. In: Takagi, T., Uesaka, M. (eds.) Applied Electromagnetics and Mechanics. JASEM Studies, pp. 1917–1927. IOS Press, Amsterdam (2001)
6. Cacciola, M., Morabito, F.C., Polimeni, D., Versaci, M.: Fuzzy characterization of flawed metallic plates with eddy current tests. Prog. Electromagn. Res. **72**, 241–252 (2007)
7. Cacciola, M., La Foresta, F., Morabito, F.C., Versaci, M.: Advanced use of soft computing and eddy current test to evaluate mechanical integrity of metallic plates. NDT E Int. **405**, 357–362 (2007). https://doi.org/10.1016/j.ndteint.2006.12.011
8. Versaci, M.: Soft computing approach to predict Intracranial pressure values. Am. J. Appl. Sci. **115**, 844–850 (2014). https://doi.org/10.3844/ajassp.2014.844.850
9. Zadeh, L.: An arithmetic approach for the computing with words paradigm. computing with words in information/intelligent systems. In: Zadeh, L.A., Kacprzyk, J., (eds.) What is Computing with Words? Wrsburg: Physica Verlag (1999)

10. Zadeh, L.: From computing with numbers to computing with words. From manipulation of measurements to manipulation of perceptions. In: IEEE Transactions on Circuits an Systems I: Fundamental Theory and Applications (1999). https://doi.org/10.1109/81.739259
11. Beliakov, G., Calvo, T., Lazaro, J.: Pointwise construction of Lipschitz aggregation operators with specific properties. Int. J. Uncertain. Fuzziness Knowl. Based Syst. **15**, 193–223 (2007)
12. Burrascano, P., Callegari, S., Montisci, A., Ricci, M., Versaci, M.: Ultrasonic nondestructive evaluation systems: Industrial application issues (2015). https://doi.org/10.1007/978-3-319-10566-6
13. Delgrado, M., Duarte, O., Requena, I.: An arithmetic approach for the computing with words paradigm. Int. J. Intell. Syst. **21**, 121–142 (2006)
14. Pavo, J., Gasparics, A., Sebestyen, I., Vertesy, G., Darczi, C.S., Miya, K.: Eddy Current Testing with Fluxset Probe. App. Electromagn. Mech. JSAEM (1996)
15. Tversky, A.: A features of similarities. IPsychol. Rev. **844**, 327–352 (1977)
16. Versaci, M.: Fuzzy approach and Eddy currents NDT/NDE devices in industrial applications. Electron. Lett. **52**, 943–945 (2016)
17. Chaira, T., Ray, A.K.: Fuzzy Image Processing and Application with MatLab. CRC Press, Taylor & Francis Group (2010)

Chapter 18
Appraisal of Enhanced Surrogate Models for Substrate Integrate Waveguide Devices Characterization

Domenico De Carlo, Annalisa Sgrò and Salvatore Calcagno

Abstract Nowadays the use of surrogate models (SMs) is becoming a common practice to accelerate the optimization phase of the design of microwave and millimeter wave devices. In order to further enhance the performances of the optimization process, the accuracy of the response provided by a SM can be improved employing a suitable output correction block, obtaining in this way a so-called enhanced surrogate model (ESM). In this paper a comparative study of three different techniques for building ESMs, i.e. Kriging, Support Vector Regression Machines (SVRMs) and Artificial Neural Networks (ANNs), applied to the modelling of substrate integrated waveguide (SIW) devices, is presented and discussed.

18.1 Introduction

Over the last few decades, computer aided design (CAD) techniques have played a fundamental role to analyze and design innovative microwave and millimeter wave devices. However, despite their great predictive accuracy, current CAD simulations are computationally demanding [1]. This poses a serious bottleneck in the device design cycle, especially during the optimization phase where a great number of CAD simulations are required [1]. To overcome this problem, the employment of surrogate models (SMs) is becoming a common practice among the designers [1]. Roughly speaking, the basic idea behind the concept of SM is the development of a

D. De Carlo (✉) · A. Sgrò
TEC Spin-In - DICEAM, University Mediterranea, via Graziella - Loc. Feo di Vito,
89121 Reggio Calabria, Italy
e-mail: domenico.decarlo@unirc.it

A. Sgrò
e-mail: annalisa.sgro@unirc.it

S. Calcagno
DICEAM, University Mediterranea, via Graziella - Loc. Feo di Vito,
89121 Reggio Calabria, Italy
e-mail: calcagno@unirc.it

© Springer International Publishing AG, part of Springer Nature 2019   201
A. Esposito et al. (eds.), *Neural Advances in Processing Nonlinear
Dynamic Signals*, Smart Innovation, Systems and Technologies 102,
https://doi.org/10.1007/978-3-319-95098-3_18

computationally tractable and reasonably accurate model able to replace the original time consuming physical model (also known as high-fidelity model) embodied in a CAD program. Despite of the fact that the SM output response is always obtained in a faster way than that provided by the related high-fidelity model, very often it is not enough close to this latter, being so mandatory improving its closeness. During the years many techniques have been proposed to this aim, but the most effective it has been proved that based on the use of a suitable correction block placed in cascade to the output of the SM. In this way a new SM, called *enhanced surrogate model* (ESM), is obtained [2]. In this paper the performances of three different techniques to build ESMs, i.e. Neural Networks, Support Vector Regression Machines and Kriging, are validated considering the characterization of a suitable substrate integrated waveguide (SIW) device: a circular cavity SIW resonator [3].

18.2 Improved Surrogate Modelling Techniques

As stated in the introduction, an ESM can be obtained by adding a correction block to the output of a SM. This block is usually built by means of a interpolation or regression process applied on the set of residuals obtained evaluating the difference between the output values provided by the high-fidelity model and by the SM, respectively, in correspondence of the same set of input values. In this section an overview of the three techniques employed in this work to assemble the correction block i.e., Kriging, Support Vector Regression Machines (SVRMs) and Artificial Neural Networks (ANNs), is given.

18.2.1 Kriging

In the Kriging model for regression [4, 5], the observed responses $\{y(\mathbf{x}^{(i)})\}_{i=1}^N \in R$ are related to the input vectors $\{\mathbf{x}^{(i)}\}_{i=1}^N \in R^n$ by means of a stochastic process of the form [6]

$$y(\mathbf{x}^{(i)}) = f(\mathbf{x}^{(i)}) + \nu(\mathbf{x}^{(i)}) \tag{18.1}$$

where $f(\cdot)$ describes the input-output relationship trend, and $\nu(\cdot)$ is a stochastic process modelling the differences between $f(\cdot)$ from $y(\cdot)$, which is characterized by a mean $\mu = 0$, a variance σ^2 and by covariance $\mathbf{C} = \sigma^2 \mathbf{P}$, where $\mathbf{P}$ is the correlation matrix. The elements $p_{i,j}$ of $\mathbf{P}$ are of the form

$$p_{i,j} = \exp\left(-\sum_{l=1}^N \theta_l |x_l^{(i)} - x_l^{(j)}|^2\right) \tag{18.2}$$

In (18.2) the terms $x_l^{(i)}$ and $x_l^{(j)}$ are the l-th components of the i-th and j-th input vectors $\mathbf{x}^{(i)}$, $\mathbf{x}^{(j)}$, respectively, and $\theta_l, l \in \{1, \ldots, N\}$ are N unknown parameters describing the degree of correlation among the components of these vectors. Once that the optimum coefficients $\hat{\theta}_l, l \in \{1, \ldots, N\}$ have been evaluated by solving a maximization problem specified on the data, the output response $y(\mathbf{x}^{(N+1)})$ at a new input $\mathbf{x}^{(N+1)}$ is given by

$$y(\mathbf{x}^{(N+1)}) = \hat{\mu} + \mathbf{r}'\mathbf{P}^{-1}\left(\mathbf{y} - \mathbf{1}\hat{\mu}\right) \tag{18.3}$$

where $\hat{\mu}$ is the mean value evaluated using the previous coefficients $\hat{\theta}_l$, $\mathbf{y}$ is the vector of the observed outputs, $\mathbf{r}'$ is the transpose vector containing the correlation between $\mathbf{x}^{(N+1)}$ and the input vectors $\{\mathbf{x}^{(i)}\}_{i=1}^N$, and $\mathbf{1}$ is the vector which components are equal to the unity.

18.2.2 Support Vector Regression Machines

A SVRM is a heuristic structures designed to solve regression problems [7]. In this context, the observed responses $\{y(\mathbf{x}^{(i)})\}_{i=1}^N \in R$ are related to the input vectors $\{\mathbf{x}^{(i)}\}_{i=1}^N \in R^n$ by using a linear relation of the form

$$f(\mathbf{x}) = \mathbf{w}\Phi(\mathbf{x}) + b \tag{18.4}$$

Equation (18.4) defines the *support vector regression estimation* function, in which $\Phi(\cdot)$ denotes a nonlinear mapping from R^n to the *feature space* (which is a higher dimensional vector space than R^n). The goal is to search for the values of the parameters $\mathbf{w}$ that minimize the following functional

$$\Phi(\mathbf{w}, \xi) = \frac{1}{2}||\mathbf{w}||^2 + C\sum_i(\xi^+ - \xi^-) \tag{18.5}$$

where ξ^+, ξ^- are slack variables representing upper and lower constraints on the output data and C is a parameter that determines the penalties on the estimation errors. The value of C is selected in a way to obtain the right compromise between penalties on the errors and SVRM generalization ability.

18.2.3 Artificial Neural Networks

An ANN can be defined as a nonlinear statistical data modeling device able to transform its internal configuration during the learning phase in conformance with the information that flows through itself [8–12]. The basic element of an ANN is the artificial neuron, which emulates the behavior of a human brain neuron. In an ANN the artificial neurons are interconnected among them and arranged in input, output and hidden layers. The performances of an ANN depends on its topology which is defined by the pattern of connections among its artificial neurons [13, 14]. Among the different ANNs architectures developed in literature in what follows the feed forward multi-layered perceptron neural network (MLPNN) will be considered. A MLPNN models the relationship between the observed responses $\{y(\mathbf{x}^{(i)})\}_{i=1}^{N} \in R$ and the input vectors $\{\mathbf{x}^{(i)}\}_{i=1}^{N} \in R^n$ as [11]

$$y(\mathbf{x}^{(i)}) = \mathbf{f}\left(\mathbf{x}^{(i)}, \mathbf{W}\right) \tag{18.6}$$

where $\mathbf{W}$ is the *weight matrix*. During the supervised learning training phase T_p for the network the weights $w_{ij} \in \mathbf{W}$ are changed until a suitable cost function $E(\mathbf{W})$

$$E(\mathbf{W}) = \sum_{p \in T_p} E_p(\mathbf{W}) \tag{18.7}$$

results minimized. In (18.7) the term $E_p(\mathbf{W})$ represents the least square error related to the p-th tuple in T_p

$$E_p(\mathbf{W}) = \left[\frac{1}{2}\sum_{k=1}^{M}(\hat{y}_{pk} - \mathbf{y}_k(\mathbf{W}, \hat{\mathbf{x}}_p))^2\right]^{\frac{1}{2}} \tag{18.8}$$

In this last relation, the term $\mathbf{y}_k(\mathbf{W}, \hat{\mathbf{x}}_p)$ represents the k-th output of MLPNN related to the input $\hat{\mathbf{x}}_p$ and $\hat{y}_{pk}$ is the kth element of the output vector $\hat{\mathbf{y}}_p$. The nonlinear least square optimization problem defined by (18.7) is usually solved by using different approaches all based on the following *update rule*

$$\mathbf{W}(t) = \mathbf{W}(t - 1) + \Delta\mathbf{W}(t - 1) \tag{18.9}$$

where $\Delta\mathbf{W}(t - 1)$ is updating matrix computed at the step $t - 1$ of the optimization process.

18.3 SIW Devices Modelling: The Case of the Lossy Circular SIW Cavity

Nowadays passive devices operating from microwaves to THz frequency band are often built by means of the technology of the substrate integrated waveguide (SIW). In particular, circular SIW cavities are intensively exploited to build many of them, as filters, antennas and so on [3, 15]. Accordingly, the proper evaluation of the fundamental resonant frequency f_r and of the related quality factor Q_r of SIW cavities can be considered of paramount importance for a correct design process of the devices developed by using this technology.

18.3.1   The Physical Model

The geometry of a circular SIW cavity is shown in Fig. 18.1. Inside a SIW structure the field scattered by the metallic vias can be formulated as [3]

$$\mathbf{H}_{cyl}^{s}(\mathbf{r}) = \sum_{l}\sum_{n,m}\left[\mathbf{M}_{n}(k_{\rho_m}, k_{z_m}, |\boldsymbol{\ae} - \boldsymbol{\ae}_l|, z)A_{m,n,l}^{TE} + \right.$$

$$\left. \mathbf{N}_{n}(k_{\rho_m}, k_{z_m}, |\boldsymbol{\ae} - \boldsymbol{\ae}_l|, z)A_{m,n,l}^{TM}\right] \qquad (18.10)$$

a series of outgoing cylindrical vector wave functions $\mathbf{M}_n$, $\mathbf{N}_n$ having coefficients $A_{m,n,l}^{TM}$, $A_{m,n,l}^{TE}$, computed by solving the following matrix system [3]

$$\left[\mathbf{L}^{TM,TE}\right]\mathbf{A}^{TM,TE} = \mathbf{G}^{TM,TE} \qquad (18.11)$$

arising from the discretization via Method of Moments of the relevant scattering operator [3]. Resonances f_r are the real part of the complex frequencies $\bar{f}_r$ for which

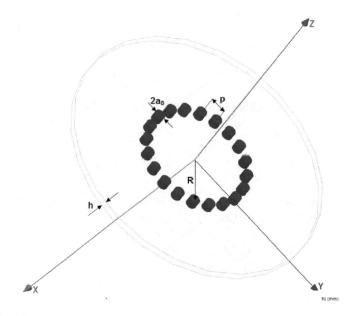

Fig. 18.1 A lossy circular SIW cavity

Eq. (18.11) has a nontrivial solution for $\mathbf{G}^{TM,TE} = 0$ (see [16, 17]). The related quality factors Q are computed as [18]

$$Q_r = \frac{Re(\bar{f}_r)}{2Im(\bar{f}_r)} \tag{18.12}$$

18.3.2 The Surrogate Model

A SM able to evaluate the resonant frequency f_r and the quality factor Q_r of the fundamental TM_{010} mode can be obtained by means of the well established relations valid for a metallic circular cavity resonator [19], i.e.

$$f_r = \frac{\chi_{10} c_0}{2\pi \sqrt{\varepsilon_r} R_{eq}} \tag{18.13}$$

$$Q_r = \left[\frac{2\sqrt{\varepsilon_r} r_s \left(1 + \frac{R_{eq}}{h}\right)}{120\pi \chi_{10}} + tan(\delta) \right]^{-1} \tag{18.14}$$

(where $\chi_{10} = 2.405$ is the first zero of the Bessel function of first kind and order zero, c_0 is the velocity of light in the vacuum, ε_r is the SIW substrate dielectric relative permittivity, $tan(\delta)$ is its dielectric loss tangent, and r_s is the surface impedance for unit length of the metallic plates) where it is used an empirical *equivalent radius* R_{eq}, defined as

$$R_{eq} = R \left[1 - \alpha_1 \left(\frac{p}{R}\right)^2 - \frac{\alpha_0}{R} \right] \tag{18.15}$$

in which the values of coefficients α_0, α_1 are given in [19].

18.4   Numerical Results

In this section we compare the performances of the three different ESMs obtained by adding to the SM described in Sect. 18.3.2 a correction block formed by exploiting an ANN, a SVRM and Kriging, respectively. These three ESMs will be denoted in the following as ESM_A, ESM_S and ESM_K. To this aim, firstly, in the range $\mathscr{D}$ of physical and geometrical parameters of interest to designers (see [18]), we have computed a database of residuals $(\hat{f}_r, \hat{Q}_r)$ obtained as the difference between the values provided by the SM (see relations (18.13) and (18.14)) and those computed by means of a full wave code based on the physical model described in Sect. 18.3.1 in

Table 18.1 Percentage normalized mean square error

Model	PNMSE N_{tp}				
	55	110	195	230	390
ESM_A (%)	7.36	3.95	2.57	1.99	1.54
ESM_S (%)	10.91	6.57	4.68	3.09	2.11
ESM_K (%)	6.50	3.34	1.95	0.97	0.54

correspondence of the same input values. In order to select these values in $\mathscr{D}$ the Latin hypercube sampling has been exploited [1]. For the numerical results reported in this work a variable number N_{sp} of sample points ($N_{sp} \in$ [55, 110, 195, 230, 390]) has been exploited to built the database of residuals. Concerning the ESM_A surrogate model, it was assembled using a MLPNN architecture having three layers: an input layer composed by six neurons, one hidden layer composed by nine neurons and an output layer composed by two neurons. All the neurons have been characterized by using a sigmoid transfer function. As learning rule, it was employed the resilient backpropagation rule. As for the ESM_S surrogate model, it was assembled using a SVRM having a two degree polynomial kernel with $C = 10$ and $\epsilon = 0.01$. Finally, to develop the ESM_k model the DACE Kriging toolbox has been employed [20]. The performances of the three ESMs models have been evaluated in term of normalized mean square error (PNMSE) as a function of N_{sp} by using a fixed number of test points $N_{tp} = 150$ randomly selected in $\mathscr{D}$. As it can be noticed by observing the results reported in Table 18.1, despite that the PNMSE decreases when N_{tp} increase for all the ESMs models considered in this study, the best performances are always offered by the ESM_K surrogate, followed by ESM_S and ESM_A surrogate, respectively.

18.5 Conclusions

In this paper three different techniques to obtain ESMs, i.e. Neural Networks, Support Vector Regression Machines and Kriging, applied to model substrate integrated waveguide (SIW) devices, have been investigated. The case of the lossy SIW circular resonator has been considered. ESMs performances have been compared in term of the PNSME as a function of the number of sample points used to assemble the residuals database. The numerical results shown the superiority of the ESM based on Kriging respect to the other methods considered in this study.

References

1. Koziel, S., Leifsson, L. (eds.): Surrogate-Based Modeling and Optimization. Springer, Berlin (2010)
2. Couckuyt, I., Koziel, S., Dhaene, T.: Surrogate modeling of microwave structures using kriging, co-kriging, and space mapping. Int. J. Numer. Model. Electron. Devices Field (2012). https://doi.org/10.1002/jnm.1833
3. Amendola, G., Angiulli, G., Arnieri, E., Boccia, L.: Computation of the resonant frequency and quality factor of lossy substrate integrated waveguide resonators by method of moments. Prog. Electromagn. Res. Lett. **40**, 107–117 (2013)
4. Pebesma, E.J.: The role of external variables and GIS databases in geostatistical analysis. Trans. GIS (2006). https://doi.org/10.1111/j.1467-9671.2006.01015.x
5. Tomislav, H., et al.: A generic framework for spatial prediction of soil variables based on regression-kriging. Geoderma (2014). https://doi.org/10.1016/j.geoderma.2003.08.018
6. Armstrong, M.: Basic Linear Geostatistic. Springer (1998)
7. Deng, N., Tian, Y., Zhang, C.: Support Vector Machines: Optimization Based Theory: Algorithms and Extensions. CRC (2012)
8. Cacciola, M., Morabito, F.C., Polimeni, D., Versaci, M.: Fuzzy characterization of flawed metallic plates with eddy current tests. Prog. Electromagn. Res. (2007). https://doi.org/10.2528/PIER07031301
9. Cacciola, M., La Foresta, F., Morabito, F.C., Versaci, M.: Advanced use of soft computing and eddy current test to evaluate mechanical integrity of metallic plates. NDT E Int. (2007). https://doi.org/10.1016/j.ndteint.2006.12.011
10. Mammone, N., Inuso, G., La Foresta, F., Versaci, M., Morabito, F.C.: Clustering of entropy topography in epileptic electroencephalography. Neural Comput. Appl. (2011). https://doi.org/10.1007/s00521-010-0505-2
11. Rojas R.: Neural Networks: A Systematic Introduction. Springer (1996)
12. Versaci, M.: Fuzzy approach and Eddy currents NDT/NDE devices in industrial applications. Electron. Lett. (2016). https://doi.org/10.1049/el.2015.3409
13. La Foresta, F., Mammone, N., Morabito, F.C.: PCA-ICA for automatic identification of critical events in continuous coma-EEG monitoring. Biomed. Signal Process. Control (2009). https://doi.org/10.1016/j.bspc.2009.03.006
14. La Foresta, F., Morabito, F.C., Azzerboni, B., Ipsale, M.: PCA and ICA for the extraction of EEG dominant components in cerebral death assessment. In: Proceedings of the International Joint Conference on Neural Networks (2005). https://doi.org/10.1109/IJCNN.2005.1556301
15. Angiulli, G.: Design of square substrate waveguide cavity resonators: compensation of modelling errors by support vector regression machines. Am. J. Appl. Sci. **9**, 1872–1875 (2012)
16. Amendola, G., Angiulli, G., Arnieri, E., Boccia, L.: Resonant frequencies of circular substrate integrated resonators. IEEE Microw. Wirel. Compon. Lett. **18**, 239–241 (2008)
17. Angiulli, G., Arnieri, E., De Carlo, D., Amendola, G.: Fast nonlinear eigenvalues analysis of arbitrarily shaped substrate integrated waveguide (SIW) resonators. IEEE Trans. Magn. **45**, 1412–1415 (2009)
18. Amendola, G., Angiulli, G., Arnieri, E., Boccia, L., De Carlo, D.: Characterization of lossy SIW resonators based on multilayer perceptron neural networks on graphics processing unit. Progress in Electromagnetic Research C. **42**, 1–11 (2013)
19. Amendola, G., Angiulli, G., Arnieri, E., Boccia, L., De Carlo, D.: Empirical relations for the evaluation of resonat frequency and quality factor of the TM_{101} mod of circular substrate integrated waveguide (SIW) resonators. Prog. Electromagn. Res. C **43**(165), 273 (2013)
20. Lophaven, S., et al.: DACE-A MATLAB Kriging Toolbox. Technical University of Denmark (2002). http://www2.imm.dtu.dk/projects/dace/

Chapter 19
Improving the Stability of Variable Selection for Industrial Datasets

Silvia Cateni, Valentina Colla and Vincenzo Iannino

Abstract Variable reduction is an essential step in data mining, which is able effectively to increase both the performance of machine learning and the process knowledge by removing the redundant and irrelevant input variables. The paper presents a variable selection approach merging the dominating set procedure for redundancy analysis and a wrapper approach in order to achieve an informative and not redundant subset of variables improving both the stability and the computational complexity. The proposed approach is tested on different datasets coming from the UCI repository and from industrial contexts and is compared to the exhaustive variable selection approach, which is often considered optimal in terms of system performance. Moreover the novel method is applied to both classification and regression procedures.

19.1 Introduction

Variable Selection, also known as feature selection, is the procedure of selecting the subset of the input variables which mostly affect a given process or phenomenon, where the terms *variable* and *feature* are often used as synonyms. Actually, the term *variable* indicates the "raw" initial variable, while *feature* refers to a variable that could also be derived by pre-processing and eventually combining some input variables. In most contexts, the two words can both be used, as the same algorihms are applied for selection in both cases [21]. Variable selection is a necessary stage especially when working with industrial datasets, for instance when the number of input variables is high with respect to the number of available samples [7]. The topic of variable selection has been deeply studied in literature for several tasks; for example

S. Cateni (✉) · V. Colla · V. Iannino
Scuola Superiore Sant' Anna - TeCIP Institute, Via Alamanni 13B, 56010 Pisa, Italy
e-mail: silvia.cateni@santannapisa.it

V. Colla
e-mail: valentina.colla@santannapisa.it

V. Iannino
e-mail: vincenzo.iannino@santannapisa.it

© Springer International Publishing AG, part of Springer Nature 2019
A. Esposito et al. (eds.), *Neural Advances in Processing Nonlinear Dynamic Signals*, Smart Innovation, Systems and Technologies 102,
https://doi.org/10.1007/978-3-319-95098-3_19

for prediction of a continuous target [10, 26, 28], for classification purposes [7, 9, 11, 16, 17] and finally for clustering tasks [30]. An important topic to be studied when treating with variables reduction methods is the *stability*. A variable reduction procedure is defined instable when exploiting different training data set, the algorithm provides different variable subsets.

The paper presents a variable selection approach merging the dominating set procedure for redundancy analysis and a wrapper approach in order to achieve an informative and not redundant subset of variables improving both the stability and the computational complexity of the overall selection procedure.

The paper is organised as follows: Sect. 19.2 provides a brief literature survey; in Sect. 19.3 a description of the proposed method is provided; the obtained results are then shown in Sect. 19.4 and finally in Sect. 19.5 some concluding remarks are given.

19.2   State of the Art

Variable reduction and selection is a significant preliminary phase of the development of Artificial Neural Network (ANN)-based models. An inadequate selection of the input variables can worsen the performance of ANN in the training phase [14]. In common thinking ANNs are able to recognize redundant and noisy variables during the training phase and this belief leads the developers to include a large number of variables thinking that if the size of the ANN is high, more information can be included. However, the number of variables which are fed as inputs to the ANN affects the computational burden as well as the training time. Moreover the occurrence of redundant variables introduce a disturbance, which can hide or mask the real input-output association [25]. Another important aspect, which is largely discussed in literature, is the so-called *curse of dimensionality* [2]: as the dimensionality of a model enhance linearly, the total volume of the modelling process domain enhance exponentially. ANN architectures are very sensitive to this phenomenon, because the number of connection weights increases with the number of variables. The variable selection techniques can be divided in three main categories: filters, wrappers and embedded approaches. **Filter approaches** can be seen as a pre-processing stage, as they are independent on the developed learning algorithm. The variables subset are generated by valuing the association between input and output and the input variables are ranked on the basis on their relevancy to the target by executing statistical tests. The main advantage of the filter approaches lies in their simplicity, which makes them fast and also appropriate for dealing with large and complex databases [8]. **Wrapper approaches**, that were presented in [23], evaluate the performance of the learning machine in order to select a subset of variables on the basis on their predictive power. Wrappers use the learning algorithm as a black box making these approaches remarkably universal. The most common wrapper approach is the *exhaustive search*, also known as *brute force method*, which considers all combinations of available variables. The exhaustive approach is considered the best ones in terms of accuracy of the model, but its main limit lies in the fact that when we deal with large dataset it becomes impractical. Considering k potential input variables, there are 2^k possible candidate variable sets to test and 2^k training procedures to execute in order to find the best one. Another popular wrapper approach is the so called *Greedy Search* strategy, which progressively generates the variables subset by addition or removal of single variables from an initial set. Greedy search in fact works into two directions: *Sequential Forward Selection* (SFS) and *Sequential Backward Selection* (SBS). SFS begins from an empty set of variables and the other variables are iteratively added until a fixed stopping condition is achieved. SBS works in the opposite direction: the procedure starts with all features and progressively eliminates the least significant ones. A variable is considered significant if the performance of the learning machine tends to worsen when it is removed. Greedy search strategies require at maximum $k(k+1)/2$ training procedures. Others variable selection procedures based on genetic algorithms are proposed in literature in order to obtain a good performance in a reasonable time. Wrappers are usually heavier from the computational point of view than filters, due to the fact that the induction algorithm is trained for each tested subset; on the other side, they are more efficient in terms of accuracy, as they exploit the adopted machine learning procedure [19]. **Embedded methods** perform the variable selection as portion of the learning phase and are typically specific of a precise learning machine. Typical examples of embedded approaches comprise classification trees, random forests [6] and other methods based on regularization approaches [5]. The main advantage of embedded methods is their inclusion in the learning algorithm which considers the relevance of the variables. Recently several hybrid variable selection approaches have been suggested in order to exploit their advantages by overcoming their shortcomings. An example of hybrid approach is proposed in [9, 12] where the set of initial variables is initially reduced with a combination of filter selection methods and then the exhaustive search is executed to achieve a sub-optimal set of variables in a reasonable time. Another hybrid algorithm is proposed in [12], where a combination of four popular filter methods with sequential selection method is performed in order to provide a more informative subset in a reasonable time. Finally, an important aspect to be considered dealing with variable selection is the stability of the solution. The stability of a variable selection approach is generally defined as the robustness of the variables selected related to modifications in the training sets generated from the same creating distribution [22]. In the last years, the stability issue has received increasing interest in the context of variable selection [13, 15, 24, 27].

19.3 Proposed Method

The proposed method is the combination of two steps: firstly, the proposed method reduces the dataset performing a redundancy analysis, in order to work with a dataset composed by variables which are almost uncorrelated to each other; then, a wrapper variable selection based on Genetic Algorithm (GA) is applied in order to select

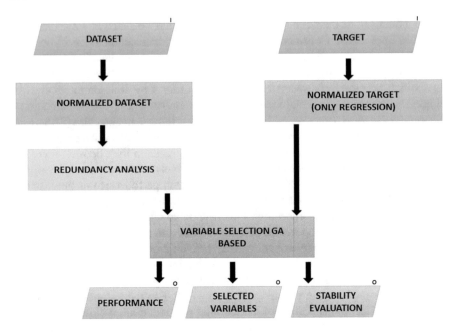

Fig. 19.1 Scheme of the proposed approach

only the relevant non-redundant variables in a reasonable time. This approach does not need a priori assumption about the considered process, which is important when treating with real-word applications. The procedure is repeated in order to evaluate its stability in providing the variables that mostly affect the considered target as well as the effects on the performance of the predicition model. A general scheme of the method is shown in Fig. 19.1.

Dataset and Target

Let us consider a multivariate dataset containing N instances, M variables and a target to predict, which is a binary or a continuous vector. If the target is binary then the fitness function will optimize the performance of the adopted binary classifier, while if the target is a continous vector then the fitness function will tend to reduce the mean average relative error of the regression model.

Normalization

Dataset and target (only in the case of continuous target) are normalized in order to obtain a value in the range [0, 1] as follows:

$$NormVariable(i) = \frac{v(i) - \mu[v(i)]}{\max[v(i)] - \min[v(i)]} \quad (19.1)$$

where $v(i)$ is the i_{th} variable, $\mu[v(i)]$ represents the mean value of the i_{th} variable and finally $\max[v(i)]$ and $\min[v(i)]$ are, respectively, the maximum and minimum value assumed by the i_{th} variable.

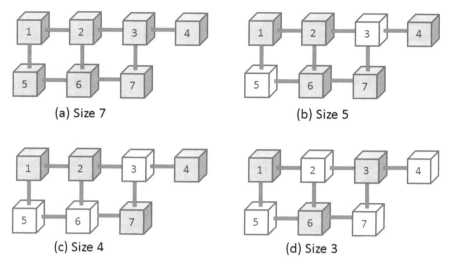

Fig. 19.2   Example of dominating set algorithm

Redundancy Analysis

Redundant variables are those variables that are highly correlated or "redundant" with one another. If so, any one of these variables should be adopted as a proxy for all the others. Removal of redundant variables provides improvement in interpretability. In the proposed approach redundant variables are identified by applying the so–called *dominating set algorithm*, which derives from the graph theory [3, 4]. Such algorithm considers redundant the variables with a high predictive capability with respect to at least another variable belonging to the same dataset. In particular, each variable is used as input to a feedforward ANN in order to predict each of remaining variables. If the prediction error is less than a fixed threshold then one of the two variables is considered as redundant. Then the algorithm identifies the minimal dominating set of the graph corresponding to the set with the minimal number of vertexes: redundant variables are those variables of the graph do not falling inside the minimal dominating set. An example of dominating set algorithm is given in Fig. 19.2 where 1–7 are the initial input variables and if a variable is connected to another one means that the two variables can be considered correlated.

A simple way to identify the correlation between two variables is the evaluation of the Pearson coefficient [31]; however. this approach identify only the linear dependence, while using the prediction based on ANN we are able to identify also non linear dependence between input variables. In the above examples (a), (b), (c), (d) are dominating set of different sizes where each vertex can be reached by selected verteces. The dominating set algorithm selects the minimal dominating set, i.e. case (d), the dominating set having the minimal size (three in the present example). The performed neural network is a feedforward network with one hidden layer. The neurons in the hidden layer has been evaluated according to an empirical formula, largely used in literature, which takes into account the number of free parameters.

GA-based Variable Selection
Once the redundant variables are removed from dataset, a GA-based variable selection procedure is applied. The chromosomes are represented by binary vectors whose length is equal to the number of non-redundant input variables: each gene corresponds to a variable. If the gene assumes an unitary value it states that the associated variable is selected to be fed as input to the model, otherwise the associated variable is discarded. The fitness function is computed for each chromosome of the population and genetic operators (mutation and crossover) are evaluated. The mutation operator randomly select a binary gene of the chromosome switching it, while the crossover operator establish each gene of the son chromosome by randomly extracting genes from the two parent chromosomes. The dataset is divided in two subsets: 75% of the data are used for the network training, the remaining 25% are adopted as validation set. The objective function to minimize is the average relative error, in case of a regression model, or the classification error, computed on the basis of the confusion matrix, in case of a classifier [20]. The developed neural model is a traditional feedforward ANN with one hidden layer and a fixed number of neurons. The number of neurons is automatically calculated based on the number of input-output samples included in the training set. The GA stops when a fixed maximum number of iterations is achieved or a plateau of the fitness function is reached. The winner chromosome outputs the subset of input variables associated to the best value of the fitness, i.e. to the minimum prediction/classification error. An important limitation of the GA is represented by the fact that the initial population is randomly selected, thus a variable can sometimes be included in the winner chromosome even if it is not correlated with the target to be predicted. For this reason, the GA is run several times and the winner chromosomes are stored in a matrix. The occurrence frequency of each variable in the winning chromosomes is evaluated and finally a variable is considered as selected only if it is present in more than the 80% of the five winning chromosomes [10, 11].

Stability Evaluation
The stability concept was introduced by Turney [29] in 1995 and represents an important aspect, mainly when variable selection is applied not only to enhance the system performance but also as a data mining technique, in order to improve the knowledge of the variables which mostly affect a particular process or phenomenon. In effect, a satisfactory variable selection algorithm should output a stable selection also when varying the training data sets [13]. Wrapper approaches can often be affected by instability problems. Therefore within the proposed approach, the *Tanimoto distance* [18] is evaluated in the different GA iterations, in order to measure the stability of the variable selection method, as it calculates the similarity between two binary vectors. As described above, a subset of input variables can be represented by a binary vector. The Tanimoto distance between two binary vectors $\mathbf{b}_1$ and $\mathbf{b}_2$ is computed as:

$$T(\mathbf{b}_1, \mathbf{b}_2) = \frac{|\mathbf{b}_1 \cdot \mathbf{b}_2|}{|\mathbf{b}_1| + |\mathbf{b}_2| - |\mathbf{b}_1 \cdot \mathbf{b}_2|} \tag{19.2}$$

where $|\cdot|$ is the norm and $\cdot$ is the scalar product.

The Tanimoto distance metric evaluates the amount of overlap between two sets of arbitrary cardinality. T includes values in the range $[0,1]$ where null value means that there is no overlap between the two sets while unitary value identifies two identical sets.

19.4 Experimental Results

The proposed variable reduction approach has been tested on different applications by exploiting different datasets included in UCI learning repository [1] as well as some datasets coming from industrial applications. The method has been applied to both classification and regression tasks. A description of the main characteristics of the exploited datasets is given in Table 19.1.

The industrial datasets, which refer to binary classification applications, derive from an industrial context related to fault detection problems and quality check issues in the metal industry. The obtained results have been compared to the exhaustive method, which is considered the optimal one in terms of accuracy, as it considers an tests all the combinations of input variables. Results related to the exhaustive approach are shown in Table 19.2. The first column represent the analysed dataset, the second column shows the error (in terms of relative error for regression tasks or misclassification rate for the classification tasks), the third column provides the length of the final selected subset and the last column is the stability index in terms of Tanimoto distance. Table 19.3 shows the results provided by the proposed approach. In this table another column has been added, which represents the improvement of the stability index with respect to the exhaustive approach.

Table 19.1 Datasets description

Dataset	#Instances	#Variables	Regression (R), Classification (C)	Source
Baseball	337	16	R	UCI
Computer hardware	209	7	R	UCI
Yacht hydrodynamics	308	6	R	UCI
Ind-I	107	15	R	Industrial
Breast cancer	699	9	C	UCI
Heart	270	13	C	UCI
PimaDiabete	768	8	C	UCI
Ind-II	1915	10	C	Industrial
Ind-III	3756	6	C	Industrial

Table 19.2 Obtained results with exhaustive method

Dataset	Error	Subset length	Stability
Baseball	0.018	4	0.38
Computer hardware	$1.4 * 10^{-3}$	4	0.43
Yacht hydrodynamics	0.003	4	0.44
Ind-I	0.056	5	0.48
Breast cancer	0.003	4	0.46
Heart	0.005	5	0.42
PimaDiabete	0.003	3	0.48
Ind-II	0.001	5	0.48
Ind-III	$5.3 * 10^{-4}$	3	0.35

Table 19.3 Obtained results with the proposed method

Dataset	Error	Subset length	Stability	Stability improvement (%)
Baseball	0.019	4	0.48	26
Computer hardware	$1.5 * 10^{-4}$	4	0.9	109
Yacht hydrodynamics	0.002	2	1	127
Ind-I	0.057	5	0.65	35
Breast cancer	0.002	3	0.56	22
Heart	0.005	4	0.49	17
PimaDiabete	0.002	3	0.52	8
Ind-II	0.001	4	0.62	30
Ind-III	$5 * 10^{-4}$	2	0.56	60

19.5   Conclusions

A novel approach for variable reduction is proposed. The main idea is the combination
of the dominating set approach for removing redundant variables before applying
a GA-based variable selection method aiming at removing irrelevant variables. The
proposed approach provides a more informative and stable subset in a reasonable
time and it can be applied to all kind of datasets without any a priori assumption
on the data. Moreover it is suitable to different purposes, such as classification or
prediction tasks. The proposed approach has been successfully tested on different
datasets coming from a public repository and on some datasets coming from industrial
contexts.

Future work deals with the application of this approach to different machines
learning; i.e. other classifiers in order to prove that the proposed method is able to
improve the stability with respect to the brute force approach, independently from
the machine learning performed, by preserving the performance of the final system.

Acknowledgements The work presented in this paper was developed within the project entitled "Piattaforma Integrata Avanzata per la Progettazione di Macchine e Sistemi Complessi" (PRO-MAS), which was co-funded under Tuscany POR FESR 2014–2020.

References

1. Asuncion, A., Newman, D.: UCI machine learning repository (2007). http://archive.ics.uci.edu/ml/datasets.html
2. Bellman, R.: Adaptive Control Processes: A Guided Tour. Princeton University Press (1961)
3. Biggs, N., Lloyd, E., Wilson, R.: Graph Theory. Oxford University Press (1986)
4. Bondy, J.A., Murty, U.: Graph Theory. Springer (2008). ISBN 978-1-84628-969-9
5. Breiman, L., Friedman, J.H., Olshen, R.A., Stone., C.J.: Classification and Regression Trees. Wadsworth and Brooks (1984)
6. Breiman, L.: Random forests. Mach. Learn. **45**, 5–32 (2001)
7. Cateni, S., Colla, V., Vannucci, M., Vannocci, M.: A procedure for building reduced reliable training datasets from realworld data. In: 13th IASTED International Conference on Artificial Intelligence and Applications, AIA 2014, Innsbruck, Austria, pp. 393–399 (2014)
8. Cateni, S., Colla, V., Vannucci, M.: A fuzzy system for combining filter features selection methods. Int. J. Fuzzy Syst. (2016)
9. Cateni, S., Colla, V., Vannucci, M.: A hybrid feature selection method for classification purposes. In: 8th European Modeling Symposium on Mathematical Modeling and Computer simulation EMS 2014, Pisa, Italy, vol. 1, pp. 1–8 (2014)
10. Cateni, S., Colla, V., Vannucci, M.: General purpose input variable extraction: a genetic algorithm based procedure give a gap. In: 9th International Conference on Intelligence Systems Design and Applications, ISDA 2009, pp. 1307–1311 (2009)
11. Cateni, S., Colla, V., Vannucci, M.: Variable selection through genetic algorithms for classification purpose. In: IASTED International Conference on Artificial Intelligence and Applications, AIA 2010, pp. 6–11 (2010)
12. Cateni, S., Colla, V.: A hybrid variable selection approach for NN-based classification in industrial context. In: Smart Innovation, Systems and Technologies (in press)
13. Cateni, S., Colla, V.: Improving the stability of sequential forward and backward variables selection. In: 15th International Conference on Intelligent Systems Design and Applications, ISDA 2015, pp. 374–379 (2016)
14. Cateni, S., Colla, V.: The importance of variable selection for neural networks based classification in an industrial context. In: International Workshop on Neural Networks, WIRN 2015, Smart Innovation, Systems and Technologies, vol. 54, pp. 363–370 (2016)
15. Cateni, S., Colla, V.,: Improving the stability of wrapper variable selection applied to binary classification. Int. J. Comput. Inf. Syst. Ind. Manag. Appl. **8**, 214–225 (2016)
16. Cateni, S., Colla, V., Vannucci, M.: A genetic algorithm based approach for selecting input variables and setting relevant network parameters of som based classifier. Int. J. Simul. Syst. Sci. Technol. **12**(2), 30–37 (2011)
17. Cateni, S., Colla, V., Vannucci, M.: A method for resampling imbalanced datadata in binary classification tasks for realworld problems. Neurocomputing **135**, 32–41 (2014)
18. Duda, R., Hart, P., Stork, D.: Pattern Classification. Wiley, New York (2001)
19. Fiasché, M.: A quantum-inspired evolutionary algorithm for optimization numerical problems. Lecture Notes in Computer Science (Including Subseries Lecture Notes in Artificial Intelligence and Lecture Notes in Bioinformatics), Part 3. LNCS, vol. 7665, pp. 686–693 (2012)
20. Fiasché, M.: SVM tree for personalized transductive learning in bioinformatics classification problems. Smart Innov. Syst. Technol. **26**, 223–231 (2014)
21. Guyon, I., Elisseeff, A.: An introduction to variable and feature selection. Mach. Learn. **3**, 1157–1182 (2003)

22. Kalousis, A., Prados, J., Hilario, M.: Stability of feature selection algorithms: a study on high-dimensional spaces. Knowl. Inf. Syst. **12**, 95–116 (2007)
23. Kohavi, R., John, G.: Wrappers for feature selection. Artif. Intell. **97**, 273–324 (1997)
24. Loscalzo, S., Yu, L., Ding, C.: Consensus group stable feature selection. In: Proceedings of ACM SIGKDD International Conference on Knowledge Discovery and Data Mining, vol. 1, pp. 567–575. ACM (2009)
25. May, R., Dandy, G., Maier, H.: Review of input variable selection methods for artificial neural networks. Artif. Neural Netw. Methodol. Adv. Biomed. Appl. (2011)
26. Mitchell, T., Toby, J., Beauchamp, J.: Bayesian variable selection in linear regression. J. Am. Stat. Assoc. **83**, 1023–32 (1988)
27. Novovicova, J., Somol, P., Pudil, P.: A new measure of feature selection algorithms stability. In: IEEE International Conference Data Mining Workshops, vol. 1, pp. 382–387 (2009)
28. Sun, Y., Robinson, M., Adams, R., Boekhorst, R., Rust, A.G., Davey, N.: Using feature selection filtering methods for binding site predictions. In: Proceedings of 5th IEEE International Conference on Cognitive Informatics (ICCI 2006) (2006)
29. Turney, P.: Techncal note: bias and the quantification of stability. Mach. Learn. **20**, 23–33 (1995)
30. Wang, S., Zhu, J.: Variable selection for model-based high dimensional clustering and its application on microarray data. Biometrics **64**, 440–448 (2008)
31. Yu, L., Liu, H.: Feature selection for high-dimensional data: a fast correlation based filter solution. In: Proceedings of the 20th International Conference on Machine Learning, ICML, vol. 1, pp. 856–863 (2003)

Chapter 20
Cause and Effect Analysis in a Real Industrial Context: Study of a Particular Application Devoted to Quality Improvement

Silvia Cateni, Valentina Colla, Antonella Vignali and Jens Brandenburger

Abstract This paper presents an analysis of the occurrence of ripple defects during Hot Deep Galvanising of flat steel products, with a focus on the study on thick coils having low zinc coating. Although skilled personnel can manage ripples defects through particular operations, for instance wiping nitrogen instead of air in air blades, the real effects of each process parameter variation is unknown. Therefore, the study of these phenomena can improve the quality of coils, by decreasing reworked or scrapped material and reducing costs related to a redundant use of nitrogen. An accurate pre-processing procedure has been performed and then the analysis focused on the possible causes of ripples occurrences. In particular, the attention is focused on the development of a model capable to identify process variables with a stronger impact on the presence or absence of ripples, by expressing such effect through an appropriate relationship.

S. Cateni (✉) · V. Colla · A. Vignali
Scuola Superiore Sant' Anna, TeCIP Institute, Via Alamanni 13B, 56010 Pisa, Italy
e-mail: silvia.cateni@santannapisa.it

V. Colla
e-mail: valentina.colla@santannapisa.it

A. Vignali
e-mail: vincenzo.iannino@santannapisa.it

J. Brandenburger
VDEh-Betriebsforschungsinstitut GmbH (BFI), Dusseldorf, Germany
e-mail: Jens.Brandenburger@bfi.de

219

20.1 Introduction

Hot-Dip Galvanizing (HDG) is one of the most used processes for coating steel with a layer of zinc. This process consists in a passage of the steel strip through molten zinc at a temperature of $860\,^\circ$F in order to obtain zinc carbonate ($ZnCO_3$). Zinc carbonate is a strong material that provides sacrificial protection against corrosion over the steel surface [16, 17]. The present work is devoted to the analysis of the occurrence of ripple defects during HDG, focusing on the study on thick coils having low zinc coating. Ripples are diffused coating ruffles, appearing on the coil surface in the form of transversal lines [2], and are identified by an Automatic Surface Inspection System (ASIS). The occurrence and gravity of this type of defect is affected by several process variables. Although skilled personnel can manage ripple defects through particular operations, for instance using nitrogen as wiping medium in air blades, the real effects of each process parameter variation is unknown. Hence, the study of these phenomena can improve the quality of coils, by decreasing reworked or scrapped material and reducing costs related to a redundant use of nitrogen. The paper is organised as follows: Sect. 20.2 provides a description of the industrial dataset analysed; in Sect. 20.3 the proposed method is described; the obtained results are then depicted in Sect. 20.4 and finally in Sect. 20.5 some concluding remarks are provided.

20.2 Dataset Construction and Description

The phenomenon under consideration has been studied by analysing process data coming from one of the HDG lines of an Italian company. Currently the production of high quality steel is supported by modern measuring systems collecting an increasing amount of high resolution (HR) quality and process data along the whole flat steel production chain. Process data sampling is based on length, meaning that many process variables are sampled every 10 m of coil, while some of them are sampled every 1 m. It is important to notice that each measurement is related to a longitudinal position on the coil, while defects, detected by ASIS, are represented by a two dimensional spatial position on the coil and by the involved area. Therefore, the coil surface has been divided into disjoint tiles, each of them representing a portion of the coil. Since the number of tiles depends on the adopted resolution, different resolution levels, named stages, have been defined and are shown in Table 20.1

Specifically, each tile of each stage has a single TileID and is thus identified by the couple Stage and TileID. Since process variables are 1D-continuous, collected measure data have been aggregated on each coil cross section by computing the mean value. Coil cross sections are called *slices* and are obtained by aggregating tiles along the coil width. On the other hand, defect areas inside a tile have been summed and the obtained value has been then normalized towards the tile area in order to find the percentage of area occupied by defects inside a tile. Furthermore, in order to develop

Table 20.1 Grid definitions

Stage	Tiles x-axis	Tiles y axis	Number of tiles
0	1	2	2
1	2	4	8
2	4	8	32
3	8	16	128
4	16	32	512
5	32	64	2048
6	64	128	8192
7	128	256	32768
8	256	512	131072

a cause and effect analysis between process variables and defects, the defect tiles are aggregated on slices by means of a sum operation. In this way a sequence of coil *slices* has been created and the measure (1D) can be correlated to defects (2D) [3]. The target of the analysis has been defined as the percentage of area occupied from defects inside a slice. The analysis is devoted to the occurrence of ripple defects on thick material and low zinc coating coils. In particular the analysed coils show the following characteristics:

- thickness ≥ 1.5 mm
- 50 g/m^2 $\leq$ Zinc coating weight ≤ 71 g/m^2

The final number of analysed coils is 356 while the process variables, which are 20, have been selected by the skilled personnel of the company and can be associated to four mainclasses:

1. Air-knife (including Nitrogen, pressure, distance and angle)
2. Temperatures (zones situated before and after zinc bath, top-roll, water bath cooling)
3. Line speed
4. Fan coolers along the cooling tower.

Some of suggested variables have been removed because highly correlated with other ones and therefore redundant. In order to study the two different cases based on the use of air or nitrogen in air blades, the coils have been divided into two groups: the first group includes 179 coils using air blowing, while the second one is made of 177 coils exploiting nitrogen blowing. Therefore, two different datasets were prepared for air and nitrogen blowing and each of them is represented by a matrix, where rows correspond to slices of coils, while columns correspond to the suggested process variables.

20.3 Proposed Method

After the dataset construction, an accurate pre-processing procedure has been performed. Firstly, a plausibility analysis has been applied in order to create prediction models starting from process parameters with acceptable values. To this aim, a plausibility range or plausible values have been defined for each process parameter with the help of technical personnel, in order to check process variable measurements during data pre-processing. Another check concerns the possibility to have missing values on some process variable on some slices of the coils, due to measurement errors: all rows or coil slices where at least one value is missing or not plausible are removed. Moreover, if more than 30% of the samples of a process variable are not plausible, that variable is eliminated. Subsequently, a fully automatic fuzzy-based outlier detection has been applied. An outlier is defined as an observation that deviates from the rest of the available data. Outliers can be caused by measurement errors or by a drastic alteration of the considered phenomenon. Thus, on the basis of the application, an outlier can be a sample to be discarded or, on the contrary, it can represent a relevant although anomalous case to be investigated. Outlier detection is an important step of data mining as well as it is a useful pre-processing stage in many applications belonging to different contexts such as financial analysis, fraud detection, network robustness analysis, network intrusion detection [24, 27, 31]. Traditional outlier detection approaches can be categorized into four main classes: distribution-based [23], distance-based [28], density-based [30] and clustering-based [26]. The approach which is proposed here calculates a feature belonging to each class and then they are combined through a Fuzzy Inference System (FIS) [8–10, 22]. The distribution-based approach is usually addressed in the field of statistics. According to this approach an outlier is defined as an object that does not fit well with a standard distribution considering that the distribution is not a priori known. A largely used method belonging to this approach is the Grubbs test for outliers which detects outliers approximating the distribution of such data by a Gaussian function. The distance based approach was presented in [26], where the following definition is provided: *An object x in a dataset T is a DB(p, D)-outlier if at least fraction p of the objects in T lie a distance greater than D from x.* In order to identify a definition of distance, the Mahalanobis distance is adopted as metric. The density-based method for outliers detection was introduced in 2000 [6]. This approach assigns to each datum a degree of outlierness considering the density of its neighbourhood called Local Outlier Factor (LOF). Finally the clustering based method defines an outlier as a point that does not belong to any cluster after an appropriate clustering operation. In this context the adopted clustering method is the Fuzzy C-means [19]. The four features are fed in input to the FIS [29] that outputs an outlierness degree belonging to the range [0, 1], that if it is close to one the corresponding pattern is labeled as outlier [8]. In this industrial context, outlier detection has been applied on each coil independently from other coils, being some process settings depending on coil basic features. After the pre-processing phase, the analysis has been focused on the identification of possible causes of ripples occurrences. In particular, the attention is focused on the development of a model capable to identify process variables with a stronger impact on the presence/absence of ripples by expressing such effect through an appropriate relationship. To this aim, a binary classification based on a Decision Tree (DT) has been adopted, which considers two classes:

- **Class 0**: slice without ripples defects.
- **Class 1**: highly defective slices, with defects exceeding a threshold automatically calculated and equal to the 95th quantile of the empirical cumulative distribution of the percentage area of defects.

In the considered context, the focus is on the identification of class 0 samples in the air-blowing case and of class 1 samples in the nitrogen-blowing case [1, 18, 32]. The performance of the developed classifier has been evaluated in terms of average accuracy, also known as Balanced Classification Rate [25], which is suitable also for imbalanced dataset [12, 15] that are quite frequent in industrial contexts. The formula of Balanced Classification Rate (BCR) is defined as in Eq. 20.1:

$$BCR = \frac{1}{2}(\frac{TP}{TP + FN} + \frac{TN}{TN + FP}) \qquad (20.1)$$

where:

- **TP** True Positive, number of unitary samples correctly classified.
- **TN** True Negative, number of null samples correctly classified.
- **FP** False Positive, number of null samples incorrectly classified.
- **FN** False Negative, number of unitary samples incorrectly classified.

Furthermore, the ratios $\frac{TP}{TP+FN}$ and $\frac{TN}{TN+FP}$ are also known as *sensitivity* and *specificity*, respectively. The main advantage on the use of a DT-based classifier lies in the fact that it is a white box model and it is understandable by unskilled people, as it appears as a chain of simple if-then rules [4, 5, 20]. Each node of the DT is associated to a process variable, a branch corresponds to a range of values and finally leaf nodes are associated to the two classes. Firstly, data are randomly shuffled and then are divided into a training set, represented by the 75% of the data, and a validation set corresponding to the remaining 25% of data. 10 different training procedures have been performed in order to ensure a general validity of results. The performance of the DT has been evaluated by computing the BCR on the validation set and Afterward the average accuracy on different trainings has been computed. Finally an appropriate procedure to determine the most important variables, i.e. variables which mostly affect the classification [7, 10–14], has been adopted in order to provide a measure of the impact of each input variable on the occurrence or absence of defects [12]. This procedure computes an estimate of variable importance for a decision tree by summing variations in the risk caused by splits on each variable. One *importance degree* is then returned for each variable in the data adopted to train the tree. For each node a *risk* is evaluated as the node error value weighted by the node probability. The

Variable importance degree $Imp(v)$ is calculated as the proportion of observations from the initial data that fulfill the conditions for the considered node, as follows:.

$$Imp(v) = \frac{1}{\text{Nodes}} \sum_{N \in N_v} \frac{E_N - (E_R + E_L)}{\text{instances}} \qquad (20.2)$$

where N_v is the set of all nodes associated to the variable v, *nodes* is the number of nodes, *instances* is the number of available samples, E_N represents the number of errors in the node N and finally, E_R and E_L correspond to the number of errors of the right and left children, respectively. The error of a node is represented by the declassification probability for that node. The variable importance has been computed for each attribute present in the DT, while a null value is associated to the attributes that are missing in the DT. Afterwards, each variable importance degree has been normalized with respect to the maximum value in order to obtain an importance value in the range [0, 1]. The unitary value corresponds to the most important variable. The procedure is repeated 10 times and an average importance degree is determined for each variable.

20.4 Obtained Results

A plausibility range has been provided for each input variable and samples out of range have been discarded. Moreover, the above presented outlier fuzzy-based approach has been applied in order to obtain a clean and reliable dataset [15]. Tables 20.2 and 20.3 show the number of input samples, the number of not plausible data, the number of detected outliers and finally the percentage of outliers in both case studies: air and nitrogen blowing.

Outlier detection has been applied coil per coil and it has been omitted for lower stages (0–3), since the number of samples for each coil is too low. Results on the

Table 20.2 Pre-processing results concerning air-blowing

Stage	Samples	Not plausible	Outliers	Outliers (%)
0	358	0	–	–
1	716	0	–	–
2	1432	0	–	–
3	2864	0	–	–
4	5728	0	216	3.77
5	11456	0	207	1.81
6	22912	0	228	1
7	45024	0	237	0.52
8	91648	0	253	0.28

Table 20.3 Pre-processing results concerning Nitrogen-blowing

Stage	Samples	Not plausible	Outliers	Outliers (%)
0	354	0	–	–
1	708	0	–	–
2	1416	0	–	–
3	2832	0	–	–
4	5664	0	236	4.17
5	11328	1	220	2.02
6	22656	3	214	1.23
7	45312	3	215	1.19
8	90624	3	214	1.18

Table 20.4 Balanced classification rate

Blowing type	Accuracy (%)		
	overall	class 0	class 1
Air	99.34	99.78	98.90
Nitrogen	97.65	98.92	96.37

pre-processing phase show that in both cases (air or nitrogen blowing) a low number of samples have been discarded. The two datasets, obtained with air or nitrogen blowing, are individually given in input to a DT-based classification algorithm. The average accuracy of the binary classification has been evaluated in terms of Balanced Classification Rate (BCR) and Afterward the mean accuracy on ten iterations on different training sets has been calculated. The results for both cases (air and nitrogen blowing) are shown in Table 20.4.

The obtained results not only show good classification performances in terms of average accuracy, but also provide very precise rules, inferred by DTs, with a simple syntax that allows to easily act on the affecting process variables in order achieve the desired quality level. This result, from an industrial point of view, is clearly very satisfactory, as it provides an effective way to support decisions and to inform operators on the actions to take in order to avoid defects [21]. Furthermore, the most relevant variables have been identified. Tables 20.5 and 20.6 show the mean importance degree of the relevant variables in a decreasing order in both studied cases: air blowing and nitrogen blowing.

The variables that are listed in the two tables are in line with the expectations of industrial experts. In addition, all variables involved in this analysis except for the *Water bath cooling temperature* can be adjusted in line in order to tune the process in "ripples safe" conditions. The *Water bath cooling temperature* represents the temperature of the water quench located towards the end of the coating process and it is a consequence of other parameters, such as the residual strip heat, speed or thickness, and consequently the operator cannot modify it.

Table 20.5 Mean variable importance in the case of air blowing

Process variables	Mean importance degree
Distance air-blade	1
Tunnel zone temperature	0.69
Speed process section	0.29
Pressure air-blade	0.25
Fans reference speed	0.15
Top-roll zone temperature	0.08
Water bath cooling temperature	0.07
Strip temperature	0.02

Table 20.6 Mean variable importance in the case of Nitrogen blowing

Process variables	Mean importance degree
Water bath cooling temperature	1
Distance air-blade	0.45
Hot briddle zone temperature	0.39
Speed process section	0.14
Pressure air-blade	0.14
Fans reference speed	0.10
Height air knife	0.05
Top-roll zone temperature	0.02

20.5 Conclusions

The analysis of the occurrence of ripple defects during HDG in the production of flat steel products has been presented, with a focus on the study on thick coils having low zinc coating. An accurate pre-processing procedure has been applied and then a DT-based classification model has been developed. The proposed method is generic and does not require any a-priori assumptions, thus it can be applied to other real applications such as the study of the occurrence of different defects type. The obtained results demonstrate the effectiveness of the proposed method (average accuracy more than 97% in the nitrogen blowing case and an average accuracy even greater than 99% in the air blowing case), moreover another important result lies in the fact that DTs approach provide defined rules with simple sintax becoming an effective tool to support decisions. Finally, the approach outputs a mean importance degree of relevant variables giving a deeper understanding of the process that can lead to improved coil quality, by reducing the reworked or discarded material and reducing the costs associated with an excessive use of the nitrogen.

References

1. Ali, S., Smith, K.A.: On learning algorithm selection for classification. Appl. Soft Comput. **6**(2), 119–138 (2006)
2. Borselli, A., Colla, V., Vannucci, M., Veroli, M.: A fuzzy inference system applied to defect detection in flat steel production. In: IEEE World Congress on Computational Intelligence, WCCI (2010)
3. Brandenburger, J., Colla, V., Nastasi, G., Ferro, F., Schirm, C., Melcher, J.: Big data solution for quality monitoring and improvement ion flat steel production. In: 17th Symposium on Control, Optimization and automation in Mining, Mineral and Metal Processing, Vienna (2016)
4. Breiman, L.: Random forests. Mach. Learn. **45**, 5–32 (2001)
5. Breiman, L., Friedman, J.H., Olshen, R.A., Stone, C.J.: Classification and Regression Trees. Wadsworth and Brooks (1984)
6. Breunig, M.M., Kriegel, H.P., Ng, R.T., Sander, J.: Lof: identifying density-based local outliers. In: ACM Sigmod Record, vol. 29, pp. 93–104. ACM (2000)
7. Cateni, S., Colla, V.: Improving the stability of wrapper variable selection applied to binary classification. Int. J. Comput. Inf. Syst. Ind. Manag. Appl. **8**, 214–225 (2016)
8. Cateni, S., Colla, V., Nastasi, G.: A multivariate fuzzy system applied for outliers detection. J. Intell. Fuzzy Syst. **24**(4), 889–903 (2013)
9. Cateni, S., Colla, V., Vannucci, M.: A fuzzy logic based method for outliers detection. Proceedings of the IASTED International Conference on Artificial Intelligence and Applications AIA **2007**, 561–566 (2007)
10. Cateni, S., Colla, V., Vannucci, M.: General purpose input vvariable extraction: a genetic algorithm based procedure give a gap. In: 9th International Conference on Intelligence Systems design and Applications ISDA09, pp. 1307–1311 (2009)
11. Cateni, S., Colla, V., Vannucci, M.: Variable selection through genetic algorithms for classification purpose. In: IASTED International Conference on Artificial Intelligence and Applications AIA2010, pp. 6–11 (2010)
12. Cateni, S., Colla, V., Vannucci, M.: A genetic algorithm based approach for selecting input variables and setting relevant network parameters of som based classifier. Int. J. Simul. Syst. Sci. Technol. **12**(2), 30–37 (2011)
13. Cateni, S., Colla, V., Vannucci, M.: A hybrid feature selection method for classification purposes. In: 8th European Modeling Symposium on Mathematical Modeling and Computer simulation EMS2014, Pisa, Italy, vol. 1, pp. 1–8 (2014)
14. Cateni, S., Colla, V., Vannucci, M.: A fuzzy system for combining filter features selection methods. Int. J. Fuzzy Syst. (in Press)
15. Cateni, S., Colla, V., Vannucci, M., Vannocci, M.: A procedure for building reduced reliable training datasets from realworld data. In: 13th IASTED International Conference on Artificial Intelligence and Applications AIA 2014, Innsbruck (Austria), pp. 393–399 (2014)
16. Colla, V., Valentini, R., Bioli, G.: Mechanical properties prediction for aluminium-killed and interstitial-free steels. Revue de Métalurgie **Special Issue JSI**, 100–101 (2004)
17. Colla, V., Vannucci, M., Fera, S., Valentini, R.: Ca-treatment of al-killed steels: inclusion modification and application of artificial neural networks for the prediction of clogging. In: Proceedings of the 5th European Oxygen Steelmaking Conference EOSC'06, vol. 1, pp. 387–394 (2006)
18. Duda, R., Hart, P., Stork, D.: Pattern Classification. Wiley, New York (2001)
19. Dunn, J.C.: A fuzzy relative of the isodata process and its use in detecting compact well-separated clusters (1973)
20. Fiasché, M.: Svm tree for personalized transductive learning in bioinformatics classification problems. Smart Innov. Syst. Technol. **26**, 223–231 (2014)
21. Fiasché, M., Cuzzola, M., Irrera, G., Iacopino, P., Morabito, F.: Advances in medical decision support systems for diagnosis of acute graft-versus-host disease: molecular and computational intelligence joint approaches. Front. Biol. China **6**(4), 263–273 (2011)

22. Fiasché, M., Ripamonti, G., Sisca, F., Taisch, M., Tavola, G.: A novel hybrid fuzzy multi-objective linear programming method of aggregate production planning. Smart Innov. Syst. Technol. **54**, 489–501 (2016)
23. Grubbs, F.E.: Procedures for detecting outlying observations in samples. Technometrics **11**(1), 1–21 (1969)
24. Hawkins, D.M.: Identification of Outliers. Chapman and Hall, London (1980)
25. Kalousis, A., Gama, J., Hilario, M.: On data and algorithms: understanding inductive performance. Mach. Learn. **54**(3), 275–312 (2004)
26. Knox, E.M., Ng, R.T.: Algorithms for mining distancebased outliers in large datasets. In: Proceedings of the International Conference on Very Large Data Bases, pp. 392–403. Citeseer (1998)
27. Koc, L., Carswell, A.D.: Network intrusion detection using a hnb binary classifier. In: 17th UKSIM-AMSS International Conference on Modelling and Simulation (2015)
28. Mahalanobis, P.C.: On the generalized distance in statistics. Proc. Natl. Inst. Sci. (Calcutta) **2**, 49–55 (1936)
29. Mamdani, E., Assilian, S.: An experiment in linguistic synthesis with a fuzzy logic controller. Int. J. Man-Mach. Stud. **7**, 1–13 (1975)
30. Ramaswamy, S., Rastogi, R., Shim, K.: Efficient algorithms for mining outliers from large data sets. In: ACM Sigmod Record, vol. 29, pp. 427–438. ACM (2000)
31. Shetty, M., Shekokar, N.M.: Data mining techniques for real time intrusion detection systems. Int. J. Sci. Eng. Res. **3**(4), 1–7 (2012)
32. Theodoridis, S., Koutroumbas, K.: Pattern Recognition. Academic Press (1999)

Chapter 21
An Improved PSO for Flexible Parameters Identification of Lithium Cells Equivalent Circuit Models

Massimiliano Luzi, Maurizio Paschero, Antonello Rizzi and Fabio Massimo Frattale Mascioli

Abstract Nowadays, the equivalent circuit approach is one of the most used methods for modeling electrochemical cells. The main advantage consists in the beneficial trade-off between accuracy and complexity that makes these models very suitable for the State of Charge (SoC) estimation task. However, parameters identification could be difficult to perform, requiring very long and specific tests upon the cell. Thus, a more flexible identification procedure based on an improved Particle Swarm Optimization that does not require specific and time consuming measurements is proposed and validated. The results show that the proposed method achieves a robust parameters identification, resulting in very accurate performances both in the model accuracy and in the SoC estimation task.

21.1 Introduction

Battery Management Systems (BMSs) are one of the most critical and influencing devices in the technological evolution concerning energetic efficiency and sustainable mobility. In fact, they are of extreme importance for monitoring and protecting any Energy Storage System (ESS) used in Smart Grids and microgrids [1–3], or in hybrid and electric vehicles [4, 5]. The main role of any BMS is to ensure that each electrochemical cell composing the ESS remains in its safety operating area; second, BMS has to estimate the State of Charge (SoC) of each cell, giving thus information about the amount of residual stored energy. In particular, SoC estimation is the most

M. Luzi (✉) · M. Paschero · A. Rizzi · F. M. F. Mascioli
University of Rome "La Sapienza", Via Eudossiana 18, 00184 Rome, Italy
e-mail: massimiliano.luzi@uniroma1.it

M. Paschero
e-mail: maurizio.paschero@uniroma1.it

A. Rizzi
e-mail: antonello.rizzi@uniroma1.it

F. M. F. Mascioli
e-mail: fabiomassimo.frattalemascioli@uniroma1.it

© Springer International Publishing AG, part of Springer Nature 2019
A. Esposito et al. (eds.), *Neural Advances in Processing Nonlinear Dynamic Signals*, Smart Innovation, Systems and Technologies 102,
https://doi.org/10.1007/978-3-319-95098-3_21

critical task of BMS, since only an accurate information about this quantity allows to maximize the life, the efficiency and the effectiveness of ESSs.

Several methods for SoC estimation have been proposed in the literature [6]. Among them, the techniques based on nonlinear state observers, such as Extended Kalman Filters (EKF) or Unscented Kalman Filters (UKF), are showing the most robust and promising estimation accuracies [7–9]. In these methods, a suitable model of electrochemical cells in which the SoC quantity belongs to the state space vector is used for estimating SoC, together with other system state components. Thus, it is clear that the effectiveness of state observer methods is strictly dependent on the accuracy of the considered cell model.

Among the several approaches proposed in the literature [10–12], equivalent circuit models has showed a promising trade-off between accuracy and computational demand [13, 14]. Thus, they are often used for SoC estimation. In these models, a connection of bipolar lumped elements is adopted in order to emulate the voltage response of the cell to the input current. However, the parameters identification procedure could be often stiff and/or inaccurate. For example, in [13] the identification procedure has to be performed offline and it requires a very long and specific charging and discharging test in order to get accurate values for the electric components composing the model. Alternatively, in [15, 16] a Gray Wolf Optimization (GWO) and a Particle Swarm Optimization (PSO), respectively, have been used for identifying the model parameters by fitting the datasheet curves of the cell. Nevertheless, these approaches can result in a loss of accuracy because the cells can work differently with respect to the datasheet due to manufacturing deviations or aging.

In this work, a flexible parameters identification procedure based on an improved PSO is proposed. Instead of considering the generic data of the datasheet, the proposed method searches for the best parameters set resulting in the best fitting of data directly measured on the cell under test. Consequently, this approach results in a more flexible identification procedure that does not require any specific test over the cells, allowing a possible application in real time during the actual usage of ESS. Moreover, it brings to a more accurate identification of the considered cell.

In the following section, the architecture of the equivalent circuit model is presented. In Sect. 21.3, a detailed description both of the proposed identification procedure and of the improvements implemented on the PSO are discussed. Then, the performed tests are discussed and commented in Sect. 21.4. Finally, Sect. 21.5 contains the concluding remarks.

21.2 Equivalent Circuit Model

A new equivalent circuit model based on a mechanical analogy has been proposed in [13, 14]. In these works, the authors showed that the generic voltage response V_{out} of any cell can be thought as the superimposition of three contributions related to different timescales, namely the *instantaneous*, the *dynamic* and the *quasi-stationary*

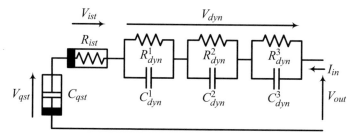

Fig. 21.1 Equivalent circuit model of an electrochemical cell

contributions. More precisely, the instantaneous term V_{ist} models the memoryless relationship between the input current I_{in} and the output voltage V_{out}; the dynamic term V_{dyn} models the low pass transient response of the cell; the quasi-stationary term V_{qst} models the voltage contribution due to the amount of currently stored charge. In particular, V_{qst} is equivalent to the Open Circuit Voltage (OCV) of the cell. The model proposed in [13] is shown in Fig. 21.1. Its main contribution consists in the use of nonlinear circuit components, such as nonlinear resistors and capacitors, in order to model the nonlinear response to the input current. This allows to obtain a model closer to the physical behavior of the cell, avoiding any kind of mathematical artifices such as SoC dependent resistors.

In this circuit, V_{qst} is modeled with a nonlinear capacitor whose governing function is the OCV-SoC curve typical of any cell, whereas V_{ist} and V_{dyn} are modeled with a nonlinear resistor and the series of three first order RC low pass filters, respectively. The input and the output of the model are the current I_{in} and the voltage V_{out}, respectively, whereas the state variables are SoC and each voltage V_{dyn}^i related to the RC filters. Thus, the following state form equations can be derived:

$$
\begin{cases}
\begin{Bmatrix} SoC[k] \\ V_{dyn}^1[k] \\ V_{dyn}^2[k] \\ V_{dyn}^3[k] \end{Bmatrix} =
\begin{Bmatrix} SoC[k-1] + \eta \frac{T_s}{C_n} I_{in}[k] \\ e^{-T_s/\tau_{dyn}^1} V_{dyn}^1[k-1] + (1 - e^{-T_s/\tau_{dyn}^1}) R_{dyn}^1 I_{in}[k] \\ e^{-T_s/\tau_{dyn}^2} V_{dyn}^2[k-1] + (1 - e^{-T_s/\tau_{dyn}^2}) R_{dyn}^2 I_{in}[k] \\ e^{-T_s/\tau_{dyn}^3} V_{dyn}^3[k-1] + (1 - e^{-T_s/\tau_{dyn}^3}) R_{dyn}^3 I_{in}[k] \end{Bmatrix} \\
V_{out}[k] = \tilde{V}_{qst}(SoC[k]) + \sum_{i=1}^{3} V_{dyn}^i[k] + \tilde{R}_{ist}(I_{in}[k]) I_{in}[k]
\end{cases} \quad (21.1)
$$

In the above formula, SoC has been evaluated with the Coloumb Counting approach [17], where T_s is the sampling time, η is the Coloumbic efficiency and C_n is the nominal capacity expressed in Ah, that is the total amount of charge that the cell can store between the maximum and minimum allowed SoC. The nonlinear function $\tilde{V}_{qst}(.)$ is the OCV-SoC curve modeling the behavior of the nonlinear capacitor C_{qst}, whereas $\tilde{R}_{ist}(.)$ is the nonlinear function modeling the nonlinear resistor R_{ist}. Consequently, the parameters to be identified are the nonlinear functions $\tilde{V}_{qst}(.)$ and

$\tilde{R}_{ist}(.)$ with respect to the quasi-stationary and the instantaneous terms, respectively, and the values of R^i_{dyn} and $\tau^i_{dyn} = R^i_{dyn} C^i_{dyn}$ for each RC dipole modeling the dynamic contribution.

In [13] the authors has proposed a very accurate parameters identification procedure for this model. Its main advantage is that it identifies separately the parameters related to the different timescale contributions. However, in order to achieve this separated identification, the method requires to test the cell with a very specific current profile. Moreover, this test cycle must have a very long time duration and it could last some days in order to allow the identification of the dynamic parameters related to the slowest transient response. As a consequence, this identification procedure is very rigid and expensive to perform, because it requires to disassemble the cell from ESS and to test it offline, with a high accuracy and for a very long time.

21.3 Flexible Identification Procedure

In order to facilitate the parameters identification, the model of Fig. 21.1 has been simplified by modeling V_{ist} with a linear resistor in place of the nonlinear one of the original model. Thus, all the parameters have been expressed by only one real number, except for those of the quasi-stationary contribution. In fact, the identification of the OCV-SoC curve is more complicated since it requires to retrieve the shape of the nonlinear function $\tilde{V}_{qst}(.)$. In order to do that, the same procedure discussed and proposed in [14] has been considered here. In this procedure, a limited number L of samples equally distributed in the function domain are used as representatives of the nonlinear function to identify. Then, the considered samples are interpolated with a cubic polynomial in order to recover the overall shape of the function. Thus, the identification procedure is in charge to find only the values of the L samples, searching for the best sampling representing $\tilde{V}_{qst}(.)$. Summarizing, the parameters set θ to be identified is given by the following vector:

$$\theta = \left\{ R_{ist}, \tau^1_{dyn}, R^1_{dyn}, \tau^2_{dyn}, R^2_{dyn}, \tau^3_{dyn}, R^3_{dyn}, V^1_{qst}, V^2_{qst}, \dots, V^L_{qst} \right\} \qquad (21.2)$$

where V^i_{qst} indicates the i-th sample of the OCV-SoC curve representation.

In this work, it has been adopted an approach similar to those proposed in [15, 16]. More precisely, the identification procedure has been formulated as a fitting problem solved with swarm intelligence algorithms. However, instead of considering the curves of the datasheet, the fitting problem has been applied to generic data measured directly on the cell. This procedure has allowed to generalize the parameters identification, to make it more flexible, and to tailor it on the cell under test. Thus, given any measured sequence of voltage V_{out} and current I_{in}, the optimization algorithm will search for the optimal parameters set θ_{opt} that minimizes the error between the estimated voltage and the measured one. Thus, being $\bar{V}^\theta_{out}$ the estimated voltage given the parameters set θ, the identification procedure consists in solving the following optimization problem:

$$\theta_{opt} = \underset{\theta}{\mathrm{argmin}} \left\{ \mathrm{MSE} \left(V_{out}, \bar{V}_{out}^{\theta} \right) \right\} \tag{21.3}$$

where MSE(.) indicates the mean square error evaluated between V_{out} and $\bar{V}_{out}^{\theta}$.

In this work, the optimization problem (21.3) has been solved by means of a PSO. In particular, in order to improve both the convergence to the global minimum and the robustness of the parameters identification, an improved implementation of PSO, called Hybrid Genetic-PSO (HG-PSO), has been developed. Three kind of improvements have been introduced with the aim of reducing the stagnation effect, to increase the exploration and exploitation capability, and to avoid the convergence to local minima.

The main improvement has regarded the hybridization of PSO with the Genetic Algorithm (GA). This hybridization has aimed at enhancing the exploitation capability of the algorithm, facilitating it in escaping from local minima. At each iteration, a certain number of new particles are generated by applying the genetic operators of crossover and mutation. In particular, a K-tournament algorithm is considered for determining the particles affected by the genetic operators. Once the new individuals have been generated, these will substitute a set of particles randomly chosen from the worst half of the original swarm.

The second improvement has been a multi-swarm implementation of PSO. Herein, the entire swarm is split in a number $G > 1$ of sub-swarms, each one characterized by having its own global best. In this topology, each particle belongs to the sub-swarm related to the closest global best, and consequently it updates its position with respect to that global best. The advantage of using a multi-swarm implementation consists in an improved exploration capability that allows to search in different areas of the solution space, and to avoid a premature convergence to local minima.

The final improvement was the implementation of the *Guaranteed Convergence* update rule [18]. In this version, the velocity of each particle close to the global best is updated with expression (21.4) in place of the standard rule of PSO:

$$v[n] = \omega v[n-1] + g_{best} - \theta[n-1] + \delta(1 - 2r) \tag{21.4}$$

where v, θ and g_{best} are the velocity, the current particle position and the global best, respectively, ω is the inertial coefficient, whereas $r \in [0, 1]$ is a random number chosen from the uniform distribution, and δ is a scaling factor. This update rule introduces a more random behavior of the particles close to the global best, allowing to avoid stagnation and to improve the exploitation around the current global best.

21.4 Validation Procedure

The proposed method has been validated by performing the parameters identification
procedure upon a real Li-ion cell. In order to highlight the achieved flexibility, the
Randomized Battery Usage Dataset [19] collected by the *NASA Ames Research
Center* has been considered for building the Training Set (TrS) and the Test Sets
(TsS). This dataset is composed of the measurements performed by testing a Li-ion
cell *model 18650* with a sequence of random charging/discharging current pulses.
Thus, the considered data for the parameters identification is totally generic and not
chosen a priori. The TrS and the TsS are shown in Fig. 21.2.

In order to analyze the robustness of the proposed procedure, the parameters
identification has been performed ten times, considering a different random initialization of the swarm for each run. Then, a statistical analysis of the results has been
performed considering the values of the 1-st, 2-nd and 3-rd quartiles; note that the
2-nd quartile coincides with the median value. The configuration of HG-PSO is
shown in Table 21.1, where ω, c_p and c_g are the inertial coefficient and the accel-
eration coefficients related to the personal and global bests, respectively. Moreover,
the identification of the nonlinear function $\tilde{V}_{qst}(.)$ has been performed considering
$L = 15$ and a total of 1000 interpolation points.

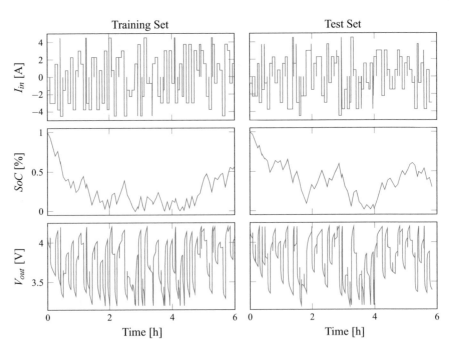

Fig. 21.2 Dataset related to the *Randomized Battery Usage Dataset*

Table 21.1 HG-PSO configuration

Particles	ω	c_p	c_g	Mutations	Crossovers	K	Sub-swarms	δ
50	0.7298	1.4962	1.4962	5	5	5	2	0.1

Table 21.2 Comparison between HG-PSO and PSO

	Median		Q1		Q3		Q3-Q1	
	HG-PSO	PSO	HG-PSO	PSO	HG-PSO	PSO	HG-PSO	PSO
MSE TrS	2.49e-4	3.08e-4	2.49e-4	2.81e-4	2.50e-4	3.34e-4	9.84e-7	5.32e-5
MSE TsS	7.46e-4	8.24e-4	7.44e-4	7.73e-4	7.50e-4	9.62e-4	6.00e-6	1.90e-4
R_{ist} [mV]	78.92	85.64	78.81	83.68	79.12	86.07	0.31	2.39
R_{dyn}^1 [mV]	20.35	26.21	20.19	22.62	20.54	27.65	0.36	5.03
τ_{dyn}^1 [h]	7.58e-3	2.50e-2	7.29e-3	1.87e-2	7.95e-3	2.90e-2	6.57e-4	1.02e-2
R_{dyn}^2 [mV]	21.34	16.89	21.11	15.11	21.43	50.62	0.33	35.51
τ_{dyn}^2 [h]	8.58e-2	5.14e-1	8.23e-2	1.51e-1	9.14e-2	4.66	9.05e-3	4.51
R_{dyn}^3 [mV]	16.63	27.39	6.93	12.89	69.87	54.36	62.94	41.47
τ_{dyn}^3 [h]	5.43	6.26	4.31	4.04	10.75	7.81	6.44	3.77

Q1 = 1st Quartile; Q3 = 3rd Quartile

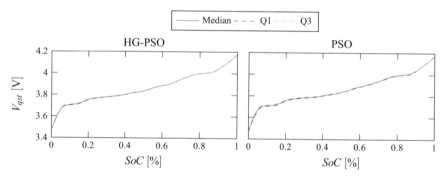

Fig. 21.3 Comparison between the $\tilde{V}_{qst}$ identified by the HG-PSO and the PSO

With the aim of highlighting the improvement gained with the customization discussed in Sect. 21.3, the same identification procedure has been performed with a standard PSO sharing the same configuration of HG-PSO, except for the genetic and multi-swarm setup. The results related to the performed MSE and the values of the statistical analysis of R_{ist}, R_{dyn}^i and τ_{dyn}^i are shown in Table 21.2, whereas the statistical analysis of the $\tilde{V}_{qst}$ function is shown in Fig. 21.3.

Both HG-PSO and PSO has achieved a great accuracy with the former performing a median MSE of 2.49e-4 and 7.46e-4 for the TrS and the TsS, respectively, whereas PSO has performed a median MSE of 3.08e-4 and 8.24e-4. The similar MSE performed in the TrS and the TsS proves the good generalization capability of

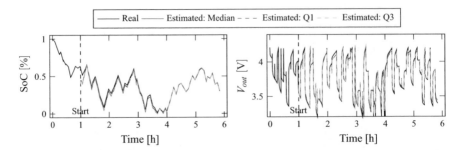

Fig. 21.4 Comparison between real and estimated SoC and V_{out} performed by executing the SR-UKF over the models built considering the parameters got from the HG-PSO procedure

the proposed method. Moreover, beside the better MSE performed both in the TrS and in the TsS, the most relevant result of HG-PSO is related to the significantly better robustness of the identified parameters. In fact, the differences between the third and the first quartiles performed by HG-PSO are almost always two order of magnitude lower than those of PSO. This is because the three implemented upgrades has allowed HG-PSO to converge every time to very close solutions, and thus to find almost always the same parameters set. On the contrary, PSO often has stuck in local minima, resulting in a greater deviation both in the identified parameters and in the performed MSE.

In addition to the robustness analysis, the ten models obtained by executing the HG-PSO procedure have been used into a Square Root UKF (SR-UKF) [7] for performing SoC estimation upon the TsS data. In order to avoid a trivial test in which the states of the cell are known, SoC estimation has been performed considering a temporal offset of $t = 1$ h. This way, the cell is in a non-stationary condition in which each dynamical state variable is totally unknown. The obtained estimations of SoC and V_{out} are shown in Fig. 21.4, where SR-UKF has been initialized considering a generic state with $SoC = 0.5$ and $V_{dyn}^i = 0$ for each RC filter. It can be seen that SR-UKF succeeds in correcting the state estimation in a few minutes, achieving then an accurate SoC estimation along the entire length of the TsS sequence. In particular, it has performed a median MSE between the real SoC and the estimated one equal to 6.13e-4. Moreover, it is noticeable also the robustness of the estimation with the median almost always overlapped with both the first and the third quartiles evaluated on the SoC estimated by the ten models.

21.5 Conclusions

In this work, a flexible procedure for performing the parameters identification related to an equivalent circuit model of an elecrochemical cell has been addressed. The identification procedure has been formulated as a fitting problem in which a PSO is in charge to find the best parameters set resulting in the minimization of the MSE between the measured voltage and the estimated one. In order to increase the robustness of the algorithm, three improvements to the standard PSO have been implemented: the genetic hybridization and the guaranteed convergence has allowed to improve exploitation and to avoid the stagnation on local minima; the multi-swarm implementation has improved the exploration capability. The main advantage of the proposed approach is that it does not require any specific test on the cell, and it is able to identify the model parameters starting from generic measurements.

The proposed method has been validated by performing parameters identification upon the *Randomized Battery Usage Dataset* of the *NASA Ames Research Center*. In particular, the procedure has been performed ten times and a statistical analysis of the results has been made. Moreover, the effectiveness of the obtained models has been analyzed by using them in a SR-UKF for the SoC estimation. The results show very promising performances, with the identified models succeeding not only in achieving a good estimation accuracy of V_{out} and SoC, but also a very robust identification of the model parameters.

Future efforts will focus on using and testing the proposed procedure for performing an online parameters identification. Furthermore, it will be investigated a closed loop architecture in which the SR-UKF and the HG-PSO work together for estimating the SoC and updating the model parameters at the same time.

References

1. Leonori, S., Paschero, M., Rizzi, A., Frattale Mascioli, F.: An optimized microgrid energy management system based on FIS-MO-GA paradigm. In: 2016 IEEE International Conference on Fuzzy Systems (FUZZ-IEEE) (2017)
2. Leonori, S., De Santis, E., Rizzi, A., Frattale Mascioli, F.M.: Optimization of a microgrid energy management system based on a fuzzy logic controller. In: IECON 2016-42nd Annual Conference of the IEEE Industrial Electronics Society, pp. 6615–6620 (2016)
3. Leonori, S., De Santis, E., Rizzi, A., Frattale Mascioli, F.M.: Multi objective optimization of a fuzzy logic controller for energy management in microgrids. In: 2016 IEEE Congress on Evolutionary Computation (CEC), pp. 319–326 (2016)
4. Ehsani, M., Gao, Y., Emadi, A.: Modern electric, hybrid electric, and fuel cell vehicles: fundamentals, theory, and design. CRC Press (2009)
5. Xie, S., Zhong, W., Xie, K., Yu, R., Zhang, Y.: Fair energy scheduling for vehicle-to-grid networks using adaptive dynamic programming. IEEE Trans. Neural Netw. Learn. Syst. **27**(8), 1697–1707 (2016)
6. Chang, W.Y.: The state of charge estimating methods for battery: a review. ISRN Appl. Math. (2013)
7. Luzi, M., Paschero, M., Rossini, A., Rizzi, A., Frattale Mascioli, F.M.: Comparison between two nonlinear kalman filters for reliable SoC estimation on a prototypal BMS. In: IECON 2016-42nd Annual Conference of the IEEE Industrial Electronics Society, pp. 5501–5506 (2016)
8. Sun, F., Hu, X., Zou, Y., Li, S.: Adaptive unscented kalman filtering for state of charge estimation of a lithium-ion battery for electric vehicles. Energy **36**(5), 3531–3540 (2011)
9. Plett, G.L.: Extended kalman filtering for battery management systems of LiPB-based HEV battery packs: Part 1-2-3. J. Pow. Sour. **134**(2), 252–292 (2004)

10. Fan, G., Pan, K., Storti, G.L., Canova, M., Marcicki, J., Yang, X.G.: A reduced-order multi-scale, multi-dimensional model for performance prediction of large-format li-ion cells. J. Electrochem. Soc. **164**(2), A252–A264 (2017)
11. Paschero, M., Di Giacomo, V., Del Vescovo, G., Rizzi, A., Frattale Mascioli, F.M.: Estimation of Lithium Polymer cell charachteristic parameters through genetic algorithms. In: Proceedings of the ICEM 2010-International Conference on Electrical Machines (2010)
12. Du, J., Liu, Z., Wang, Y.: State of charge estimation for li-ion battery based on model from extreme learning machine. Control Eng. Pract. **26**, 11–19 (2014)
13. Paschero, M., Storti, G.L., Rizzi, A., Frattale Mascioli, F.M., Rizzoni, G.: A novel mechanical analogy-based battery model for SoC estimation using a multicell EKF. IEEE Trans. Sustain. Energy **7**(4), 1695–1702 (2016)
14. Luzi, M., Paschero, M., Rizzi, A., Frattale Mascioli, F.M.: A PSO algorithm for transient dynamic modeling of lithium cells through a nonlinear RC filter. In: 2016 IEEE Congress on Evolutionary Computation (CEC), pp. 279–286 (2016)
15. Sangwan, V., Kumar, R., Rathore, A.K.: Estimation of battery parameters of the equivalent circuit model using grey wolf optimization. In: 2016 IEEE 6th International Conference on Power Systems (ICPS), pp. 1–6 (2016)
16. Wang, Y., Li, L.: Li-ion battery dynamics model parameter estimation using datasheets and particle swarm optimization. Int. J. Energy Res. **40**(8), 1050–1061 (2016). ER-15-5937.R2
17. Ng, K.S., Huang, Y.F., Moo, C.S., Hsieh, Y.C.: An enhanced coulomb counting method for estimating state-of-charge and state-of-health of lead-acid batteries. In: INTELEC 2009-31st International Telecommunications Energy Conference, pp. 1–5 (2009)
18. Peer, E.S., van den Bergh, F., Engelbrecht, A.P.: Using neighbourhoods with the guaranteed convergence PSO. In: Proceedings of the 2003 IEEE Swarm Intelligence Symposium, 2003. SIS 2003, pp. 235–242 (2003)
19. Bole, B., Kulkarni, C., Daigle, M.: Randomized battery usage data set. Technical Report, NASA Ames Prognostics Data Repository, NASA Ames Research Center, Moffett Field, CA (2014). http://ti.arc.nasa.gov/project/prognostic-data-repository

Part IV
Intelligent Tools for Decision Making in Economics and Finance

Chapter 22
Yield Curve Estimation Under Extreme Conditions: Do RBF Networks Perform Better?

Alessia Cafferata, Pier Giuseppe Giribone, Marco Neffelli and Marina Resta

Abstract In this paper we test the capability of Radial Basis Function (RBF) networks to fit the yield curve under extreme conditions, namely in case of either negative spot interest rates, or high volatility. In particular, we compare the performances of conventional parametric models (Nelson–Siegel, Svensson and de Rezende–Ferreira) to those of RBF networks to fit term structure curves. To such aim, we consider the Euro Swap–EUR003M Euribor, and the USDollar Swap (USD003M) curves, on two different release dates: on December 30th 2004 and 2016, respectively, i.e. under very different market situations, and we examined the various ability of the above–cited methods in fitting them. Our results show that while in general conventional methods fail in adapting to anomalies, such as negative interest rates or big humps, RBF nets provide excellent statistical performances, thus confirming to be a very flexible tool adapting to every market's condition.

22.1 Introduction

The yield curve represents a relationship between the spot rates of zero coupon bonds and their respective maturities and provides a way of understanding whether the economy will be strong or weak. Understanding the evolution of the yield curve is therefore an important issue in finance, especially for assets pricing, financial risk management and portfolio allocation.

A. Cafferata · M. Neffelli · M. Resta (✉)
DIEC, University of Genova, via Vivaldi 5, 16126 Genova, Italy
e-mail: marina.resta@economia.unige.it

A. Cafferata
e-mail: alessia.cafferata@economia.unige.it

M. Neffelli
e-mail: marco.neffelli@edu.unige.it

P. G. Giribone
Banca Carige, Financial Engineering and Pricing, Genova, Italy
e-mail: piergiuseppe.giribone@carige.it

© Springer International Publishing AG, part of Springer Nature 2019
A. Esposito et al. (eds.), *Neural Advances in Processing Nonlinear Dynamic Signals*, Smart Innovation, Systems and Technologies 102,
https://doi.org/10.1007/978-3-319-95098-3_22

During the past forty years considerable research efforts have been devoted to this task. The two main research tracks refer to equilibrium models, as pioneered by Vasicek [26], and statistical models. Here we are mainly concerned with discussing those latter, as the underlying approach is strongly related to our research question: can Radial Basis Function nets provide a suitable environment to fit the yield curve under extreme conditions? Statistical contributions embrace a wide range of techniques, including the smoothed bootstrap by Bliss and Fama [13], and the parametric approaches by Nelson and Siegel [21], Svensson [24], and de Rezende and Ferreira [10]. These techniques deal with in–sample estimation of the yield curve, while the forecasting issue has been addressed in a more recent literature track. Diebold and Li [11], [12] pioneered the field with a dynamic version of the Nelson–Siegel model (NSm), and [2] modified the NSm by way of a procedure based on ridge regression to avoid collinearity issues. Furthermore, approaches employing Machine Learning (ML) paradigms have been already explored by Ait Sahalia [1], Cottrell et al. [8], Tappinen [25] to cite some, who proposed nonparametric models, with no restriction on the functional form of the process generating the structure of interest rates. More recently Bose et al. [5], Joseph et al. [19], Rosadi et al. [22], Sambasivan and Das [23] explained the behavior of the yield curve with various neural architectures, while Barunik and Malinska [3] employed artificial neural networks to fit the term structures of crude oil future prices. However, so far very little attention has been devoted to the practice of fitting the yield curve in extreme conditions within the ML framework: to the best of our knowledge, Gogas et al. [15] is the only attempt of forecasting recession from a variety of short (treasury bills) and long term interest rate bonds applying Support Vector Machines [7] for classification. We therefore think that there is enough room for contributing with Radial Basis Function (RBF) Networks. Indeed RBF nets have been already widely employed in financial applications: for some examples one can refer to [16, 20]; however, we intend to explore how much this technique can be effective in providing in–sample matching to the yield curve under conditions of stress, and we are going to compare RBF nets performances to those of traditional statistical models.

What remains of the paper is organized as follows; Sect. 22.2 introduces the theoretical framework with a brief discussion concerning both conventional parametric techniques and Radial Basis Function Networks. Section 22.3 contains simulation settings and the results discussion. Section 22.4 concludes.

22.2 Theoretical Framework

In this section we provide an overview on the estimation models generally employed to fit the yield curve: starting from the parametric models in Sect. 22.2.1, we then describe the RBF Networks in Sect. 22.2.2.

22.2.1 Thirty Years of Parametric Estimation Models of the Yield Curve

We are mainly concerned on the Nelson and Siegel model and on the extensions discussed by Svensson [24] and by de Rezende and Ferreira [10]: for other variants the reader can refer to [9]. Nelson and Siegel suggested to model the yield curve in the following way:

$$y_{NS}(t, \boldsymbol{\beta}, \tau) = \beta_0 + \beta_1 \frac{\tau \left[1 - \exp(-t/\tau)\right]}{t} + \beta_2 \frac{\tau \left[1 - \exp(-t/\tau)\right] - \exp(t/\tau)}{t}$$

(22.1)

where the dependent variable y_{NS} represents the zero rate to be determined, t is the time to maturity, $\boldsymbol{\beta} = [\beta_0 \ \beta_1 \ \beta_2]'$ is the parameters vector, with β_0 representing the impact of long–run yield levels, β_1 and β_2 expressing the short–term and the mid–term components, respectively; finally τ is the decay factor. By properly estimating the parameters value, (22.1) makes possible to explain the different shapes the yield curve can assume: flat, humped or S–shaped.

The extension suggested by Svensson in 1994 introduced the possibility to model a second hump in the yield curve:

$$y_{SV}(t, \boldsymbol{\beta}, \tau) = \beta_0 + \beta_1 \frac{\tau_1 \left[1 - \exp(-t/\tau_1)\right]}{t} + \beta_2 \frac{\tau_1 \left[1 - \exp(-t/\tau_1)\right] + - \exp(t/\tau_1)}{t} +$$
$$+ \beta_3 \frac{\tau_2 \left[1 - \exp(-t/\tau_2)\right] - \exp(t/\tau_2)}{t}$$

where $\boldsymbol{\beta} = [\beta_0 \ \beta_1 \ \beta_2 \ \beta_3]'$, with β_0, β_1, β_2 likewise in (22.1), while β_3 is the parameter associated to the second hump. Moreover, we now have: $\tau = [\tau_1 \ \tau_2]'$ representing the decay factors associated to earlier three parameters (τ_1) and to β_3 (τ_2), respectively.

Finally, de Rezende and Ferreira [10] in 2011 discussed a five parameters extension of the Nelson–Siegel model, allowing to insert a third hump in the yield curve:

$$y_{dRF}(t, \boldsymbol{\beta}, \tau) = \beta_0 + \beta_1 \frac{\tau_1 \left[1 - \exp(-t/\tau_1)\right]}{t} + \beta_2 \frac{\tau_1 \left[1 - \exp(-t/\tau_1)\right] - \exp(t/\tau_1)}{t} +$$
$$+ \beta_3 \frac{\tau_2 \left[1 - \exp(-t/\tau_2)\right] - \exp(t/\tau_2)}{t} + \beta_4 \frac{\tau_3 \left[1 - \exp(-t/\tau_3)\right] - \exp(t/\tau_3)}{t}$$

where $\boldsymbol{\beta} = [\beta_0 \ \beta_1 \ \beta_2 \ \beta_3 \ \beta_4]'$, with β_4 being the parameter associated to the third hump, and $\tau_3 \in \tau = [\tau_1 \ \tau_2 \ \tau_3]'$ representing the decay factor associated to β_4.

Clearly both the Svensson (SV) and de Rezende–Ferreira (dRF) variants of the original Nelson–Siegel (NS) model are more complex to manage than the NS model, but they generally improve the desired fitting, as rising the number of parameters the related SSE, and MSE decrease and R^2 increases. In all the examined cases, the estimation of parameters can be performed by way of quasi–Newton methods like the Broyden–Fletcher–Goldfarb–Shanno algorithm–BFGS–[4] or with an optimization heuristic, as in [14]. However, while this latter solution seems to be capable of reliably solving the models, it fails (likewise the BFGS) in assuring the stability of estimated parameters under certain conditions, namely under small perturbations of the data. This motivated us to explore a non–parametric alternative, represented by RBF networks.

22.2.2 Radial Basis Function Networks

Radial Basis Function Networks–RBF–[6] are a kind of neural architecture generally organized into three layers, as illustrated in Fig. 22.1.

Here $x_1, x_2, \ldots, x_m$ represent the components of the input vector $\mathbf{x}$ that are transmitted to the first layer nodes (in the same number as the input elements), by acting a linear transformation. The signal then moves to the hidden layer where it is interpreted by a number of radial functions $\phi_j(\cdot)$, $j = 1, \ldots, n$, whose number n is decided by the user, being:

$$\phi_j(\mathbf{x}) = \exp\left(-\frac{||\mathbf{x} - \mathbf{c}_j||}{r_j}\right), \quad j = 1, \ldots, n \tag{22.2}$$

where $\mathbf{c}_j$ and r_j are the center and the radius of the function, respectively. The characteristic feature of those functions $\phi_j(\cdot)$ is that their response decreases monotonically with distance from a central point. Finally the output neuron generates a weighted sum of the information processed by the hidden layer units:

$$F(\mathbf{x}) = \sum_{j=1}^{n} \omega_j \phi_j(\mathbf{x}) \tag{22.3}$$

The output signal $F(x)$ is then compared to the observed value, and the weights ω_j $(j = 1, \ldots, n)$ are adjusted accordingly, by way of an iterative process, until a stopping criterion is reached. It is a common practice to initialize the number n of nodes in the hidden layer to a small value, iteratively inserting an additional node if the desired tolerance is not fullfilled.

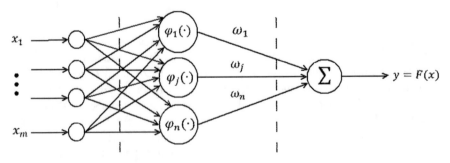

Fig. 22.1 The traditional Radial Basis Function Network

22.3   Simulation Settings and Results Discussion

The goal of our work is to asses the capability of RBF nets to fit the yield curve in situations of stress, likewise in case of extreme humps or when interest rates turn being negative. To such aim, we collected end–of–month data from Bloomberg sheets containing the bid and ask par rates for the Euro Swap–EUR003M Euribor, and the USDollar Swap–USD003M Curves, both observed on two different release dates, on December 30th 2004 and 2016, respectively. The average between bid and ask par rates was computed for each tenor and employed to derive the zero rates for each curve. We were therefore able to manage four curves whose name is provided in Table 22.1.

Our choice can be easily motivated: the credits crunch in 2007–2008 and the Eurozone sovereign debt crisis in 2009–2012 have changed the fixed income market, fostering the emergence of the s.c. multiple curve issue [18] and altering traditional connections between interest rates and zero coupon bond prices [17]; our rationale is then to consider curves in both pre–crisis and crisis times to check the different fitting ability of conventional interpolation techniques against RBF nets. Figure 22.2 shows the dynamics of the yield curves under examination.

At the first glance, the 3 Months Euribor Curves (both Eur003MOld and Eur003M) appear being a bit more tricky to fit than the 3 Months USD Swap curve: starting from the graphs in the left–hand side, in fact, the USD003MOld curve is sensitively smoother than the EUR003MOld; moving to the right–hand side of Fig. 22.2, the actual EUR003M profile shows singularities and slowdowns to negative values, while the USD003M is still quite flat, apart from a hump at short maturities.

Figures 22.3, 22.4, 22.5 and 22.6 show the graphical comparison among the interpolations obtained with the various methods.

Figures are self–explaining: the parametric methods work well, at the same level of the RBF net in interpolating the EUR003M old yield curve; the performance, however, is declining, at least for what is concerning the de Rezende–Ferreira approximation model, in the fitting of the USD003MOld; this is probably due to the known problems (already discussed in Sect. 22.2.2) of precision in the parameters estimation procedure. Things go worst when we turn to examine the estimation performed on actual data. In this case, the fitting to the observed yield curves is quite poor for all the examined parametric techniques; on the contrary the RBF net performances maintain stable. This evidence is also confirmed by analysing the main statistics of

Table 22.1   Yield curves employed in this work

ID	Extended name	Inception date	Length
EUR003MOld	3–Months Euribor	12/30/2004	22
EUR003M	3–Months Euribor	12/30/2016	22
USD003MOld	3–Months USDollar Swap	12/30/2004	22
USD003M	3–Months USDollar Swap	12/30/2016	22

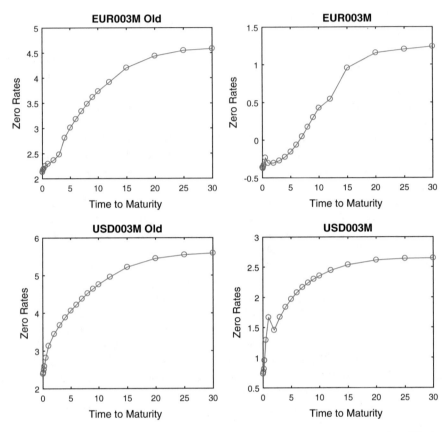

Fig. 22.2 The dynamics of the examined yield curves: time to maturity is on the x–axis, while the value of zero rates appears on y–axis. On top: the 3 Months Euribor Curve (release date Dec. 2004 on the left, release date Dec. 2016 on the right), on the bottom: the 3 Months USD Swap Curve (release date Dec. 2004 on the left, release date Dec. 2016 on the right)

the various methods, given in Table 22.2 for the parametric techniques, including parameters estimation, and in Table 22.3 for the RBF network.

The values in Tables 22.2 and 22.3 support the graphical evidence: quite surprisingly the de Rezende–Ferreira model is the worst, in terms of R^2 and RMSE, in three over four cases (namely, in approximating the EUR003M, USD003MOld and USD003M curves). The remaining parametric techniques (Nelson–Siegel and Svensson) maintain satisfying values of the R^2, but the related RMSE is sensitively higher than in the case of RBF net interpolation.

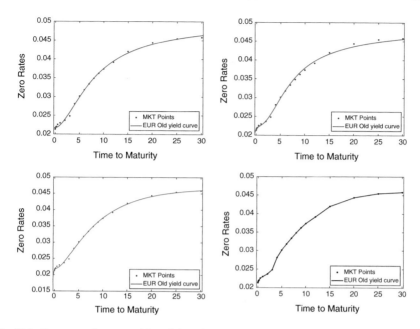

Fig. 22.3 From top to bottom and from left to right: interpolation of the EUR003M Old yield curve with the Nelson–Siegel (NS), Svensson (SV), de Rezende–Ferreira (dRF) models and with the RBF network

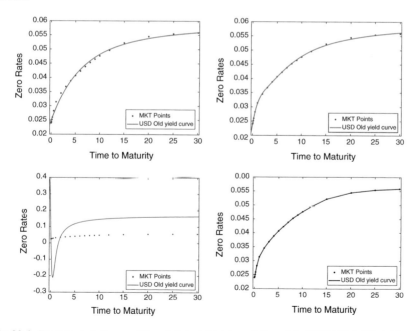

Fig. 22.4 From top to bottom and from left to right: interpolation of the USD003M Old yield curve with the NS, SV, dRF models and with the RBF network

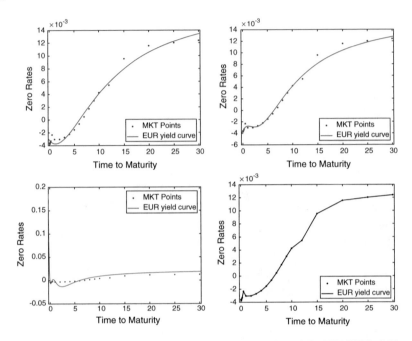

Fig. 22.5 From top to bottom and from left to right: interpolation of the EUR003M yield curve with the NS, SV, dRF models and with the RBF network

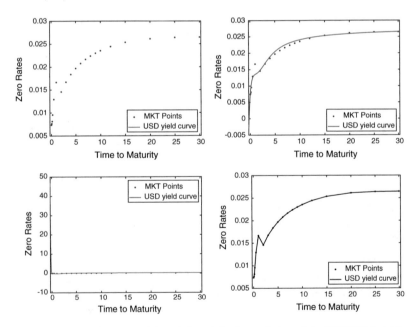

Fig. 22.6 From top to bottom and from left to right: interpolation of the USD003M yield curve with the with the NS, SV, dRF models and with the RBF network

Table 22.2 Estimated coefficients for the parametric techniques

Coeff.	Eur003M Swap Curve					
	NS_{2004}	SV_{2004}	dRF_{2004}	NS_{2016}	SV_{2016}	dRF_{2016}
β_0	0.051205	0.049531	0.023256	0.019413	0.017525	0.023256
β_1	−0.029522	−0.028132	−0.024378	−0.022607	−0.021132	−0.024378
β_2	−0.029323	0.00111	−17.834098	−0.029211	−0.006591	−17.83410
β_3	–	−0.05842	17.874810	–	−0.053531	17.87481
β_4	–	–	−0.157251	–	–	−0.157251
τ_1	2.471681	0.740510	0.230981	3.4671849	0.556850	0.230980
τ_2	–	1.687435	0.232039	–	2.3897270	0.232039
τ_3	–	–	1.048890	–	–	1.048890
R^2	0.998042	0.997884	0.998594	0.986832	0.992417	0.314388
$RMSE$	0.000423	0.000467	0.000408	0.000701	0.000564	0.005736
Coeff.	USD003M Swap Curve					
	NS_{2004}	SV_{2004}	dRF_{2004}	NS_{2016}	SV_{2016}	dRF_{2016}
β_0	0.060161	0.0606401	0.023256	0.028773	0.027744	0.023256
β_1	−0.035384	−0.0368601	−0.024378	−0.020578	−0.020762	−0.024378
β_2	0.003668	−0.0145134	−17.83410	0.002111	0.105636	−17.834098
β_3	–	−0.0512392	17.87481	–	−0.122490	17.874810
β_4	–	–	−0.157251	–	–	−0.157251
τ_1	4.034936	0.5146503	0.230981	2.641094	0.583050	0.230981
τ_2	–	2.0924553	0.232039		0.704717	0.232039
τ_3	–	–	1.048890	–	–	1.048890
R^2	0.994607	0.9996403	−93.09985	0.965927	0.931953	−397.747844
$RMSE$	0.000905	0.0002480	0.1355922	0.001413	0.001091	0.173339

Table 22.3 RBF nets settings for the observed yield curves

	Eur003MOld	Eur003M	USD003MOld	USD003M
Max Nr Neur.	100	100	100	100
RMSE	7.32293E-11	1.30206E-10	9.38584E-11	1.96893E-10

22.4   Conclusion

In this work we performed a comparison between parametric techniques and RBF networks to fit the yield curve under conditions of stress. The issue is of actual interest in this challenging time of high volatility and negative interest rates, because the yield curve is an important tool in finance, especially for assets pricing, financial risk management and portfolio allocation. We therefore investigated the capability of various methods to interpolate the yield curve under such extreme conditions of instability; to such aim, we considered the Euro Swap Euribor (EUR003M), and the

USDollar Swap (USD003M) curves, on two different release dates (December 30th 2004 and 2016), corresponding to very different market situations, and we examined the various ability of the above–cited methods in fitting them. The results confirm that RBF nets can reach very satisfying results to manage anomalies such as extreme humps or negative interest rates. Besides, our opinion is that the results pave the way to a research trail more focused on the use of Machine Learning methods to provide a integrate model of in–sample fitting and forecasting that actually is under the study of our research group.

References

1. Ait-Sahalia, Y.: Nonparametric pricing of interest rate derivative securities. Econometrica **64**, 527–560 (1996)
2. Annaert, J., Claes, A.G., De Ceuster, M.J.K., Zhang, H.: Estimating the spot rate curve using the Nelson–Siegel model. a ridge regression approach. Int. Rev. Econ. Financ. 27, 482–496 (2013)
3. Barunik, J., Malinska, B.: Forecasting the term structure of crude oil futures prices with neural networks. Appl. Energy **164**, 366–379 (2016)
4. Bonnans, J.F., Gilbert, J.Ch., Lemarchal, C., Sagastizbal, C.A.: Numerical Optimization, Theoretical and Numerical Aspects. 2nd edn. Springer (2006)
5. Bose, S.K., Sethuraman, J., Raipet, S.: Forecasting the term structure of interest rates using neural networks. In Kamruzzaman, J., Begg, R.K., Sarker, R.K. (eds.) Artificial Neural Networks in Finance and Manufacturing, chap. 8, pp. 124–138. IGI Global (2006)
6. Broomhead, D.S., Lowe, D.: Multivariate function interpolation and adaptive networks. Complex Syst. **2**, 321–355 (1988)
7. Cortes, C., Vapnik, V.: Support-vector networks. Mach. Learn. **20**(3), 273–297 (1995)
8. Cottrell, M., de Bodt, E., Grégoire, P.: Interest rates structure dynamics: a non parametric approach. In: Refenes, A.N., Burgess, A.N., Moody, J.E. (eds.) Decision Technologies for Computational Finance: Proceedings of the Fifth International Conference Computational Finance, pp. 259–266. Springer US, Boston (1998)
9. De Pooter, M.: Examining the Nelson–Siegel class of term structure models. Technical Report 043/4, Tinbergen Institute (2007)
10. de Rezende, R.B., Ferreira, M.S.: Modeling and forecasting the yield curve by an extended nelson-siegel class of models: a quantile autoregression approach. J. Forecast. **32**, 111–123 (2013)
11. Diebold, F., Li, C.: Forecasting the term structure of government bond yields. J. Econ. **130**(2), 337–364 (2006)
12. Diebold, F., Rudebusch, G.D.: The Dynamic Nelson–Siegel Approach to Yield Curve Modeling and Forecasting. Princeton University Press (2013)
13. Fama, E., Bliss, R.R.: The information in long maturity forward rates. Am. Econ. Rev. **77**(4), 680–692 (1987)
14. Gilli, M., Grosse, S., Schumann, E.: Calibrating the Nelson-Siegel-Svensson model. Technical report, COMISEF (2010)
15. Gogas, P., Papadimitriou, T., Matthaiou, M., Chrysanthidou, E.: Yield curve and recession forecasting in a machine learning framework. Comput. Econ. **45**, 635–645 (2015)
16. Golbabai, A., Ahmadian, D., Milev, M.: Radial basis functions with application to finance: american put option under jump diffusion. Math. Comput. Model. **55**, 1354–1362 (2012)
17. Grbac, Z., Runggaldier, W.J.: Interest Rate Modeling: Post-Crisis Challenges and Approaches. Springer (2014)

18. Henrard, M.: Interest Rate Modelling in the Multi–curve Framework. Palgrave McMillan (2014)
19. Joseph, A., Larrain, M., Singh, E.: Predictive ability of the interest rate spread using neural networks. Procedia Comput. Sci. **6**, 207–212 (2011)
20. Larsson, E., Gomes, S.M., Heryudono, A., Safdari-Vaighani, A.: Radial basis function methods in computational finance. In: Proceedings of the 13th International Conference on Computational and Mathematical Methods in Science and Engineering, CMMSE 2013 (2013)
21. Nelson, C., Siegel, A.F.: Parsimonious modeling of yield curves. J. Bus. **60**, 473–489 (1987)
22. Rosadi, D., Nugraha, Y.A., Dewi, R.K.: Forecasting the Indonesian government securities yield curve using neural networks and vector autoregressive model. Technical report, Department of Mathematics, Gadjah Mada University, Indonesia (2011)
23. Sambasivan, R., Das, S.: A statistical machine learning approach to yield curve forecasting. Technical report, Chennai Mathematical Institute (2017)
24. Svensson, L.E.O.: Estimating the term structure of interest rates for monetary policy analysis. Scand. J. Econ. **98**, 163–183 (1996)
25. Tappinen, J.: Interest rate forecasting with neural networks. Technical Report 170, Government Institute for Economic Research (1998)
26. Vasicek, O.: An equilibrium characterization of the term structure. J. Financ. Econ. **5**, 177–188 (1977)

Chapter 23
New Challenges in Pension Industry: Proposals of Personal Pension Products

Valeria D'Amato, Emilia Di Lorenzo and Marilena Sibillo

Abstract Within the current post-crisis economic environment, characterized by low growth and low interest rates, retirement and long-term saving represent a crucial challenge. Furthermore, the expansion of life expectancies modifies the demand of pension products and insurers and pension providers have to guarantee the sustainability and competitiveness of their products, in spite of the economic stagnation. Within the context of the personal pension products, in the paper we propose a new contract with profit participation, which consists in a deferred life annuity with variable benefits changing according with two dynamic financial elements: the periodic financial result of the invested fund year by year and the first order financial technical base checked at the beginning of predefined intervals all along the contract life. A numerical implementation explains the forecasted trend of the inflows and outflows connected to the contract under financial and demographic stochastic assumptions.

Keywords Pension · Variable annuity · Participating profit

23.1 Introduction

An ever-growing interest in the development of funded private pension plans can be observed in every advanced economy, where private pension systems are regarded as tools for economic development, as well as means to fuel social security systems

---

V. D'Amato · M. Sibillo (✉)
Department of Economics and Statistics, Campus Universitario,
University of Salerno, 84084 Fisciano, SA, Italy
e-mail: msibillo@unisa.it

V. D'Amato
e-mail: vdamato@unisa.it

E. Di Lorenzo
Department of Economic and Statistical Sciences, via Cintia, University
of Naples Federico II, 80126 Naples, Italy
e-mail: diloremi@unina.it

© Springer International Publishing AG, part of Springer Nature 2019
A. Esposito et al. (eds.), *Neural Advances in Processing Nonlinear
Dynamic Signals*, Smart Innovation, Systems and Technologies 102,
https://doi.org/10.1007/978-3-319-95098-3_23

(cf. [12]). This is because funded private pension plans can easily represent a funding source for enterprises and long-term investments (cf. [2]). As highlighted in [2], then, pension funds can be seen as meaningful tools to the aim of funding productive enterprises. This can be achieved, for example, by steering pension funds investments toward listed companies, thereby triggering positive effects not only with regards to the individual enterprises (both in terms of funding and of control on strategic decisions), but also the overall dynamics and functioning of the market.

It is no surprise, therefore, that pension funds in Italy, among other investors (including private ones), have recently been granted exemption from income taxes, should they operate medium-and long-term investments in the real economy (cf. [15]).

Private pension plans' strategic role, therefore, appears to have a two-fold value: these represent, in fact, a driving force for enterprises activities, on the one hand, and an effective means to safeguard households investments, on the other. The centrality of these aspects has systematically been acknowledged by European authorities; among the various initiatives on the matter, the European Insurance and Occupational Pension Authority (EIOPA) promoted a consultation, taking place in October 2016, dealing with personal pension products (PPP), aiming at triggering an integration process among member states (cf. [8]). This consultation, which academics, individuals, specialists and consumers associations took place in, attempted to overcome PPPs normative and contractual fragmentation, as these products constitute the core structure of the third pension pillar. Additionally, it endeavored to lay down the foundation of a discussion platform concerned with Europe's pension future.

Public pension systems' fragility, coupled with elderly populations growing needs in member States, as well as the prolonged episode of low interest rates and low economic growth (cf. [9]), let EIOPA to shelve the Pan-European Personal Pension product (PEPP), namely a plan for individual pension products, standardized and homogenized at a European level. These products ought to entail common funds, policies and other financial products, which can be defined in relation to consumers pension needs and their risk profile. One of the foreseen advantages protecting individual consumers is represented by a significant reduction of costs connected with the sales network; from this perspective, it can be argued that the ambitious aim of amalgamating retirement and long-term saving may well revivify capitals market, even thought the harmonization of fiscal aspects still represents the Achilles heel of the entire project.

The European Commission nonetheless highly prioritized the debate on individual pension products in 2017, aiming to strike a balance between consumer protection and a good uptake at EU level of the PEPP product, as illustrated as part of the Update by the European Commission in [11]. The Organization for Economic Co-Operation and Development (OECD) estimated private pension assets to have globally exceeded USD 38 trillion in 2015 (cf. [11]); furthermore, returns were positive in most countries. Yet, despite all this, the number of pension funds has sensibly decreased in several countries, probably due to competition reasons. An investigation led by the OECD in 20 countries, where assets invested by pension funds represent

65% of global pension assets (cf. [12]) revealed that there is no significant correlation between the number of pension funds and the real net rate of returns.

The interest of the market in retirement and annuity products in the recent years (see [13, 17]) is reflected in the current actuarial literature, where several products are proposed focusing on specific guarantee structures.

Among the various proposals, we recall the contractual forms with minimum guaranteed benefits (see, for example, [14]), the annuities with benefits linked to actual mortality experience (see [16]), as well as the annuities with guaranteed minimum benefits and participation in insurers' surpluses (cf. [10]). Denuit et al. [4] propose that annuitants bear the nondiversifiable mortality risk and successively (cf. [5]) they develop this idea considering longevity-contingent deferred life annuities. Bravo et al. [1] aim at sharing both longevity and investment risks.

In general the overall suggestions aims at designing products which include a risk transfer system between annuitants and provider (cf. [3]). Within this framework an interesting analysis is developed by Weal et al. (cf. [20]), who compare traditional indexed annuities with annuities where the payout rates are linked to differences between expected and actual mortality rates of the specific annuitants cohort.

The contribution we provide in this paper is framed in a life annuity market experiencing at present new vivacity and increasing tangible and potential demand. Starting from the consideration that the current financial context offers heavily low interest rates, the life annuity market, as most of the financial activities personal-saving oriented, is affected by the scanty desirability of its products. The guaranteed interest rate, known as the first order technical financial base, is fixed at the issue time and, due to the financial market present dynamics, has to be very low. Even in case of profit sharing schemes (cf. [6, 7]), when the benefits due in case of life are linked to the periodic financial result of the invested fund, the technical financial base applied in the contract constitutes the hard tie. The low level of the guaranteed interest rate is actually a strong constraint in particular if it is highly probable an increasing trend, as it seems currently possible. Even more meaningful this consideration is if you take into account the marked high length in average of the life annuity contracts.

In light of the above mentioned recalls, we introduce an innovative structure, against the proposals provided in the current literature, within the life annuity contractual scheme, consisting of a kind of dynamic profit participation. The point is to remove the hypotheses that the first order technical financial base is fixed at the issue time and to allow it to increase if the financial general conditions so allow. It consists in a system of periodic upwards adjustments of the first order financial technical base, structured considering the prudential point of view of the pension provider together with the insureds chance of benefiting by settled better market conditions. We propose a variable deferred life annuity, where the structure of the profit participation is based on the periodic financial result of the invested fund year by year as well as on periodic adjustments according to the spreads between the current trend of the first order financial base and the financial guarantees (planned at the issue time). This contractual scheme allows providers' sustainability and insureds' profitability.

In this paper, in Sect. 23.2 the new product is outlined and its mathematical structure is presented. In Sect. 23.3, after a quantitative representation of the financial and
demographic scenario, we investigates a case study, providing a numerical implementation in order to forecast the behavior of inflows and outflows related to the
contract.

23.2 The Contract. Profit Participation Annuities with Periodic Adjusted Guaranties

The new contract we are proposing is placed within the framework of the life insurance products with profit participation [7]. It consists in a variable deferred life
annuity issued on an individual aged x, with constant premiums paid at most all
along the deferment period; the benefits are due to the insured after the deferment
interval in case of life.

We suppose the benefits are variable according with two dynamic financial elements: they are linked to the periodic financial result of the invested fund year by
year and can be also adjusted at the beginning of predefined intervals according to
the trend the first order financial base is following with respect to that one guaranteed
to the insured at the issue time.

Indicating by d the deferment period, p the constant time interval between two
guarantee adjustments and T the number of periods p constituting the whole benefit payment length, the following financial equivalence describing the contractual
equilibrium holds:

$$P\left\{E\sum_{h=0}^{\tau}v(0,h)_h p_x\right\} = E\left\{\sum_{h=0}^{T-1}\tilde{B}_h\sum_{k=d+hp+1}^{d+(h+1)p}v(0,h)_h p_x\right\} \qquad (23.1)$$

where P is the level premium paid at the beginning of each year until the time τ,
$\tau < d$, and $\tilde{B}_h$ the annual variable installment paid at the end of each year ($h = d +
1, d + 2, \omega - x - 1$), both due in case of life of the insured; $v(0, h)$ is the stochastic
value in $t = 0$ of a monetary unit in t.

In the benefit settlement we insert an embedded option involving the insureds
participation in the fund annual financial profit, if, after paying the fund management
expenses, it is strictly positive and according to a predefined participation rate. The
profit rate assignable to the insured is that one arising from the difference between
the Cash Net Profit Rate (CNP) and the first order technical base, guaranteed inside
the contract to the insured, both referred to the same payment period of one year. The
rate CNP of the fund in which the premiums are invested is the profit rate calculated
year by year on the cash-flow representing the increase in the net assets from the
operations attributable to the contract-holders (cf. [7]).

The new characteristic we insert in the contract is added to the aim of relaxing the strong conditioning due to the setting of the first order technical financial base at the issue time. Especially when the interest rates are very low, the competitiveness of the life product can be strongly improved making the guaranteed interest rate flexible, able to match the interest rate market behavior. This aspect is particularly relevant if the interest rates are forecasted on an upward trend and in all the cases, as the life products mainly are, characterized by long durations. In light of these considerations, we assume that the insurer will be able to calibrate the profit participation of the insureds also taking into account the trend of the current first order technical financial base. We pose he can resort to these adjustments at the beginning of each of the T periods of p years each.

We indicate by ρ the participation rate and by i the initial first order annual technical base, fixed at the issue time. The benefit $\tilde{B}_h$ paid to the insured at the end of the h-th year, is given by:

$$\tilde{B}_h = B_h + [\rho(CNP_h - i_h + s_h)^+]\left[\sum_{j=0}^{\tau-1} P_j(1+i)^{h-j}\right] \mathbf{1}_{T \leq t < |K(x) \geq t}. \qquad (23.2)$$

In formula (23.2), s_h is the spread the insurer assesses at the beginning of each of the T periods of p years, suitably quantified according to the annual interest rates i_j, $j = d, d+p, d+2p, \ldots, d+(T-1)p$, the first order financial technical base detectable at the beginning of each period p. We suppose that s_h is constant during the period p. The indicator function takes the value 1 if the event at subscript happens, otherwise takes the value 0, $K(x)$ is the random curtate lifetime of an annuitant aged x at the issue time, $(CNP_h - i_h + s_h)^+$ is the maximum between $(CNPh - i_h + s_h)$ and 0.

23.3 Numerical Application and Forecasting Evidences

The goal of the section is to present the empirical outcomes of the model implementation.

In order to define the contract financial details, we need to consider two important issues in the risk management evaluation: on the one hand the demographic projection and on the other hand the financial forecasting.

23.3.1 The Demographic Scenario

About the stochastic mortality rates, we refer to a Poisson Lee Carter model, which works well in terms of goodness of fitting according to [18]. The Lee-Carter model in the Poisson setting is characterized by the following expressions:

$$D_{xt} \approx Poisson(E_{xt}\mu_{xt}) \tag{23.3}$$

where the force of mortality is assumed to have the log-bilinear form:

$$ln(\mu_{xt}) = \alpha_x + \beta_x k_t. \tag{23.4}$$

E_{xt} represents the exposures to the risk of death (in other words, the number of person years from which D_{xt} occurs), α_x means the main age effect, k_t maps out the mortality changes as function of time, β_x represents the impact of k_t on the population (cf. [18]). They are subject to the constraints:

$$\sum_t k_t = 0; \quad \sum_x \beta_x = 1.$$

The dataset is represented by Male US population collected by 1985–2007, related to an age range from 0 to 110. We assume the inception date in January 2017, the expiration date in January 2037.

Following [18], we fit the model by a likelihood methodology. The resulting parameter estimates are shown in Fig. 23.1. The diagnostics of residuals is illustrated by Fig. 23.2 and it looks realistically random, showing a good model performance.

Figure 23.3 is referred to the parameter forecasts and in particular to the stochastic time-varying parameter k_t, which depicts mortality changes over times (cf. [18]), projected till the expiration date, to the aim of getting the mortality rates.

23.3.2 The Financial Scenario

As regards the financial scenario, we describe either the future evolution of the technical base either the periodic financial result of the invested fund by means of two suitable stochastic models. Starting from the first one, in order to study the investment fund performance to the aim of forecasting the profit sharing behavior, we need to estimate the future evolution of the rate of return. Our example is based on the choice of a balanced investment, where the equity component is represented by 30% on US market and fixed-income securities are divided into 40% Treasury bills and 30% Treasury bonds. Different investment strategies would imply different stochastic description and obviously different financial outcomes. In this specific case we choose to implement a Vasicek model, described by the following equation:

$$dr_t = a(b - r_t)dt + \sigma dW_t \tag{23.5}$$

where r_t is the interest rate, a the speed of reversion, b the long term mean, σ the instantaneous volatility, W_t a Wiener process (cf. [19]).

The parameter estimation has been performed on the dataset composed by time series from 1928 to 2006 (http://www.stern.nyu.edu/_adamodar/New_Home_Page/data.html).

In Fig. 23.4 the forecasted trend is shown, fixing the issue time at the beginning of 2017.

Figure 23.5 provides a comparison between the active funds related to contract duration; the dashed one referred to the guaranteed rate at level of $i = 0.015$ and the black one to the forecasted rates. The bandwidth between the two cases highlights the different trends.

Passing to the financial technical base projection, we implement a CIR model, described by the following equation:

$$dr_t = a(b - r_t)dt + \sigma \sqrt{r_t} dW_t \qquad (23.6)$$

where the parameters has the same meaning as in formula (23.5).

In this case, the parameter estimation has been developed on the dataset of Treasury bonds downloaded by Federal Reserve (www.fedgov.org) collected from 1980 (January) to 2016 (December). More specifically Fig. 23.6 presents the outcomes, also in this case fixing the initial time in $t = 0$.

23.3.3 Experiment Results

We consider the case of a temporary deferred life annuity issued on a policyholder aged $x = 60$, the deferment period being equal to 5 years. The insurer promises a guaranteed rate (financial first order technical base) $i = 0.015$ and the guaranteed benefit payable at the end of each year in case of the insureds life is equal to 1. The contract, as described in Sect. 23.2, is based on the variability of the benefits according to two financial elements: the profit participation of the insured year by year and the improvement, if it happens, of the financial technical base.

We pose to proceed with the adjustment every 3 years starting from the end of the deferment, that is year 5. In our exemplification, we pose to fix the new financial technical base as the average of the interest rates referred to the three years of interest. The adjustment will be put into effect if this average is greater than the guaranteed one (1.5%).

Figure 23.6 presents the benefits outlook all along the contract duration. The benefits vary according to the profit participation mechanism, as in formula 23.2, and the guaranteed interest rate adjustment at periods 5, 8, 11, 14, 17.

The plot in Fig. 23.6 clearly puts in evidence the profitability of the proposed contract in particular in the insured's perspective. The policyholder can take advantage not only from the good behavior of the invested fund, but also from the increasing trend of the financial technical base. This virtuous mechanism happens by means of a level premium of 2.61, calculated applying formula (23.1) to the contract structure, paid at the beginning of each year during the deferment period.

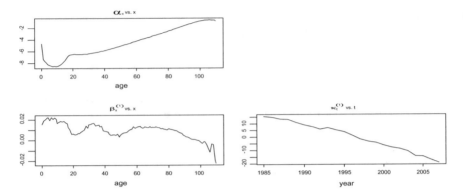

Fig. 23.1 Poisson Lee Carter model fitting on Male US population

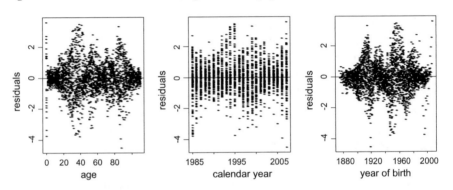

Fig. 23.2 Analysis of residuals on Poisson Lee Carter model fitted on Male US population

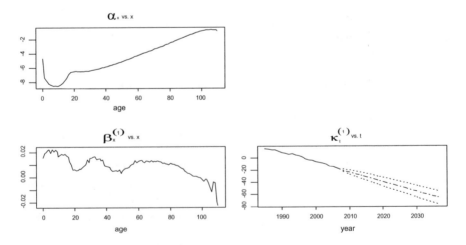

Fig. 23.3 Forecasting Poisson Lee Carter model on Male US population

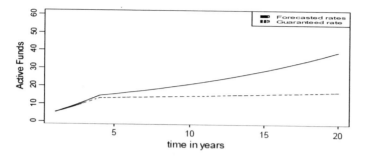

Fig. 23.4 Active portfolio funds, contract duration

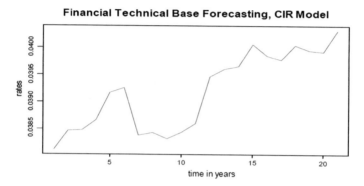

Fig. 23.5 Financial technical base forecasting, CIR model

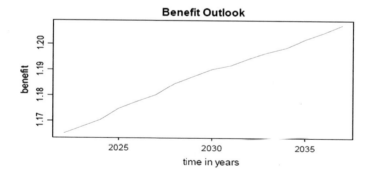

Fig. 23.6 Insured's inflow performance forecasting

23.4 Conclusions

In the paper we proposed a personal pension product, which consists in a variable deferred life annuity with profit participation and improvable guarantees.

The benefits are linked to the periodic financial result of the invested fund and are calculated year by year at the payment time. Moreover, at the beginning of predefined intervals, they also can vary according to the dynamics in time of the first order financial base, with respect to that one guaranteed to the insured at the issue time. In the contract the first order technical financial base set at the issue time constitutes the floor of the yield assigned to the insured. Such a contract involves embedded options with expiry dates at each benefits payment time for what concerns the profit participation and at the beginning of each predefined period for what concerns the adjustable guarantees.

We simulated the cash flows of the proposed Personal Pension Product (PPP), to show its performance under an illustrative scenario of the fund in which the premiums are invested, according to a specific investment policy.

Future research could explore the effectiveness of the calibration between the product competitiveness, due to the additional participating benefits, and the consumers' perception of the profitability, due, among others, to costs, returns and taxation, as these are filtered according to the personal risk aversion attitude and collective behavior in financial/insurance markets.

References

1. Bravo, J. M., El Mekkaoui de Freitas N.: Valuation of longevity-linked life annuities. Insur. Math. Econ. In press, Available online 28 September 2017 (2017). https://www.sciencedirect.com/science/article/pii/S016766871730450X
2. CONSOB: Il finanziamento dell'economia reale e il ruolo dei fondi pensione (Audizione del Direttore generale G. Caputi) 16 Aprile 2014 (2014). http://www.consob.it/documents/46180/46181/Audizione_caputi_20140416.pdf/69243d76-c4c3-4716-a20b-9debbcbf667f
3. Denuit, M., Haberman, S., Olivieri, A., Pitacco, E.: Modelling Longevity Dynamics for Pensions and Annuity Business. Oxford University Press (2009)
4. Denuit, M., Haberman, S., Renshaw, A.: Longevity-indexed life annuities. NAAJ 15(1), 97111 (2011)
5. Denuit, M., Haberman, S., Renshaw. A: Longevity-contingent deferred life annuities. J. Pension Econ. Financ. 14(03), 1–13 (2015)
6. D'Amato, V., Di Lorenzo, E., Orlando, A., Sibillo, M.: Life annuity portfolios: risk-adjusted valuations and suggestions on the product attractiveness. In: Skiadas H.C. (ed) Proceedings of Stochastic Modeling Techniques and Data Analysis International Conference with Demographic Workshop, 1–4 June 2016, Valletta (Malta), pp. 123–131 (2016). https://onedrive.live.com/redir?resid=CB6060F40BD0FF92!348&authkey=!AAQ6_VpgAZuZB4s&ithint=file%2cpdf
7. DAmato, V., Di Lorenzo, E., Orlando, A., Sibillo, M., Tizzano, R.: Profit-sharing and personal pension products: a proposal. In: Pensions: Global Issues, Perspectives and Challenges, pp. 97–111. Nova Science Publishers, New York (2017)

8. EIOPA-CP-16/001: Consultation Paper on EIOPAs Advice on the Development of an EU Single Market for Personal Pension Products (PPP) (2016). https://eiopa.europa.eu/Pages/Consultations/EIOPA-CP-16-001-Consultation-Paper-on-EIOPA%E2%80%99s-advice-on-the-development-of-an-EU-Single-Market-for-personal-pension-product.aspx
9. International Monetary Fund: Global Financial Stability Report, April 2017 (2017). http://www.imf.org/~/media/Files/Publications/GFSR/2017/April/ch02-4thproofs-040517-v2.ashx?la=en
10. Maurer, R., Rogalla, R., Siegelin, I.: Participating payout life annuities. Lessons from Germany. ASTIN Bull. **43**(02), 159187 (2013)
11. Occupational Pensions Stakeholder Group (OPSG): Fifth meeting: Summary of Conclusions, 28 February 2017-04-18 (2017). https://eiopa.europa.eu/Publications/Meetings/EIOPA-OPSG-17-05_Summary_of_Conclusions28.02.17_mtg.pdf#search=PEPP%20consultation%20April%202017
12. OECD: Pension Markets in Focus (2016). http://www.oecd.org/daf/fin/private-pensions/Pension-Markets-in-Focus-2016.pdf
13. Olivieri, A., Pitacco, E.: Introduction to Insurance Mathematics. Technical and Financial Features of Risk Transfers. Springer (2011)
14. Penga, J., Leungb, K.S., Kwokc, Y.K.: Pricing guaranteed minimum withdrawal benefits under stochastic interest rates. Quant. Financ. **12**(6), 933–941 (2012)
15. Piazza M.: Premiate le Casse che investono in aziende. Il Sole 24ore, 30 Ottobre 2016 (2016). http://www.ilsole24ore.com/art/notizie/2016-10-30/premiate-casse-che-investono-aziende-104348.shtml?uuid=ADt37ulB&refresh_ce=1
16. Richter, A., Weber, F.: Mortality-indexed annuities. Managing longevity risk via product design. NAJ **4**(1), 212236 (2011)
17. Rocha, R., Vittas, D., Rudolph, H.P.: Annuities and other retirement products. Designing the payout phase The World Bank, Washington D.C. (2011). http://www.iopsweb.org/researchandworkingpapers/48266689.pdf
18. Renshaw, A.E., Haberman, S.: Lee-Carter mortality forecasting with age-specific enhancement. Insur. Math. Econ. **33**(2), 255–272 (2003)
19. Vasicek, O.: An equilibrium characterization of the term structure. J. Financ. Econ. **5**, 177–188 (1977)
20. Weal, M., van de Ven, J.: Variable annuities and aggregate mortality risk. Natl. Inst. Econ. Rev. (2016). http://journals.sagepub.com/doi/abs/10.1177/002795011623700117

Chapter 24
A PSO-Based Framework for Nonsmooth Portfolio Selection Problems

Marco Corazza, Giacomo di Tollo, Giovanni Fasano and Raffaele Pesenti

Abstract We propose a Particle Swarm Optimization (PSO) based scheme for the solution of a mixed-integer nonsmooth portfolio selection problem. To this end, we first reformulate the portfolio selection problem as an unconstrained optimization problem by adopting an exact penalty method. Then, we use PSO to manage both the optimization of the objective function and the minimization of all the constraints violations. In this context we introduce and test a novel approach that adaptively updates the penalty parameters. Also, we introduce a technique for the refinement of the solutions provided by the PSO to cope with the mixed-integer framework.

Keywords Portfolio selection problems · Particle swarm optimization
Mixed-integer nonlinear programming

24.1 Introduction

Particle Swarm Optimization (PSO) is an iterative bio-inspired population-based metaheuristic for the solution of unconstrained global optimization problems. Recently, this metaheuristic has been applied to the unconstrained reformulation of a realistic portfolio selection problem [3]. This portfolio selection problem is NP-

M. Corazza (✉)
Department of Economics, Ca' Foscari University of Venice,
Cannaregio 873, 30121 Venezia, Italy
e-mail: corazza@unive.it

G. di Tollo · G. Fasano · R. Pesenti
Department of Management, Ca' Foscari University of Venice,
Cannaregio 873, 30121 Venezia, Italy
e-mail: giacomo.ditollo@unive.it

G. Fasano
e-mail: fasano@unive.it

R. Pesenti
e-mail: pesenti@unive.it

© Springer International Publishing AG, part of Springer Nature 2019
A. Esposito et al. (eds.), *Neural Advances in Processing Nonlinear
Dynamic Signals*, Smart Innovation, Systems and Technologies 102,
https://doi.org/10.1007/978-3-319-95098-3_24

265

hard, nonconvex, nondifferentiable and mixed-integer. Since exact methods could be extremely time-consuming for its solution, a PSO approach was proposed for practical purposes. This problem was initially reformulated as an unconstrained optimization problem by adopting an exact penalty method where all the penalty parameters were constant. Then, PSO was used to both optimize the objective function and minimize all the constraints violations.

On this guideline, in this paper we propose a novel approach that adaptively updates the penalty parameters. We also introduce a technique for the refinement of the solution provided by the PSO to cope with the mixed-integer framework. Results from numerical experiences show that this solution approach seems preferable, with respect to a more standard one from the literature. In particular, the latter conclusion holds for costly problems, where a reduced number of PSO iterations is allowed and the refinement technique proved to have a terrific impact.

The remainder of this paper is organized as follows. In the next section, we introduce the portfolio selection problem we deal with. In Sect. 24.3, we recall the basics of standard PSO. In Sect. 24.4, we first reformulate the portfolio selection problem in terms of an unconstrained nonsmooth optimization problem, then we present our PSO approach. In Sect. 24.5, we provide some computational results coming from a series of applications to the Italian stock market. Finally, in Sect. 24.6 we draw some conclusions.

24.2 The Portfolio Selection Problem

Making an effective portfolio selection in real stock markets is not an easy task, for the following reasons. It is necessary to assess the risk by measures that satisfy appropriate theoretical properties, are able to deal with the non-normal return distributions characterizing real stock markets and are parameterized with respect to the investor's risk attitude. In addition, it is necessary to take into account several practices and rules of the portfolio management industry.

To deal with the above issues, we consider the portfolio selection problem proposed by Corazza et al. in [3]. In this problem, a coherent risk measure based on the combination of lower and upper moments of different orders of the portfolio return distribution is considered [2]. This measure both manages non-Gaussian distributions of asset returns, and takes into account the risk contained in the left tail of the distributions and the chance contained in the right one; finally, it is parameterized to model different investors' risk attitudes. The problem in [3] imposes bounds on the minimum and the maximum number of stocks to trade, and on the minimum and the maximum capital percentage to invest in each asset. This allows to easily model some of the most common professional practices and rules. The overall portfolio selection problem is formulated as follows:

$$\min_{x,z} \rho_{a,p}(r) := a\|(r - E[r])^+\|_1 + (1 - a)\|(r - E[r])^-\|_p - E[r] \qquad (24.1a)$$

$$\text{s.t. } E[r] \geq r_e \qquad (24.1b)$$

$$\sum_{i=1}^{N} x_i = 1 \qquad (24.1c)$$

$$K_d \leq \sum_{i=1}^{N} z_i \leq K_u \qquad (24.1d)$$

$$z_i d \leq x_i \leq z_i u, \qquad i = 1, \ldots, N \qquad (24.1e)$$

$$z_i \in \{0, 1\}, \qquad i = 1, \ldots, N \qquad (24.1f)$$

where: $x_i \in \mathbb{R}$, for $i = 1, \ldots, N$, is the unknown percentage of capital to invest in the i-th asset of the portfolio; N is the number of assets; $z_i \in \{0, 1\}$, for $i = 1, \ldots, N$, is the binary variable such that $z_i = 1$ if the i-th asset is included in the portfolio, $z_i = 0$ otherwise; $r = \sum_{i=1}^{N} x_i r_i$ is the random variable indicating the portfolio percentage rate of return, in which r_i, for $i = 1, \ldots, N$, is a random parameter indicating the percentage rate of return of the i-th asset; $\rho_{a,p}(r)$ is a coherent risk measure associated to r, being $a \in [0, 1]$ and $p \in [1, +\infty)$ parameters used to model different investor's risk attitudes; $E[\cdot]$ indicates the expected value of its argument; $y^- := \max\{0, -y\}$ and $y^+ := (-y)^-$; r_e is the minimum expected percentage rate of return of the portfolio desired by the investor; K_d and K_u are the minimum and the maximum numbers of stocks to trade, respectively; d and u are the minimum and the maximum percentage of capital to invest in each asset, respectively.

Note that the above nonconvex, nondifferentiable and mixed-integer portfolio selection problem (24.1a)–(24.1f) can be proven to be NP-hard.

24.3   Basics on Standard PSO

PSO is a bio-inspired methodology for the solution of global optimization problems [4], which iteratively tries to improve *swarm* candidate solutions, the so-called *particles*. Several PSO variants are proposed in the literature, both for unconstrained and constrained [1, 7] problems. Let P be the size of the swarm and $f : \mathbb{R}^n \to \mathbb{R}$ be the function to minimize such that for each $\bar{y} \in \mathbb{R}^n$ the corresponding *level set* of f

$$\mathcal{L}_f(\bar{y}) := \{y \in \mathbb{R}^n : f(y) \leq f(\bar{y})\}$$

is compact. For each particle j of the swarm, the basic PSO iteration $k \geq 0$ yields

$$\xi_j^{k+1} = \xi_j^k + v_j^{k+1}, \qquad j = 1, \ldots, P,$$

where $\xi_j^k \in \mathbb{R}^n$ is the current *position* of the j-th particle (i.e., the j-th candidate solution in the language of optimization), while $v_j^{k+1} \in \mathbb{R}^n$ is the current *velocity* of particle j (i.e., the *search direction*). Thus, ξ_j^{k+1} will be the new position of particle j in the next iteration.

Unlike the standard gradient based methods, the vector v_j^{k+1} is not necessarily a descent direction for function $f(y)$ at ξ_j^k. This suggests that at each step k the j-th particle might not provide an iterate which improves the objective function value. Indeed, the direction v_j^{k+1} is a cone combination of the vector v_j^k, which represents the *inertia* of particle j to modify its trajectory, with other two vectors. The latter two vectors affect the trajectory of the j-th particle exploiting the best solutions so far found by particle j and by the whole swarm, respectively. In particular, we have

$$v_j^{k+1} = v_j^k + \alpha_j^k \otimes (p_j^k - \xi_j^k) + \beta_j^k \otimes (p_g^k - \xi_j^k), \qquad k \geq 0, \qquad (24.2)$$

where $\alpha_j^k, \beta_j^k \in \mathbb{R}^n$ are positive vectors, the symbol '$\otimes$' indicates the entry-by-entry product between vectors, and the vector p_j^k, respectively p_g^k, is the best solution so far found by particle j, respectively by the swarm, i.e.

$$p_j^k \in \arg\min_{0 \leq h \leq k} \left\{ f(\xi_j^h) \right\}, \quad j = 1, \dots, P \quad \text{and} \quad p_g^k \in \arg\min_{\substack{0 \leq h \leq k \\ j=1,\dots,P}} \left\{ f(\xi_j^h) \right\}.$$

In (24.2), the parameters α_j^k (*cognitive parameter*) and β_j^k (*social parameter*), respectively consider the contribution to v_j^{k+1} from the history of the j-th particle (i.e. $p_j^k - \xi_j^k$), and from the history of the entire swarm (i.e. $p_g^k - \xi_j^k$). Their general expressions in the literature are respectively

$$\alpha_j^k = c_j^k r_1^k, \quad r_1^k \in U[0,1]^n, \quad \text{and} \quad \beta_j^k = c_g^k r_2^k, \quad r_2^k \in U[0,1], \quad k = 0, 1, \dots,$$

being $U[0,1]$ the uniform distribution with n entries between 0 and 1, and typically $c_j^k, c_g^k \in (0, 2.5]$. In this paper, as proposed in [5], we consider a slightly more general reformulation of PSO iteration for particle $j \in \{1, \dots, P\}$, where at any step $k \geq 0$:

$$\begin{cases} v_j^{k+1} = \chi^k \left[w^k v_j^k + \alpha_j^k \otimes (p_j^k - \xi_j^k) + \beta_j^k \otimes (p_g^k - \xi_j^k) \right], \\ \xi_j^{k+1} = \xi_j^k + v_j^{k+1}, \end{cases}$$

being χ^k the so called *constriction coefficient*, and w^k the *inertia coefficient*.

24.4 A Nonsmooth Portfolio Reformulation

Given problem (24.1), we consider the following auxiliary problem:

$$\min_{x \in \mathbb{R}^N, \, z \in \mathbb{R}^N} \quad P(x, z; \varepsilon), \tag{24.3}$$

where

$$
\begin{aligned}
P(x, z; \varepsilon) := \ & \rho_{a,p}(r) + \frac{1}{\varepsilon_0} \Bigg[\varepsilon_1 \max \Bigg\{ 0, r_e - \sum_{i=1}^{N} \hat{r}_i x_i \Bigg\} + \varepsilon_2 \left| \sum_{i=1}^{N} x_i - 1 \right| \\
& + \varepsilon_3 \max \Bigg\{ 0, K_d - \sum_{i=1}^{N} z_i \Bigg\} + \varepsilon_4 \max \Bigg\{ 0, \sum_{i=1}^{N} z_i - K_u \Bigg\} \\
& + \varepsilon_5 \sum_{i=1}^{N} \max \{ 0, z_i d - x_i \} + \varepsilon_6 \sum_{i=1}^{N} \max \{ 0, x_i - z_i u \} \\
& + \varepsilon_7 \sum_{i=1}^{N} |z_i (1 - z_i)| \Bigg]
\end{aligned}
$$

$\hat{r}_i = E[r_i]$ and $\varepsilon = (\varepsilon_0, \varepsilon_1, \ldots, \varepsilon_7)^T > 0$ is the vector of the penalty parameters.

Problem (24.3) can be seen as an unconstrained reformulation of problem (24.1). The two problems present to large extent equivalent solutions, given an opportune choice of the vector ε (see [3]). In particular, note that the terms $z_i(1 - z_i)$ allow $P(x, z; \varepsilon)$ to be continuous and represent a reformulation of the binary constraints (24.1f). Indeed, the condition $z_i \in \{0, 1\}$ can be expressed as $z_i(1 - z_i) = 0$, for $i = 1, \ldots, N$.

In the next subsections, we first use PSO to approximately solve the problem (24.3), adaptively updating the values of the penalty vector ε. Then, we use a procedure to refine the obtained approximate solution.

24.4.1 Penalty Vector Settings

In this subsection we describe our approach that adaptively updates the penalty parameter vector ε during the solution of the equivalent portfolio problem (24.3). Within this approach, vector ε is possibly updated during some iterations of PSO.

Hereinafter, we use the symbol ε^k to indicate the eight entries of vector ε at iteration k. In addition, since the problem (24.3) has $2N$ unknowns (namely the vectors x and z), we represent the best position of PSO particle p_g^k as

$$ p_g^k = \begin{pmatrix} x_g^k \\ z_g^k \end{pmatrix}. $$

For $k = 0$ the initial parameters vector ε^0 is set as

$$\varepsilon^0 = \begin{pmatrix} 10^{-4} & 1 & 1 & 1 & 1 & 1 & 1 & 1 \end{pmatrix}^T.$$

The values of the entries ε_i^0, $i = 1, \ldots, 7$ are chosen to initially impose an equal penalization for all constraints violations. Differently, the value of ε_0^0 is chosen much smaller than all the constraints violations, in order to initially privilege feasible solutions over optimal ones.

For $k \geq 1$, we update vector ε^k according with the following rules. We update ε_0^k by checking for a possible decrease of the value of $\rho_{a,p}(r_g^k)$, where $r_g^k = \sum_{i=1}^{N} (x_g^k)_i \hat{r}_i$. We update ε_i^k, $i = 1, \ldots, 7$, by checking for the violation of the constraints:

$$v_1(x_g^k, z_g^k) := \max \left\{ 0, r_e - \sum_{i=1}^{N} \hat{r}_i (x_g^k)_i \right\}$$

$$v_2(x_g^k, z_g^k) := |\mathbf{1}^T x_g^k - 1|$$

$$v_3(x_g^k, z_g^k) := \max\{0, K_d - \mathbf{1}^T z_g^k\}$$

$$v_4(x_g^k, z_g^k) := \max\{0, \mathbf{1}^T z_g^k - K_u\}$$

$$v_5(x_g^k, z_g^k) := \sum_{i=1}^{N} \max \left\{ 0, (z_g^k)_i d - (x_g^k)_i \right\}$$

$$v_6(x_g^k, z_g^k) := \sum_{i=1}^{N} \max \left\{ 0, (x_g^k)_i - (z_g^k)_i u \right\}$$

$$v_7(x_g^k, z_g^k) := \sum_{i=1}^{N} \left| (z_g^k)_i \left(1 - (z_g^k)_i \right) \right|.$$

Specifically, we adopt the following practical strategy:

- every 20 iterations of PSO we update the entry ε_0^{k+1} of ε^{k+1} as follows:

$$\varepsilon_0^{k+1} = \begin{cases} \min\{3\varepsilon_0^k, 1\} & \text{if } \rho_{a,p}(r_g^k) \geq \rho_{a,p}(r_g^{k-1}) \\ \max\{0.6\varepsilon_0^k, 10^{-15}\} & \text{if } \rho_{a,p}(r_g^k) < 0.90 \cdot \rho_{a,p}(r_g^{k-1}) \\ \varepsilon_0^k & \text{otherwise;} \end{cases} \quad (24.4)$$

- every 40 iterations of PSO we update the entries ε_i^{k+1}, $i = 1, \ldots, 7$, of ε^{k+1} as follows:

$$\varepsilon_i^{k+1} = \begin{cases} \min\{2\varepsilon_i^k, 10^4\} & \text{if } v_i(x_g^k, z_g^k) > 0.95 \cdot v_i(x_g^{k-1}, z_g^{k-1}) \\ \max\{\frac{1}{2}\varepsilon_i^k, 10^{-4}\} & \text{if } v_i(x_g^k, z_g^k) < 0.90 \cdot v_i(x_g^{k-1}, z_g^{k-1}) \\ \varepsilon_i^k & \text{otherwise.} \end{cases} \quad (24.5)$$

The choice of the coefficients in (24.4) and (24.5) is motivated by efficiency reasons, and is obtained after a coarse initial tuning over a reference portfolio selection instances.

Roughly speaking, relation (24.4) imposes that when the risk functional $\rho_{a,p}(r_g^k)$ increases, ε_0^{k+1} must increase, too. This fact in turn implies that constraints' violations are proportionally less penalized. With a similar reasoning, when $\rho_{a,p}(r_g^k)$ decreases then also ε_0^{k+1} is forced to decrease, in order to improve feasibility. As regards (24.5), if the i-th violation $v_i(x_g^k, z_g^k)$ significantly increases with respect to $v_i(x_g^{k-1}, z_g^{k-1})$ (i.e. we are possibly pursuing optimality while worsening feasibility), then the corresponding penalty parameter ε_i^{k+1} is increased accordingly. Conversely, with an opposite rationale, in case we observe a relevant improvement of feasibility (i.e. $v_i(x_g^k, z_g^k) \ll v_i(x_g^{k-1}, z_g^{k-1})$), then the parameter ε_i^{k+1} is decreased.

24.4.2 Refinement of the PSO Solution

In this subsection we present the last step of our solution approach. It is a refinement technique of the solution of the optimization problem (24.3) provided by PSO.

With reference to the constraints (24.1e) and (24.1f), the solution given by PSO generally shows slight infeasibility. For instance, the value $\hat{z}_i$ associated to the i-th stock included in the portfolio may not be 1 but very close to 1. Such small approximations do not invalidate the solution $\hat{x}$, but they risk to make it unusable for practical purposes. For this reason, we refine the solution $(\hat{x}, \hat{z})$ of problem (24.3) provided by PSO, by applying the following "feasibilization" procedure:

1. Determine the refined $\tilde{z}_i$ as:
$$\tilde{z}_i = \begin{cases} 0 & \text{if } \hat{x}_i \in (-\infty, d) \cup (u, +\infty) \\ 1 & \text{otherwise} \end{cases}$$

2. Determine the refined $\tilde{x}_i$ as:
$$\tilde{x}_i = \frac{\hat{x}_i \tilde{z}_i}{\sum_{i=1}^{N} \hat{x}_i \tilde{z}_i}.$$

This refinement procedure permits $\tilde{x}_i$ and $\tilde{z}_i$ to satisfy, not only (24.1e) and (24.1f), but also the constraint (24.1c). On the contrary, it does not guarantee the satisfaction of constraints (24.1b) and (24.1d). However, our experience, coming from several applications, suggests that the latter constraints are generally satisfied. Note also that, generally, the refined solution $(\tilde{x}, \tilde{z})$ is characterized by a fitness value which is significantly lower than the one characterizing the solution obtained by PSO before the refinement.

24.5 Numerical Experiences

In this section we compare the solution of the optimization problem (24.3), provided by the standard PSO, with the solution of the same problem provided by our PSO based framework. The purpose of this comparison is twofold: first, to analyze the differences (if any) between the behaviours over iterations of the fitness function $P(\cdot, \cdot; \cdot)$ and of the risk measure $\rho_{a,p}(\cdot)$; then, to investigate the main characteristics of the selected portfolios.

In the numerical experiences, in accordance with [2, 3], we approximate the expected values that appear in the objective function (24.1a) of our portfolio selection problem (24.1), with the associated sample means over T periods:

$$E[r_i] \approx \frac{1}{T} \sum_{t=1}^{T} r_{i,t}.$$

We consider the forty assets which compose the Italian stock index FTSE MIB. In particular, we use the time series of the daily closing returns from November 14, 2016 to April 28, 2017. Furthermore, we use the following parameter settings for problem (24.3): $a = 0.5$, $p = 2$, $r_e \in \{0.02500\%, 0.04375\%, 0.06250\%, 0.08125\%, 0.10000\%\}$, $d = 0$, $u = 1$ and $(K_d, K_u) \in \{(11, 30), (16, 25), (19, 21)\}$, for a total of 15 different settings. Note that the values of r_e ensure that a feasible solution exists for the problem (24.3), and that the values of the other parameters are consistent with those usually suggested in the literature. In particular, as regards our implementation of PSO, we set $P = 160$, that is the number of particles doubles the number of variables. We arrested PSO iterations at just the 150th iteration, to show the effectiveness of our implementation of PSO since the early iterations. Finally, given the random initialization of particles positions, we compute average results over 100 runs, for each considered parameter setting.

Hereinafter, we use the following terminology. A *portfolio* is the best solution detected by any member of a swarm; an *optimal portfolio* is the best solution detected by a swarm; a *global optimal portfolio* is the best solution detected by all the swarms with the same parameter setting during the different runs. Thus, since any swarm selects 80 portfolios, and we consider 100 runs for each of the 15 different parameter settings, our approach produces $80 \cdot 100 \cdot 15 = 120.000$ portfolios, $100 \cdot 15 = 1.500$ optimal portfolios and 15 global optimal portfolios.

Figure 24.1 represents the typical behaviours over iterations of the fitness functions and of the risk measures of the optimal portfolio. In particular, it is related to the setting $r_e = 0.10000\%$ and $(K_d, K_u) = (11, 30)$. In the upper panel, the fitness functions provided by standard PSO (continuous line) and by our implementation (dashed line) are represented; in the lower panel, the risk measure provided by standard PSO (continuous line) and by our implementation (dashed line) are represented. Figure 24.1 suggests that, since the early iterations, the optimal portfolio fitness function values provided by our implementation of PSO decrease significantly faster (apart from few exceptions) than the corresponding ones produced by standard PSO. We believe that this indicates the effectiveness of the dynamic management of the penalty parameters in our framework. On the other hand, there is not such a strong difference in terms of the optimal portfolio risk measures. Indeed, the best portfolio risk measures given by our implementation are comparable with the corresponding ones produced by standard PSO. This is consistent with the fact that, unlike the standard PSO, our solution approach has been developed to pursue both feasibility and optimality of the found portfolios.

Table 24.1 presents some results relative to some selected portfolios. In particular, columns 3 and 4 respectively report the value of the global optimal portfolio fitness function before the refinement of the solution (P_A), and after the refinement of the solution (P_B). Column 5 reports the value of the risk measure after the refinement of the global optimal portfolio (ρ_B). Column 6 reports the number of stocks to trade.

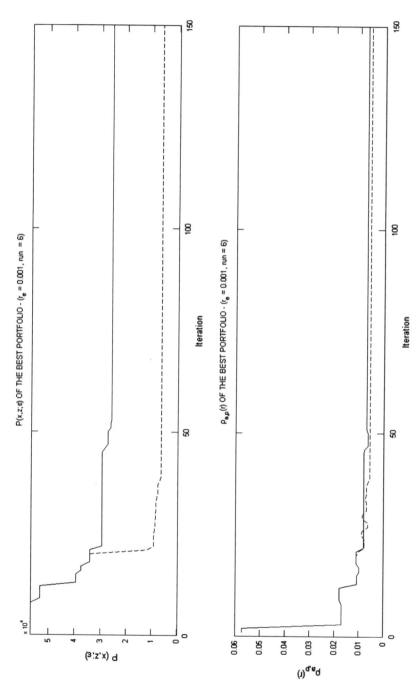

Fig. 24.1 Fitness functions and risk measures related to the setting $r_e = 0.001\%$ and $(K_d, K_u) = (11, 30)$, produced in the run number 6

Table 24.1 Results concerning the main characteristics of the selected portfolios

r_e (%)	(K_d, K_u)	P_A	P_B	ρ_B	#	$\% >$ (%)	$\% \geq$ (%)	$\%_F$ (%)
0.02500	(11, 20)	99.98767	0.00264	0.00264	19	72.00	72.00	100.00
	(16, 25)	84.73256	0.00279	0.00279	23	71.00	71.00	100.00
	(19, 21)	28731.29529	0.00288	0.00288	21	68.00	68.00	100.00
0.04375	(11, 20)	11898.72752	0.00254	0.00254	23	70.00	70.00	100.00
	(16, 25)	11.34181	0.00212	0.00212	19	69.00	69.00	100.00
	(19, 21)	361.94096	0.00279	0.00279	21	71.00	71.00	100.00
0.06250	(11, 20)	14.90991	0.00254	0.00254	24	74.00	74.00	100.00
	(16, 25)	57931.05767	0.00260	0.00260	19	69.00	69.00	100.00
	(19, 21)	30.57782	0.00239	0.00239	21	68.00	68.00	100.00
0.08125	(11, 20)	9334.44747	0.00261	0.00261	19	70.00	70.00	100.00
	(16, 25)	363.09568	0.00288	0.00288	21	66.00	66.00	100.00
	(19, 21)	398.23171	0.00189	0.00189	19	75.00	75.00	100.00
0.10000	(11, 20)	44022.38778	0.00282	0.00282	21	70.00	70.00	100.00
	(16, 25)	1001.45436	0.00239	0.00239	21	72.00	72.00	100.00
	(19, 21)	44598.40436	0.00270	0.00270	19	70.00	70.00	100.00

Columns 7 and 8 report the percentages of iterations, calculated over all the runs for all the portfolios, such that the value of the fitness function after the refinement in Sect. 24.4.2 is better ($\% >$) or not worse ($\% \geq$) than the value of the same quantity of the global optimal portfolio given by the standard PSO. Column 9 reports the percentage of runs in which the optimal portfolios produced by our implementation of PSO are feasible ($\%_F$).

Generally, our implementation of PSO works significantly better than standard PSO. Considering the behaviours of the fitness functions of the selected portfolios, results in columns 7 and 8 highlight that to large extent the refinement of the solutions in Sect. 24.4.2 is much effective. Moreover, the refinement procedure gives fitness values for all the 15 global optimal portfolios (column 4) which are lower than the corresponding values before the refinement procedure (column 3).

Then, note that for all the 15 global optimal portfolios the fact that P_B (column 4) is equal to ρ_B (column 5) indicates that all the constraints are satisfied.

Finally, to further confirm the robustness and reliability of our approach, all the 1500 optimal portfolios detected by our implementation of PSO are feasible (column 9).

24.6   Final Remarks

In this paper we have proposed a novel PSO-based scheme, for the solution of an unconstrained nonsmooth reformulation of a complex portfolio selection problem. Our original portfolio problem is a nonconvex, nondifferentiable, mixed-integer and NP-hard constrained optimization problem. The results we have obtained show that the adaptive update of the penalty parameters can play an important role for PSO-based solvers, when embedded within exact penalty frameworks.

In order to carefully detect features and drawbacks of our novel approach, in future researches further investigations are necessary with respect to different risk measures, constraints and data.

Acknowledgements The research is partially supported by the Italian Flagship Project RITMARE, coordinated by the Italian National Research Council and funded by the Italian Ministry of Education, University and Research.

References

1. Campana, E.F., Fasano, G., Pinto, A.: Dynamic analysis for the selection of parameters and initial population, in particle swarm optimization. J. Glob. Optim. **48**, 347–397 (2010)
2. Chen, Z., Wang, Y.: Two-sided coherent risk measures and their application in realistic portfolio optimization. J. Bank. Financ. **32**, 2667–2673 (2008)
3. Corazza, M., Fasano, G., Gusso, R.: Particle swarm optimization with non-smooth penalty reformulation, for a complex portfolio selection problem. Appl. Math. Comput. **224**, 611–624 (2013)
4. Kennedy, J., Eberhart, R.C.: Particle swarm optimization. In: Proceedings of the 1995 IEEE International Conference on Neural Networks, Perth, Australia, IEEE Service Center, Piscataway, IV (1995)
5. Poli, R., Kennedy, J., Blackwell, T.: Particle swarm optimisation: an overview. Swarm Intell. J. **1**, 33–57 (2007)
6. Schaerf, A.: Local search techniques for constrained portfolio selection problems. Comput. Econ. **20**, 177–190 (2002)
7. Van den Berg, F., Engelbrecht, F.: A study of particle swarm optimization particle trajectories. Inf. Sci. J. **8**, 937–971 (2005)

Chapter 25
Can PSO Improve TA-Based Trading Systems?

Marco Corazza, Francesca Parpinel and Claudio Pizzi

Abstract In this paper, we propose and apply a methodology to improve the performances of trading systems based on Technical Indicators. As far as the methodology is concerned, we take into account a simple trading system and optimize its parameters—namely, the various time window lengths—by the metaheuristic known as Particle Swarm Optimization. The use of a metaheuristic is justified by the fact that the involved optimization problem is complex (it is nonlinear, nondifferentiable and integer). Therefore, the use of exact solution methods could be extremely time-consuming for practical purposes. As regards the applications, we consider the daily closing prices of eight important stocks of the Italian stock market from January 2, 2001, to April 28, 2017. Generally, the performances achieved by trading systems with optimized parameters values are better than those with standard settings. This indicates that parameter optimization can play an important role.

Keywords Technical analysis · Trading systems · Particle Swarm Optimization
FTSE MIB

25.1 Introduction

The investors working in financial market are always looking for the philosopher's stone: a model or an algorithm which converts the huge bulk of data available in financial market to useful informations about future stock prices. As pointed out in [7], an effective information extraction is necessary to forecast direction of asset

M. Corazza (✉) · F. Parpinel · C. Pizzi
Department of Economics, Ca' Foscari University of Venice, Cannaregio 873,
30121 Venezia, Italy
e-mail: corazza@unive.it

F. Parpinel
e-mail: parpinel@unive.it

C. Pizzi
e-mail: pizzic@unive.it

© Springer International Publishing AG, part of Springer Nature 2019   277
A. Esposito et al. (eds.), *Neural Advances in Processing Nonlinear
Dynamic Signals*, Smart Innovation, Systems and Technologies 102,
https://doi.org/10.1007/978-3-319-95098-3_25

prices and this is an important task because it means to be able to elaborate profitable trading rules. The recurring financial crises clearly suggests us that none has found the philosopher's stone yet. Paper [4] presents a comprehensive review on evolutionary computation in algorithmic trading. Among them we can find the Particle Swarm Optimization (PSO) that, as well as Ant Colony Optimization, looks for the optimum of a fitness function by mimic the behaviour of a large group of animals or insects. Among the analysis method, they have identified three different analysis: fundamental, blending and technical. On one hand, the aim of fundamental analysis is to generate trading rules when the stock is undervalued or overvalued with respect to its fundamental value. On the other hand, Technical Analysis (TA) considers Technical Indicators (TIs) that are built using the time series of stock prices and volumes. Thus, the analysis of the patterns of sequence of prices or of volumes enables us to generate trading rules [2]. The blending analysis combines both ones.

TIs, used typically in trend detection, are the Simple and Exponential Moving Average (SMA and EMA, respectively), the Relative Strength Index (RSI), the Moving Average Convergence/Divergence (MACD), and the Bollinger Bands (BB). All TIs depend on one or more parameters that often assume default values. In this paper we use the swarm intelligence approach to obtain optimal values for the parameters that characterize four TIs that is SMA, RSI, MACD and BB. The next Section will be devoted to introduce the methodology we will use, and in Sect. 25.3 we will show some results of the implemented procedure. Finally, in Sect. 25.4 we draw some conclusions.

25.2 Methodology

Our idea is to investigate the improvement of the performance of a trading system based on TA tools. The evidence is that traders generally use a rule of thumbs to choose TIs' parameters, so we look for a procedure based on historical data, typically more objective. Here, we try to improve the performances of TA indicators by optimizing their parametrizations. To this aim, we consider a simple trading system constituted by four classical indicators, and optimize their parameters—namely, the time-window lengths—by the metaheuristic known as PSO. The need to use a metaheuristic as solver is justified by the fact that, as it will be explained later, the involved optimization problem is "complex". Therefore, the use of exact solution methods could be extremely time-consuming for practical purposes. In the following sections we will describe the simple trading system we will employ.

25.2.1 The Trading System

We consider a trading system based on the following TIs: EMA, RSI, $MACD$, and BB. These indicators are so well-known among academicians and practitioners that there is no need to describe them (anyway, for details we suggest to refer to [6]). Below, we first present four decisional rules based on these indicators, each of them providing one trading signal. In particular, the trading signals may be: "−1", namely *"Sell or stay short in the market"*; "0", namely *"Stay out from the market"*; "+1", namely *"Buy or stay long in the market"*. Then, we propose how to aggregate such four trading signals in order to obtain a single operational one.

Now, let us consider as trading period the discrete time interval $t = 1, \ldots, T > 1$, and let us assume that at time $t = 1$ each of the four trading signals is equal to 0. From $t = 2$ to $t = T$, the four decisional rules are:

- the one based on EMA, with $EMA_f(\cdot)$ a fast EMA and $EMA_s(\cdot)$ a slow EMA:

$$signal_{EMA}(t) = \begin{cases} -1 \text{ if } EMA_f(t) < EMA_s(t) \wedge EMA_f(t-1) \geq EMA_s(t-1) \\ +1 \text{ if } EMA_f(t) > EMA_s(t) \wedge EMA_f(t-1) \leq EMA_s(t-1); \\ signal_{EMA}(t-1) \text{ otherwise} \end{cases}$$

- the one based on RSI:

$$signal_{RSI}(t) = \begin{cases} -1 \text{ if } RSI(t) > 70 \wedge RSI(t-1) \leq 70 \\ +1 \text{ if } RSI(t) < 30 \wedge RSI(t-1) \geq 30; \\ signal_{RSI}(t-1) \text{ otherwise} \end{cases}$$

- the one based on $MACD$, where $DL(\cdot)$ and $SL(\cdot)$ are, respectively, the Differential Line and the Signal Line:

$$signal_{MACD}(t) = \begin{cases} -1 \text{ if } DL(t) < SL(t) \wedge DL(t-1) \geq SL(t-1) \\ +1 \text{ if } DL(t) > SL(t) \wedge DL(t-1) \leq SL(t-1); \\ signal_{MACD}(t-1) \text{ otherwise} \end{cases}$$

- the one based on BB, where $P(\cdot)$ is the closing price of the considered financial asset, and $BB_L(\cdot)$ and $BB_U(\cdot)$ are, respectively, the Lower Bollinger Band and Upper Bollinger Band:

$$signal_{BB}(t) = \begin{cases} -1 \text{ if } P(t) < BB_U(t) \wedge P(t-1) \geq BB_U(t-1) \\ +1 \text{ if } P(t) > BB_L(t) \wedge P(t-1) \leq BB_L(t-1). \\ signal_{BB}(t-1) \text{ otherwise} \end{cases}$$

We remember that any indicator, and consequently any decisional rule, depends on a given parametrization. In particular, in the following we will indicate by w_f and w_s, respectively, the parameters related to EMA_f and to EMA_s, by w_{RSI} the parameter related to RSI, by w_{sl}, $w_{MACD,1}$ and $w_{MACD,2}$ the three parameters related to $MACD$, and by w_{BB} the parameter related to BB. Moreover, we denote with ω

the vector of all the parameters, that is $\omega = (w_{EMA_f}, w_{EMA_s}, w_{RSI}, w_{sl}, w_{MACD,1}, w_{MACD,2}, w_{BB})$.

As regards the definition of one operational trading signal, we propose to aggregate the trading signals coming from the four decisional rules as follows:

$$signal(t) = \text{sign}(signal_{EMA}(t) + signal_{RSI}(t) + signal_{MACD}(t) + signal_{BB}(t)),$$

where $\text{sign}(\cdot)$ is the function signum. It is easy to prove that if three or all the decisional rules give the same trading signal, then the single operational trading signal is equal to it; it is also easy to prove that if two decisional rules provide the same trading signal and the other two decisional rules provide different trading signals, also between them, then the single operational trading signal is equal to the one of the two former decisional rules.

25.2.2 The Constrained Optimization Problem

There are several ways by which we can measure the performance of a trading system. In this paper we employ a simple and intuitive measure, that is the net capital at the end of the trading period, $C(T)$, where "net" means that the transaction costs are explicitly taken into account.

To determine $C(T)$, first, let us indicate by δ the transaction costs expressed in percentages, and define the net rate of return obtained by the trading system from $t - 1$ to t as follows:

$$e(t) = signal(t-1) \ln \left(\frac{P(t)}{P(t-1)} \right) - \delta \, |signal(t) - signal(t-1)| \, , t = 2, \ldots, T;$$

then, let us specify as follows the equity line produced by the trading system:

$$C(t) = C(t-1)[1 + e(t)], t = 2, \ldots, T,$$

with $C(1)$ a fixed amount.

At this point we are able to formalize the constrained optimization problem in the following way:

$$\max_{\omega} C(T)$$

$$\text{s.t.} \quad \begin{cases} w_{EMA_f} < w_{EMA_s} \\ w_{sl} < w_{MACD,1} < w_{MACD,2} \\ w_{EMA_f}, w_{EMA_s}, w_{RSI}, w_{sl}, w_{MACD,1}, w_{MACD,2}, w_{BB} \in \mathbb{N}^+ \end{cases} \tag{25.1}$$

As mentioned above, this problem is "complex" as it is formulated in terms of integer variables and its objective function is highly nonlinear and nondifferentiable. In general, this kind of problems is difficult to solve, and exact solution algorithms which are both effective and efficient are still sought at present. For these reasons, we need to use a metaheuristic, namely the PSO, as solver.

25.2.3   Particle Swarm Optimization and Its Implementation

PSO is an iterative bio-inspired population-based metaheuristic for the solution of global unconstrained continuous optimization problems. Note that, unlike, our optimization problem is global *constrained integer*. Because of this, first we will introduce the basics on standard PSO we use, then we will present the performed implementation in order to take into account the peculiarities of our optimization.

Basics on PSO The basic idea of PSO is to replicate the social behaviour of shoals of fish or flocks of birds cooperating in the search for food. To this purpose, each member of the flock explores the search area keeping memory of its best position reached so far, and it exchanges this information with the neighbors in the swarm. Thus, the whole swarm (hopefully) tends to converge towards the best global position reached by the members. In its mathematical counterpart, the paradigm of a flying flock may be formulated as follows: given a minimization problem, every member of the swarm, namely a particle, represents a possible solution of the minimization problem. Every particle is initially assigned to a random position, x_j^1, and velocity, v_j^1, which is used to determine its initial direction of movement.

For a formal description of PSO, let us consider the global optimization problem $\min_{x \in \mathbb{R}^d} f(x)$, where $f : \mathbb{R}^d \mapsto \mathbb{R}$ is the objective function. Suppose we apply PSO for its solution, where M particles are considered. At the k-th iteration of the algorithm, three vectors are associated to the j-th particle, with $j = 1, \ldots, M$:

- $x_j^k \in \mathbb{R}^d$, which is the position;
- $v_j^k \in \mathbb{R}^d$, which is the velocity;
- $p_j \in \mathbb{R}^d$, which is the best position visited so far.

Moreover, $pbest_j = f(p_j)$ is the value of the objective function in the position p_j, and $gbest = f(p_g)$ where p_g is the best position visited by the particles of the swarm. The overall algorithm, in the version with inertia weights (which is the one we will use), is reported in the following:

1. Set $k = 1$ and evaluate $f(x_j^k)$ for $j = 1, \ldots, M$. Set $pbest_j = +\infty$ for $j = 1, \ldots, M$, and $gbest = +\infty$.
2. If $f(x_j^k) < pbest_j$ then set $p_j = x_j^k$ and $pbest_j = f(x_j^k)$. If $f(x_j^k) < gbest$ then set $p_g = x_j^k$ and $gbest = f(x_j^k)$.

3. Update position and velocity of the j-th particle, with $j = 1, \ldots, M$, as

$$\begin{cases} \mathbf{v}_j^{k+1} = w^{k+1}\mathbf{v}_j^k + c_1(\mathbf{p}_j - \mathbf{x}_j^k) + c_2(\mathbf{p}_g - \mathbf{x}_j^k) \\ \mathbf{x}_j^{k+1} = \mathbf{x}_j^k + \mathbf{v}_j^{k+1} \end{cases}.$$

4. If a convergence criterion is not satisfied then set $k = k + 1$ and go to step 2.

The values of c_1 and c_2 affect the strength of the attractive forces towards the personal and the swarm best positions explored so far by the j-th particle. Thus, in order to get the convergence of the swarm, they have to be set carefully in accordance with the value of the inertia weight w^k. The parameter w^k is generally linearly decreasing with the number of steps, that is

$$w^k = w_{max} + \frac{w_{min} - w_{max}}{K}k,$$

where typical values for w_{max} and w_{min} are respectively 0.9 and 0.4, while K is usually the maximum number of iterations allowed.

The implementation As said before, the optimization problem (25.1) is a global constrained integer one, whereas PSO is a solver for global unconstrained continuous ones. Therefore, we needed to adapt properly the standard PSO algorithm for dealing with these peculiarities.

As far as the presence of integer variables is concerned, we apply one of the few approaches available in the specialized literature, the one proposed in [5], following which ≪[e]*ach particle of the swarm* [is] *truncated to the closest integer, after the determination of its new position*≫ ([5], page 1584). Note that this novel approach seems to be already effective at the current state and promising for future improvements. In particular, ≪[t]*he truncation of real values to integers seems not to affect significantly the performance of the method, as the experimental results indicate. Moreover, PSO outperforms the* [Branch and Bound] *technique for most test problems*≫ ([5], page 1583). So, the use of this approach permits to manage the constraints $w_{EMA_f}, w_{EMA_s}, w_{RSI}, w_{sl}, w_{MACD,1}, w_{MACD,2}, w_{BB} \in \mathbb{N}^+$ of our optimization problem.

As regards the presence of the other constraints, different strategies are proposed in the specialized literature to ensure that achievable positions are generated at any iterations of PSO. However, in this paper we use PSO accordingly to the original intent, that is as a tool for the solution of unconstrained optimization problems. To this purpose, we have reformulated our problem into an unconstrained one using the nondifferentiable ℓ_1 penalty function method described by [3] and recently applied in the financial context, [1]. Such an approach is known as *exact penalty method*, where the term "exact" refers to the correspondence between the minimizers of the original constrained problem and the minimizers of the unconstrained (penalized) one. So, the reformulated version of our optimization problem is

$$\max_{\omega} \ C(T) - \frac{1}{\epsilon} \Big[\max\{0, w_{EMA_f} - w_{EMA_s}\} + \max\{0, w_{sl} -$$
$$-w_{MACD,1}\} + \max\{0, w_{MACD,1} - w_{MACD,2}\} \Big], \qquad (25.2)$$

where ϵ is the so-called penalty parameter. Note that a correct choice of ϵ ensures the correspondence between the solutions of the original constrained problem (25.1) and of the reformulated unconstrained one (25.2). Note also that in the latter version of the optimization problem, the costraints w_{EMA_f}, w_{EMA_s}, w_{RSI}, w_{sl}, $w_{MACD,1}$, $w_{MACD,2}$, $w_{BB} \in \mathbb{N}^+$ do not appear as they are managed by the approach described above.

25.3 Applications

As stated above, the purpose of this paper consists in improving the performances of a simple trading system based on indicators coming from technical analysis, in particular, finding the optimal values of the parameters of such indicators by using PSO.

In this section, we compare the results coming from such a trading system with standard settings of the parameters with the results coming from the same trading system with parameters values optimized solving the optimization problem (25.2) by using the version of the PSO described in Sect. 25.2.3. The purpose of this comparison is twofold: firstly, to analyze the differences, if any, between the performances, then, to investigate the features of the optimized parameters. We apply the methodology proposed in Sect. 25.2 to the time series of closing prices of eight important stocks in the Italian stock market: BUZZI UNICEM S.p.A. (BU), ENEL S.p.A. (EE), ENI S.p.A. (EI), Generali S.p.A. (GE), INTESA SANPAOLO S.p.A. (IS), LUXOT-TICA GROUP S.p.A. (LG), STMICROELECTRONICS S.p.A. (ST) and TELECOM ITALIA S.p.A. (TE). We choose these stocks because they represent the meaningful sectors of the Italian economy. Furthermore, all of them are components of the Italian stock index FTSE MIB. The considered time series goes from January 2, 2001 to April 28, 2017 (4144 prices), while the trading period goes from March 15, 2001 to April 28, 2017 (4092 prices).[1] In all applications, we have used the following values for δ and $C(1)$: $\delta = 0.15\%$, which is a percentage transaction cost currently applied by several Italian brokerage companies, and $C(1) = 100$. As far as the trading system with standard setting is concerned, given the relevant professional literature, we have used the following values for the parameters: $w_{EMA_f} = 12$, $w_{EMA_s} = 26$, $w_{RSI} = 26$, $w_{sl} = 9$, $w_{MACD,1} = 12$, $w_{MACD,2} = 26$ and $w_{BB} = 26$. As regards the trading systems with parameters values optimized, we have used the following value for M, K, c_1, c_2 and ϵ: $M = 10$, $K = 100$, $c_1 = c_2 = 1.49618$ and $\epsilon = 0.0001$; the first two values have been determined by a trial-and error procedure, the last three

---

[1]The first 52 prices need to calculate the starting indicators.

Table 25.1 Performances achieved by the various trading systems

Stock	r (%)	$\bar{r}$ (%)	s_r (%)	$[\cdot,\cdot]_{95\%,r}$	r_{min} (%)	r_{max} (%)	$\overline{\% >}$ (%)	$\overline{\% \geq}$ (%)
BU	−1.73	15.72	4.32	[7.25%, 24.20%]	0.92	22.98	88.31	89.70
EE	−16.66	3.84	2.61	[−1.29%, 8.96%]	−6.24	8.95	88.42	89.74
EI	−16.47	1.62	4.21	[−6.63%, 9.86%]	4.21	13.50	98.36	99.65
GE	−5.79	6.22	3.67	[−0.98%, 13.42%]	−2.88	14.08	95.28	96.90
IS	−12.44	17.30	5.95	[5.62%, 28.76%]	0.00	28.76	92.38	93.86
LG	−6.49	13.32	2.97	[7.50%, 19.15%]	2.64	17.56	74.18	75.63
ST	3.93	20.77	4.06	[12.81%, 28.73%]	7.29	29.64	97.07	98.40
TE	−9.88	9.83	4.79	[0.45%, 19.22%]	−4.91	18.15	98.17	99.56

values are commonly suggested in the prominent specialized literature. Furthermore, we remember that the methodology proposed in Sect. 25.2.3 is stochastic because of the random initialization of particles positions and velocities. For these reasons, we have applied 100 times our methodology to each stock, then we have calculated mean or median values of the quantities of interest. In Table 25.1 we present the performances achieved by the various trading systems. In particular, in column 2 we report the annualized rate of return performed by the trading system with standard setting (r); in columns 3 and 4 we respectively report the average annualized rate of return performed by the trading system with parameters values optimized ($\bar{r}$) and the associated standard deviation (s_r); in column 5 we report the 95% confidence interval calculated using $\bar{r}$ and s_r ($[\cdot,\cdot]_{95\%,r}$); in columns 6 and 7 we report the minimum, respectively, the maximum, of r over the 100 applications of our methodology (r_{min} and r_{max}, respectively); in columns 8 and 9 we report the average percentages of times in which, during the trading period, the value of the equity line produced by the trading system with parameters values optimized has been greater than, and greater than or equal to, the value of the equity line produced by the trading system with standard setting ($\overline{\% >}$ and $\overline{\% \geq}$, respectively).

Below, we propose some remarks about these first results. Generally, with the only exception of the stock ST, all the annualized rates of return achieved by the trading system with standard setting (column 2) are negative, whereas all the average annualized rates of return performed by the trading system with parameters' values optimized (column 3) are far greater than the former and are all positive. It indicates that also in the case of simple TA-based trading systems, like the one considered in this paper, the parameter optimization can play an important role. Then, no annualized rates of return achieved by the trading system with standard setting (column 2) belong to the 95% confidence interval calculated using $\bar{r}$ and s_r (column 5). It indicates that, for all the investigated stocks, $\bar{r}$ is statistically different from r at the 5% significance level. Moreover, note also that, again for all the stocks, $r < r_{min}$ (r_{min}s are in column 6).

All the previous remarks concern with the performances achieved by the various trading systems in the final time instant $t = T =$ (April 28, 2017) of the trading

Table 25.2 Statistics on the optimized parameters

Stock	Parameter	$\overline{w}$	s_w	$[\cdot, \cdot]_{95\%,w}$	Median(w)	MAD(w)
EI	$w_{EMA,f}$	12.00	11.25	$[-10.06, 34.06]$	10	9.38
	$w_{EMA,s}$	28.30	19.13	$[-9.20, 65.80]$	33	17.46
	w_{RSI}	39.45	16.47	$[7.16, 71.74]$	46	10.29
	w_{sl}	4.03	6.44	$[-8.60, 16.66]$	1	3.03
	$w_{MACD,1}$	14.14	9.97	$[-5.40, 33.68]$	13	8.00
	$w_{MACD,2}$	33.21	14.95	$[3.92, 62.50]$	32	12.79
	w_{BB}	16.98	15.64	$[-13.67, 47.63]$	20	12.78
ST	$w_{EMA,f}$	10.28	5.862	$[-1.21, 21.77]$	10	4.42
	$w_{EMA,s}$	30.57	11.41	$[8.20, 52.94]$	29	9.31
	w_{RSI}	19.31	15.98	$[-12.00, 50.62]$	12	11.83
	w_{sl}	9.13	4.67	$[-0.02, 18.28]$	11	3.69
	$w_{MACD,1}$	17.23	5.41	$[6.62, 27.84]$	18	4.03
	$w_{MACD,2}$	36.76	11.14	$[14.93, 58.59]$	37	9.46
	w_{BB}	20.16	13.29	$[-5.89, 46.21]$	27	11.34

period. But the results in column 8 and 9 well highlight that for very large part of the trading period (never lower than 74.18%) the trading systems with parameters' values optimized perform better than the trading system with standard setting.

As example, in Fig. 25.1 we show the performances related to the stock GE. In particular: in the first panel, the closing price time series is represented; in the second panel, the operational trading signal is represented; in the third panel, the time series of the gross signal line produced by the trading system with parameters values optimized (dotted curve), of the net signal line produced by the same trading system (bold curve), and of the net signal line produced by the trading system with standard setting (continuous curve) are represented.

As regards the second purpose of our investigation, that is the characteristics of the optimized parameters, in Table 25.2 we present some results which concern the latter. In particular, in columns 3 and 4 we respectively report the average value of the parameter ($\overline{w}$) and the associated standard deviation (s_w); in column 5 we report the 95% confidence interval calculated using $\overline{w}$ and s_w ($[\cdot, \cdot]_{95\%,w}$); in column 6 and 7 we respectively report the median value of the parameter (median(w)) and the associated mean absolute deviation from the median (MAD(w)).

Note that we present the results related only to the stocks EI and ST, respectively the worst and the best in terms of $\overline{r}$, as their peculiarities are representative of those of all the considered stocks. Note also that $\overline{w}$, s_w, the extremes of the 95% confidence interval, and MAD(w) assume real values, which are inconsistent with the values assumed by the parameters, that are only integer. Despite that, these real-valued statistics allow us to get some useful remarks.

As we can see, generally, with few exceptions, for all the stocks the values of the medians of the optimized parameters (column 6 for stocks EI and ST) differ

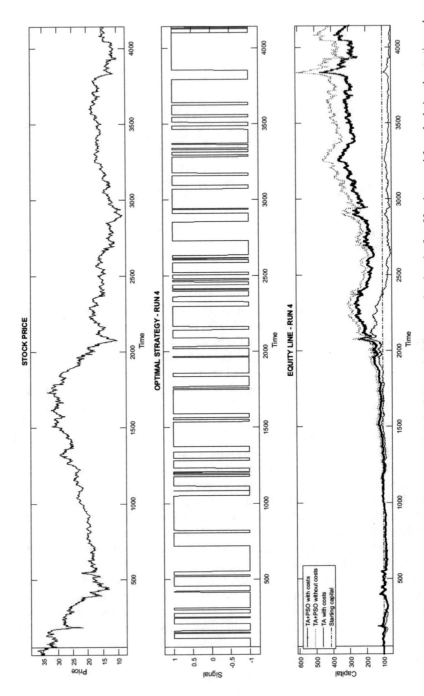

Fig. 25.1 Performances related to the stock GE. Data to the left of the vertical line are related to the first 52 prices used for calculating the starting values of the indicators

from the values of the corresponding parameters used in the trading system with standard setting. It indicates that there is room for improving the performances of standard TA-based trading systems. However, in several cases these differences are not particularly noticeable. This empirical evidence can be interpreted as follows: the swarm constituted by the human traders, in its multi ten-year acting, has reached a position in the parameter space—that is, the standard setting—which does not appear so far from the optimal setting.

Generally, with some exceptions, for all the stocks the values of the parameters used in the trading system with standard setting belong to the 95% confidence interval calculated using $\overline{w}$ and s_w (column 5 for stocks EI and ST). This may suggest the standard setting does not appear so far from the optimal setting.

Finally, for all the stocks the dispersion indices, namely s_w and $MAD(w)$ (columns 4 and 7, respectively, for stocks EI and ST), are generally decreasing when the performance of the trading system expressed in terms of $\overline{r}$ tends to increase. It likely indicates that, beyond the parameter optimization, the trading system we consider is too simple to be able to work on when applied to stock assets whose prices are characterized by particularly complex dynamics.

25.4 Conclusions

In this paper, we have proposed a PSO-based methodology to improve the performances of trading systems based on TIs coming from TA. The results, presented in Tables 25.1, 25.2 and in Fig. 25.1, show that parameter optimization can play an important role, as the results of our proposal in terms of annualized rate of return is always better than the classical TA-based one. Our future goal is to verify the proposed technique with different stocks series belonging to foreign markets. Further work will be devoted to apply the optimization and combination procedure in an *out-of-sample* framework in order to anticipate the market signals using observational data.

References

1. Corazza, M., Fasano, G., Gusso, R.: Particle swarm optimization with no-smooth penalty reformulation, for a complex portfolio selection problem. Appl. Math. Comput. **224**, 611–624 (2013)
2. Dash, R., Dash, P.K.: A hybrid stock trading framework integrating technical analysis with machine learning techniques. J. Financ. Data Sci. **2**, 42–57 (2016)
3. Fletcher, R.: Practical Methods of Optimization. Wiley, Glichester (1991)
4. Hu, Y., Liu, K., Zhang, X., Su, L., Ngai, E.W.T., Liu, M.: Application of evolutionary computation for rule discovery in stock algorithmic trading: a literature review. Appl. Soft Comput. **36**, 534–551 (2015)
5. Laskari, E.C., Parsopoulos, K., Vrahatis, M.N.: Particle Swarm Optimization for integer programming. In: Proceedings of the 2002 Congress on Evolutionary Computation, vol. 2, pp. 1582–1587 (2002)

6. Murphy, J.J.: Technical Analysis of the Financial Markets: A Comprehensive Guide to Trading Methods and Applications. New York Institute of Finance, New York (1999)
7. Wu, J., Yu, L., Chang, P.: An intelligent stock trading system using comprehensive features. Appl. Soft Comput. **23**, 39–50 (2014)

Chapter 26
Asymmetry Degree as a Tool for Comparing Interestingness Measures in Decision Making: The Case of Bayesian Confirmation Measures

Emilio Celotto, Andrea Ellero and Paola Ferretti

Abstract Bayesian Confirmation Measures are used to assess the degree to which an evidence E supports or contradicts a conclusion H, making use of prior probability $P(H)$, posterior probability $P(H|E)$ and of probability of evidence $P(E)$. Many confirmation measures have been defined till now, their use being motivated in different ways depending on the framework. Comparisons of those measures have already been made but there is an increasing interest for a deeper investigation of relationships, differences and properties. Here we focus on symmetry properties of confirmation measures which are partly inspired by classical geometric symmetries. Measures which do not satisfy a specific symmetry condition may present a different level of asymmetry: we define an asymmetry measure, some examples of its evaluation providing a practical way to appraise the asymmetry degree for Bayesian Confirmation Measures that allows to uncover some of their features, similarities and differences.

Keywords Bayesian confirmation measures · Symmetries · Asymmetry measure

26.1 Introduction

Relationships hidden in datasets are often expressed in terms of inductive rules, $E \rightarrow H$, meaning that knowledge E corroborates a conclusion H. Such rules can be supported by a dataset with different intensities that need to be assessed to rank

E. Celotto · A. Ellero (✉)
Department of Management, Ca' Foscari University of Venice, Venice, Italy
e-mail: ellero@unive.it

E. Celotto
e-mail: emilio.celotto@unive.it

P. Ferretti
Department of Economics, Ca' Foscari University of Venice, Venice, Italy
e-mail: ferretti@unive.it

© Springer International Publishing AG, part of Springer Nature 2019
A. Esposito et al. (eds.), *Neural Advances in Processing Nonlinear Dynamic Signals*, Smart Innovation, Systems and Technologies 102,
https://doi.org/10.1007/978-3-319-95098-3_26

the rules with respect to their reliability. This is of particular interest in the field of Decision Making in Economics when intelligent tools have to be considered.

A typical way in which rules are ranked is via the so-called confirmation measures: they evaluate the degree to which an evidence E supports or contradicts the conclusion, H, using prior probability $P(H)$, posterior probability $P(H|E)$ and $P(E)$, the probability of evidence E. The revealing of evidence E may change the knowledge about the occurrence of H, indeed conclusion H may be confirmed when $P(H|E) > P(H)$, or disconfirmed when $P(H|E) < P(H)$. It is therefore natural to define a measure $c(H, E)$ such that

$c(H, E) > 0$ if $P(H|E) > P(H)$ (confirmation case)
$c(H, E) = 0$ if $P(H|E) = P(H)$ (neutrality case)
$c(H, E) < 0$ if $P(H|E) < P(H)$ (disconfirmation case)

which is called a *Bayesian Confirmation Measure* (BCM). Bayesian Confirmation Measures (BCMs) have been defined and used in different contexts (see, e.g., [5, 8, 10]) often resulting in coinciding measures, which are named differently depending on the context, or with partially overlapping definitions. So, it makes sense to wonder if there is a BCM that performs better than other ones given a specific use of the measure we have in mind. The way in which BCMs are related, their similarities or differences can be clarified by means of some geometric or even visual approach, which could result particularly meaningful for the comprehension and selection of different measures (see [2, 22, 23]). In the same way, the study of analytical properties of BCMs may provide useful insights into the differences among measures (see [9, 14]).

In this paper we compare BCMs by means of their symmetry properties and, in doing that, we will consider a set of *guinea pig* measures which are well known confirmation measures:

$$d(H, E) = P(H|E) - P(H) \tag{26.1}$$

which was defined by Carnap [1],

$$F(H, E) = \frac{P(H|E) - P(H)}{P(H)} \tag{26.2}$$

defined by Finch [7],

$$G(H, E) = \log\left[\frac{P(E|H)}{P(E|\neg H)}\right] \tag{26.3}$$

which was defined by Good [12],

$$K(H, E) = \frac{P(E|H) - P(E|\neg H)}{P(E|H) + P(E|\neg H)} \tag{26.4}$$

defined by Kemeny and Oppenheim [15], and

$$Z(H,E) = \begin{cases} Z_1(H,E) = \dfrac{P(H|E) - P(H)}{1 - P(H)} & \text{in case of confirmation} \\[4mm] Z_2(H,E) = \dfrac{P(H|E) - P(H)}{P(H)} & \text{in case of disconfirmation.} \end{cases} \qquad (26.5)$$

The last measure explicitly takes into account the different confirmation and disconfirmation situations and was defined by Rescher [19] and further analysed in [5, 14].

Remark that, besides d, F and Z, with the help of some algebraic manipulation also G and K can be expressed in terms of $P(H)$ and $P(H|E)$ only:

$$G(H,E) = \log\left[\frac{P(H|E)}{P(H)} \frac{[1 - P(H)]}{[1 - P(H|E)]} \right]$$

$$K(H,E) = \frac{P(H|E) - P(H)}{P(H|E) - 2P(H|E)P(H) + P(H)}.$$

The BCMs that can be written as functions of $P(H|E)$ and $P(H)$ only, constitute the special class of IFPD (Initial Final Probability Dependence) confirmation measures. But for many BCMs (not-IFPD measures) a third variable must be included in the definition, for example the probability of the evidence $P(E)$.

For this reason, we add two more measures to our *guinea pigs* set of BCMs, considering a couple of not-IFPD measures: Mortimer's M [17] and Nozick's N [18]. They are respectively defined as

$$M(H,E) = P(E|H) - P(E) \qquad (26.6)$$

and

$$N(H,E) = P(E|H) - P(E|\neg H). \qquad (26.7)$$

The last two BCMs can be rewritten using only $P(H|E)$, $P(H)$ and $P(E)$ (but not less than three variables) as

$$M(H,E) = P(E) \frac{P(H|E) - P(H)}{P(H)}$$

and

$$N(H,E) = P(E) \frac{P(H|E) - P(H)}{P(H)(1 - P(H))}.$$

26.2   Symmetries

Symmetry properties of confirmation measures have been widely discussed in the literature (see, e.g., [11, 14]) observing that some of them should be required while some other ones should be avoided. In the following we recall the symmetry definitions that have been recently proposed in [4] (after being partially introduced in [1] and subsequently investigated in [13]).

Definition 26.1 A confirmation measure c satisfies

> *Evidence Symmetry (ES)* if $c(H, E) = -c(H, \neg E)$;
> *Hypothesis Symmetry (HS)* if $c(H, E) = -c(\neg H, E)$;
> *Evidence Hypothesis Symmetry (EHS)* if $c(H, E) = c(\neg H, \neg E)$;
> *Inversion Symmetry (IS)* if $c(H, E) = c(E, H)$;
> *Evidence Inversion Symmetry (EIS)* if $c(H, E) = -c(\neg E, H)$;
> *Hypothesis Inversion Symmetry (HIS)* if $c(H, E) = -c(E, \neg H)$;
> *Evidence Hypothesis Inversion Symmetry (EHIS)* if $c(H, E) = c(\neg E, \neg H)$.

Some of the proposed definitions refer to simple geometric symmetry properties (see [2, 3]) that, by the way, are considered desirable properties in the literature (see [4, 6]).

A geometric interpretation of some symmetries is presented in [2] referring to the so-called Confirmation Space which is based on the two dimensions $x = P(H|E)$ and $y = P(H)$, useful for confirmation measures that are IFPD measures. We define the three dimensional Extended Confirmation Space considering the variables

$$x = P(H|E), \qquad y = P(H), \qquad z = P(E)$$

and this allows to express the previously defined symmetries in terms of x, y, z, as reported in Table 26.1.

Note that the considered symmetries are defined by means of different combinations of H, E and of their negations $\neg H$ and $\neg E$, in other words by logical variations of the involved elements in the inductive rule $E \rightarrow H$. By algebraic computations and well-known probability theorems, it is then possible to reformulate their definition in terms of the probabilities $P(H|E)$, $P(H)$ and $P(E)$. For example, if we consider symmetry HS, i.e.,

$$c(H, E) = -c(\neg H, E)$$

in the Extended Confirmation Space, HS can be written as

$$c(P(H|E), P(H), P(E)) = -c(P(\neg H|E), P(\neg H), P(E))$$
$$= -c(P(1 - P(H|E), 1 - P(H), P(E))$$

that is, $c(x, y, z) = -c(1 - x, 1 - y, z)$.

26.3 Degree of Asymmetry

In general, confirmation measures do not satisfy all the above defined symmetry properties. For example, Nozick's $N(H, E) = P(E|H) - P(E|\neg H)$ satisfies only properties ES, HS and EHS, Carnap's $d(H, E) = P(H|E) - P(H)$ satisfies just HS symmetry. A BCM can also satisfy all the above considered symmetries, like Carnap's b [1] which is defined as

Table 26.1 Symmetries in the extended confirmation space

Symmetry	Definition in (H, E)	Definition in (x, y, z)
(ES)	$c(H, E) = -c(H, \neg E)$	$c(x, y, z) = -c((y - xz)/(1 - z), y, 1 - z)$
(HS)	$c(H, E) = -c(\neg H, E)$	$c(x, y, z) = -c(1 - x, 1 - y, z)$
(EHS)	$c(H, E) = c(\neg H, \neg E)$	$c(x, y, z) = c(1 - (y - xz)/(1 - z), 1 - y, 1 - z)$
(IS)	$c(H, E) = c(E, H)$	$c(x, y, z) = c(xz/y, z, y)$
(EIS)	$c(H, E) = -c(\neg E, H)$	$c(x, y, z) = -c(1 - xz/y, 1 - z, x)$
(HIS)	$c(H, E) = -c(E, \neg H)$	$c(x, y, z) = -c((1 - x)z/(1 - y), z, 1 - y)$
(EHIS)	$c(H, E) = c(\neg E, \neg H)$	$c(x, y, z) = c(1 - (1 - x)z/(1 - y), 1 - z, 1 - y)$

$$b(H, E) = P(E \wedge H) - P(E) P(H). \tag{26.8}$$

Remark that $b(H, E) = P(E) d(H, E)$, so, simply multiplying Carnap's measure $d(H, E)$ by $P(E)$, we move from a measure that satisfies only symmetry HS to a measure which satisfies all the symmetries proposed in Definition 26.1. It is this variety of behaviors that makes evaluating the degree of asymmetry of a BCM particularly interesting.

To define the degree of asymmetry of a BCM, we consider now a different way to define symmetry properties, with an approach which recalls some studies on the degree of exchangeability of continuous identically distributed random variables (see [16, 20, 21]).

Let us start by considering the Evidence Symmetry (ES). Looking at the last column of Table 26.1, we can say that a confirmation measure c satisfies ES if

$$c(x, y, z) = \hat{c}(x, y, z)$$

with

$$\hat{c}(x, y, z) = -c\left(\frac{y - xz}{1 - z}, y, 1 - z\right).$$

In a similar way, we can observe that c fulfills Hypothesis Symmetry (HS) if

$$c(x, y, z) = \hat{c}(x, y, z)$$

with

$$\hat{c}(x, y, z) = -c(1 - x, 1 - y, z).$$

More in general, a confirmation measure c satisfies symmetry σ, where

$$\sigma \in \{ES, HS, EHS, IS, EIS, HIS, EHS\}$$

if

$$c(x, y, z) = \hat{c}(x, y, z) \tag{26.9}$$

and $\hat{c}(x, y, z)$ is a suitably defined function, depending on the chosen symmetry; the functions $\hat{c}$ to be considered are reported in the right hand side term of the last column in Table 26.1. Function $\hat{c}$, in some sense, takes into account both the considered symmetry and the particular Bayesian Confirmation Measure c.

Accordingly, when condition (26.9) is not satisfied, c is called σ-asymmetric. For example, Nozick's N is IS–, EIS–, HIS– and $EHIS$–asymmetric. In the same way, Carnap's d is asymmetric with respect to ES, EHS, IS, EIS, HIS and $EHIS$.

Considering again the same two BCM's, we can deepen the investigation on their symmetry properties observing that both Nozick's N and Carnap's d are IS-asymmetric measures. The question we focus on is: what is the magnitude of their asymmetry?

With the aim of giving an answer to this question, we define

$$c_\sigma(x, y, z) = \frac{c(x, y, z) + \hat{c}(x, y, z)}{2} \qquad c_{\sigma a}(x, y, z) = \frac{c(x, y, z) - \hat{c}(x, y, z)}{2}$$

so that each Bayesian Confirmation Measure c may be written as

$$c(x, y, z) = c_\sigma(x, y, z) + c_{\sigma a}(x, y, z) \tag{26.10}$$

i.e., c can be viewed as the sum of two functions where c_σ is σ–symmetric by construction.

In fact, we observe that

$$c \text{ is } \sigma\text{-symmetric} \iff c_{\sigma a}(x, y, z) = 0 \iff c(x, y, z) = c_\sigma(x, y, z)$$

while c is σ-asymmetric otherwise.

For example, if we consider again Carnap's d and we analyse the Inversion Symmetry (IS), the confirmation measure can be decomposed into

$$d_{IS}(x, y, z) = \frac{(x - y)(y + z)}{2y}$$

and

$$d_{ISa}(x, y, z) = \frac{(x - y)(y - z)}{2y}.$$

To evaluate the degree of asymmetry of a BCM we first introduce an order of asymmetry: this definition is necessary to avoid the possibility of facing two BCMs which are inversely ordered by different asymmetry measures [21].

Definition 26.2 In the Extended Confirmation Space, a confirmation measure c_1 is called to be less σ-asymmetric than a confirmation measure c_2, written $c_1 \prec_{\sigma a} c_2$, if

$$\frac{|c_1(x, y, z) - \hat{c}_1(x, y, z)|}{|c_1(x, y, z)|} \leq \frac{|c_2(x, y, z) - \hat{c}_2(x, y, z)|}{|c_2(x, y, z)|}$$

for each feasible choice of (x, y, z).

Note that the feasibility of (x, y, z) requires that probabilities x, y and z satisfy the Total Probability Theorem (see [2]).

Observe that $\prec_{\sigma a}$ is a preorder, that is the relation is reflexive and transitive; moreover it is not antisymmetric, given that for example, both Nozick's N and Carnap's d are Hypothesis Symmetric confirmation measures. Finally, it is clearly not a total order since not all BCMs can be ordered by $\prec_{\sigma a}$: if we consider for example Carnap's d and Finch's F that are both not EIS–symmetric, when $(x, y, z) = (1/6, 2/3, 1/6)$ it is

$$\frac{|d(x, y, z) - \hat{d}(x, y, z)|}{|d(x, y, z)|} < \frac{|F(x, y, z) - \hat{F}(x, y, z)|}{|F(x, y, z)|}$$

while the inequality is reversed in $(x, y, z) = (2/5, 1/3, 1/2)$.

Finally, we can propose a general definition of asymmetry measure which is compatible with the partial order $\prec_{\sigma a}$:

Definition 26.3 A measure of σ-asymmetry is a function $\mu_{\sigma a}$ that satisfies the following conditions:

1. $\mu_{\sigma a}(c) = 0$ if and only if c is σ-symmetric;
2. if $c_1 \prec_{\sigma a} c_2$ then $\mu_{\sigma a}(c_1) \leq \mu_{\sigma a}(c_2)$.

In the following, we consider the particular class of σ-asymmetry measures defined as:

$$\mu_p(c) = \left\| \frac{c - \hat{c}}{c} \right\|_p \tag{26.11}$$

where $\| \cdot \|_p$ denotes the L^p-norm, for each $p \in [1, \infty]$. A measure μ_p is clearly compatible with the partial order $\prec_{\sigma a}$ as required in Definition 26.3.

Observe that different choices of p allow to emphasize different kinds of σ-asymmetries: for example, when $p = 1$ attention is on the average degree of σ-asymmetry, while $p = \infty$ drives the focus on the maximal degree of asymmetry that can be attained on single points.

26.4 Asymmetry Degree Comparison

In this section we provide some examples of asymmetry degree computations on BCMs; all computations were performed with Wolfram's software *Mathematica* (version 11.0.1.0).

We focus here on four symmetries, namely *ES*, *HS*, *IS*, *EIS*, and on four BCMs: *d* (which satisfies only *HS*), *F* (only *IS*), *M* (only *ES*) and *N* (which satisfies *ES* and *HS*). The asymmetry of a measures is here computed using the L_1 norm. The results are reported in Table 26.2.

Let us observe the results obtained for symmetry *ES* (see Table 26.2): *d* and *F* are asymmetric with the same asymmetry score of 0.5. While it could be not so easy to guess it *ex ante*, this common score can be explained *ex post* since *d* measures the difference between posterior and prior probabilities and *F* essentially considers their ratio. Moreover, *M* and *N* turn out to be *ES* symmetric: this fact can be proved analytically with some tricky algebraic manipulations, but the numerical computation immediately suggests the property.

With respect to symmetry *HS*, instead, Carnap's measure *d* and Nozick's measure *N* are symmetric, the corresponding asymmetry evaluations are equal to zero, while, not too surprisingly, *F* and *M* are equally asymmetric: in fact, their definitions differ only by the multiplicative factor $z = P(E)$, which does not affect the definition of symmetry *HS*.

The results concerning symmetries *IS* and *EIS*, allow to completely rank the four BCMs. Considering *IS*, measure *F* is symmetric while we find *d*, *N* and *M* with increasing asymmetry. The order changes when we consider symmetry *EIS*: the less asymmetric measure is now *M* and then with increasing asymmetry we find *d*, *F* and *N*. This result is (a priori), again unexpected.

Table 26.2 Degree of asymmetry of some BCMs (L_1 norm and μ_1)

BCM	ES	HS	IS	EIS
Carnap *d* $x - y$	5.0000E-01	0.0000E+00	3.8629E-01	3.8629E-01
Finch *F* $(x - y)/y$	5.0000E-01	9.0584E-01	0.0000E+00	5.0000E-01
Mortimer *M* $z(x - y)/y$	0.0000E+00	9.0584E-01	4.2835E+04	3.3333E-01
Nozick *N* $z(x - y)/[y(1 - y)]$	0.0000E+00	0.0000E+00	3.5118E+00	5.3809E+00

Table 26.3 Degree of asymmetry of some BCMs (L_2 norm and μ_2^2)

BCM	ES	HS	IS	EIS
Carnap d $x - y$	2.0026E+01	0.0000E+00	9.9485E+00	9.9485E+00
Finch F $(x - y)/y$	2.0026E+01	2.8020E+02	0.0000E+00	2.0026E+01
Mortimer M $z\,(x - y)/y$	0.0000E+00	2.8020E+02	3.3333E+09	7.1753E+00
Nozick N $z\,(x - y)/[y\,(1 - y)]$	0.0000E+00	0.0000E+00	3.3333E+08	1.0556E+09

26.5 Conclusions

The large variety of Bayesian Confirmation Measures made available by the literature on the subject, makes the choice of the measure to be used a rather tricky task. Ways to compare the measures are still needed, even if geometric, visual and analytical proposals have already been suggested (see [2, 22, 23]): among them, symmetry properties appear to be quite informative. Defining an asymmetry measure (Sect. 26.3) we suggest a practical way to appraise the asymmetry degree of a BCM and their ranking, also providing an example computed on a little set of *guinea pigs* measures (Sect. 26.4).

Even if the research should be extended to a larger set of BCMs and symmetries, the numerical results show how by using an asymmetry measure it is possible to easily highlight features of a BCM which are not obvious *ex ante* and, sometimes, to discover even unexpected properties, thus paving the way to the interpretation and the analytical proof of those features (see Sect. 26.4).

We remark that the definition of the asymmetry measure is not unique, in particular the choice of the norm to be used in the computations can be changed. In fact different norms (i.e. different p values) provide an evaluation of the degree of asymmetry with different sensitivity on extreme values. To illustrate this fact, we computed the degree of asymmetry not only using norm L_1 (Table 26.2) but also with norm L_2; the corresponding (squared) values are reported in Table 26.3. There is a glaring numerical difference with Table 26.2, but the asymmetry ranking of the BCMs is the same. An extension of the numerical computations to a set of other BCMs and symmetries should be made, to better understand which norms, even different from L_1 and L_2, allow a better comparison of the asymmetry degrees.

References

1. Carnap, R.: Logical Foundations of Probability. University of Chicago Press, Chicago (1950)
2. Celotto, E.: Visualizing the behavior and some symmetry properties of Bayesian confirmation measures. Data Min. Knowl. Disc. (2016). https://doi.org/10.1007/s10618-016-0487-5

3. Celotto, E., Ellero, A., Ferretti, P.: Monotonicity and symmetry of IFPD Bayesian confirmation measures. In: Torra, V., Narukawa, Y., Navarro-Arribas, G., Yañez, C. (eds.) Modeling Decisions for Artificial Intelligence. MDAI 2016. Lecture Notes in Computer Science, vol. 9880, pp. 114–125. Springer, Cham (2016)
4. Crupi, V., Tentori, K., Gonzalez, M.: On Bayesian measures of evidential support: theoretical and empirical issues. Philos. Sci. **74**(2), 229–252 (2007)
5. Crupi, V., Festa, R., Buttasi, C.: Towards a grammar of bayesian confirmation. In: Suárez, M., Dorato, M., Rédei, M. (eds.) Epistemology and Methodology of Science, pp. 73–93. Springer, Dordrecht (2010)
6. Eells, E., Fitelson, B.: Symmetries and asymmetries in evidential support. Philos. Stud. **107**(2), 129–142 (2002)
7. Finch, H.A.: Confirming power of observations metricized for decisions among hypotheses. Philos. Sci. **27**(4), 293–307 (1960)
8. Fitelson, B.: The plurality of Bayesian measures of confirmation and the problem of measure sensitivity. Philos. Sci. **66**, 362–378 (1999)
9. Fitelson, B.: Likelihoodism, Bayesianism, and relational confirmation. Synthese **156**(3), 473–489 (2007)
10. Geng, L., Hamilton, H.J.: Interestingness measures for data mining: a survey. ACM Comput. Surv. **38**(3), 1–32 (2006)
11. Glass, D.H.: Entailment and symmetry in confirmation measures of interestingness. Inform. Sci. **279**, 552–559 (2014)
12. Good, I.J.: Probability and the Weighing of Evidence. Hafners, New York (1950)
13. Greco, S., Pawlak, Z., Słowiński, R.: Can Bayesian confirmation measures be useful for rough set decision rules? Eng. Appl. Artif. Intell. **17**(4), 345–361 (2004)
14. Greco, S., Słowiński, R., Szczęch, I.: Properties of rule interestingness measures and alternative approaches to normalization of measures. Inform. Sci. **216**, 1–16 (2012)
15. Kemeny, J., Oppenheim, P.: Degrees of factual support. Philos. Sci. **19**, 307–324 (1952)
16. Klement, E.P., Mesiar, R.: How non-symmetric can a copula be? Comment Math. Univ. Carol. **47**(1), 141–148 (2006)
17. Mortimer, H.: The Logic of Induction. Prentice Hall, Paramus (1988)
18. Nozick, R.: Philosophical Explanations. Clarendon, Oxford (1981)
19. Rescher, N.: A theory of evidence. Philos. Sci. **25**, 83–94 (1958)
20. Siburg, K.F., Stoimenov, P.A.: Symmetry of functions and exchangeability of random variables. Stat. Papers **52**(1), 1–15 (2011)
21. Siburg, K.F., Stehling, K., Stoimenov, P.A., Weiß, G.N.F.: An order of asymmetry in copulas, and implications for risk management. Insur. Math. Econom. **68**, 241–247 (2016)
22. Susmaga, R., Szczęch, I.: The property of χ^2_{01}-concordance for Bayesian confirmation measures. In: Torra, V., Narukawa, Y., Navarro-Arribas, G., Megías, D. (eds.) Modeling Decisions for Artificial Intelligence. MDAI 2013. Lecture Notes in Computer Science, vol. 8234, pp. 226–236. Springer, Heidelberg (2013)
23. Susmaga, R., Szczęch, I.: Can interestingness measures be usefully visualized? Int. J. Appl. Math. Comput. Sci. **25**, 323–336 (2015)

Chapter 27
A Method Based on OWA Operator for Scientific Research Evaluation

Marta Cardin⬤, Giuseppe De Nadai⬤ and Silvio Giove⬤

Abstract This paper proposes a model for faculty evaluation based on OWA aggregation operators. Our method permits to consider interactions among the criteria in a formal way, and, at the same time, to realize an easy approach to understand, implement and apply.

Keywords Bbliometric indicator · Aggregation operators · OWA

27.1 Introduction

The evaluation of the performance of the academic activity has gained a growing interest in recent years and many Universities undertake periodic assessments of their faculty members which act as a first step for a wider performance evaluation process involving, at different levels, departments, faculties and Universities.

A vast literature is interested in the measurement of the research output of scholars and several approaches have indeed been proposed to assess the scientific production (see for example [1, 2, 4, 6, 7, 9]). The traditional method based on peer review may be accurate but subjective and very time-consuming. There are also many methods for ranking scientific publications, and as it is well known most of them are based on citations.

M. Cardin · S. Giove (✉)
Department of Economics, Ca' Foscari University of Venice,
Sestiere Cannaregio 873, Venezia, Italy
e-mail: sgiove@unive.it

M. Cardin
e-mail: mcardin@unive.it

G. De Nadai
Department of Management, Ca' Foscari University of Venice,
Sestiere Cannaregio 873, Venezia, Italy
e-mail: denadai@unive.it

© Springer International Publishing AG, part of Springer Nature 2019
A. Esposito et al. (eds.), *Neural Advances in Processing Nonlinear Dynamic Signals*, Smart Innovation, Systems and Technologies 102,
https://doi.org/10.1007/978-3-319-95098-3_27

There are many bibliometric indicators such as total number of publications, total number of citations, citations per paper, number of highly cited papers and so on. It is straightforward to note that all indicators highlight only a particular dimension of the research output.

In this paper we take into account the multidimensional nature of the considered evaluation problem, and the need of indices that provide useful information about a researcher's activity by summarizing it with a single and numerical score. A good research output indicator has to capture both the quantitative and the qualitative dimension of the research.

As it is well known, aggregation operators play an important role in several fields such as decision sciences, computer and information sciences, economics and social sciences (see [5, 8]). The most commonly used aggregation is the Weighted Averaging (WA). The ordered weighted averaging operators (OWA) developed by Yager in [14] assigns the weights to the ordered values (i.e. to the smallest value, the second smallest and so on) rather than to the specific values. Since its introduction, the OWA aggregation has been received increasingly interest in many publications (see [13]). The main reason for the success of OWA consists into the capability to take interactions among the criteria into account, so that it is a suitable tool to properly reflects the preference structure of a Decision Maker. Given its linear nature, the WA cannot include interactions between criteria, requiring the satisfaction of the Preferential Independence axiom that is rarely satisfied. This implies that WA requires complete compensativeness, i.e. a low value of one criterion, can be compensated by a high value of another one, and this can be undesiderable.

In this paper we propose a model to rank journals or authors based on OWA aggregation operators. The paper is structured as follows. Section 27.2 introduces Aggregation Operators, mainly focusing on OWA operators. In Sect. 27.3 our model is introduced while in Sect. 27.4 is reported a simulated application. The last Section concludes and presents possible extensions and future work.

27.2 OWA as Aggregation Operators

Aggregation has for purpose the simultaneous use of different pieces of information provided by several sources, in order to come to a conclusion or a decision so aggregation functions transform a finite number of inputs, called arguments, into a single output. Aggregation functions are applied in many different domains and in particular aggregation functions play an important role in different approaches to decision making. Many functions of different type have been considered, and various properties can be imposed by the nature of the considered aggregation problem. See [8] for a comprehensive overview on aggregation theory and for the characterization and the properties of Aggregation Operators.

We denote by E a non empty real interval and n represents the number of values to be aggregated an **aggregation operator** is a function $A : E^n \to E$. Two basic mathematical properties of the aggregation functions are the following:

- **Monotonicity** For all $\mathbf{x}, \mathbf{y} \in E^n$ if $x_i \leq y_i$ ($i = 1, \ldots, n$) then $A(\mathbf{x}) \leq A(\mathbf{y})$
- **Idempotence** If $x \in E$ then $A(x, \ldots, x) = x$

Moreover we consider the case in which $E = [0, 1]$ and we assume that boundary conditions $A(0, 0, \ldots, 0) = 0$, $A(1, 1, \ldots, 1) = 1$ are satisfied. These properties are very general, thus a very large family of n–-dimensional operators can be obtained, each of them with a properly own characterization and its specific set of parameters. The class of Choquet integrals is widely used in the applications, given its high generalization capability, and the same holds also for the family of Sugeno integrals. Nevertheless, both the Choquet and the Sugeno integral (see [5, 8]) requires 2^n parameters to be estimated where n is the number of the criteria.

OWA operators, introduced by Yager in [14], received particular attention due to their general representation properties and the limited number of parameters, and are defined as follows:

$$OWA^w(x_1, x_2, \ldots, x_n) = \sum_{i=1}^{n} w_i x_{\sigma(i)} \qquad (27.1)$$

where $\{x_i\}$ is the normalized value of the i-th criterion, for every i, $i = 1, 2, \ldots, n$, $\{p_i\}$ a weight, $p_i \geq 0$, $\sum_{i=1}^{n} p_i = 1$ and σ is a permutation of the index set $\{1, 2, \ldots, n\}$ such that $x_{\sigma(1)} \geq x_{\sigma(2)} \geq \ldots \geq x_{\sigma(n)}$. The weights represent the relative importance of the criteria, and need to be determined by direct assignment or using a learning procedure. The WA is *compensative* and *homogeneous* of degree 1 (see [10]). Moreover, note that OWA is linear w.r.t. the ordered values of the criteria.

As opposed to linear Aggregation Operators, as simple weighted mean, OWA operators enable a more or less significant degree of *compensativeness*, considering *synergic* or *conflicting* interactions among the criteria. That is, a low value of a criteria cannot be compensated by a high value of an other one. This property can be desirable in many applications, see for instance [5]. Different Aggregation Operators can be obtained in function of the value of the weights, see [13], moving from a completely *conjunction* behavior up to a *disjunction* behavior, and passing through the simple averaging. The MIN operator is obtained with $w_n = 1$, $w_i = 0$ for every i, $i = 1, \ldots, n - 1$ while the MAX if $w_1 = 1$, $w_i = 0$ for every i, $i = 2, \ldots, n$.

Moreover if $w_i = \frac{1}{n}$, for every i, $i = 1, \ldots, n$ we get the simple average and with suitable choice of the weights vector, we can get the median, the k-th order statistic and others.

27.3 OWA in Multi-person Decision Problems: Scientific Paper Evaluation

As we noted before when we evaluate a scholar we have to consider both the quantity and the quality of his scientific production.

A commonly used approach ranks the publications according to the Impact Factor of the Journal where they are published (other bibliometric index could also be considered, for instance considering the number of citations of the paper itself).

The *more-is-better* assumption (more citations, more papers...) is normally considered, but with care (see [12]). For instance it is not so obvious which is the better researcher if one has more citations but in lowly cited papers while the other has a number of highly cited publications. Then many indicators, like the well-known h-index, consider only a subset of an author's publications. The h-index considers the first most productive publications each having at least h citations. Although we can note that h-index is insensitive to lowly cited papers, on the other hand the it does not enlight researchers with a moderate number of research papers but with a very high impact. Moreover it does not satisfy the independence property as emphasized in the following example presented in [4].

Consider two authors; the first one has 4 papers with 4 citations each; the second has 3 papers with 6 citations each. We see that the first one is judged better than the second one, according to the h-index, since the h-index for the first author is equal to 4, whereas the h-index for the second author is 3. Suppose now that each of the two authors publish together an additional paper with 6 citations. Now the two authors are judged as equivalent, because the h-index is equal to 4 for both.

We recall also the approach proposed by [3, 11] which is based on the concept of *scoring rules*. This approach considers summation-based rankings and therefore authors are ranked according to the sum over all their publications, where each paper is evaluated by some partial scores. It is interesting to note that scoring rule approach satisfies independence.

Anywise, in some real-world situations, for instance to evaluate the scientific performances of a University Department member, a critical debate regards the *optimal* number of publications to be considered inside a fixed period. For what above said, too few number of publications (only one in an extreme case) characterizes a strong *non compensatory* behavior, i.e. clearly the *excellence* is considered, given that only the *best* one is taken into account. On the other side, a more *production*-prone behavior tends to consider all the publications, not only the *best* ones, but every assessable publication, in the spirit of the additive rule [11]. Clearly, different choices can be adopted among these two extreme situations, depending on the decision making attitude. In fact, if we ask to some Department members, we get different answers about. In our model, we choose as Decision Makers a suitable selected subset of the full time researchers of the Department. To each of them we asked for the most "right" number of publications that has to be considered for the evaluation, and elaborated all the numerical answers, see formulas (27.2) and (27.3) below, obtaining a set of OWA weights which respect as most as possible all the furnished answers. Despite to other commonly used methods, like the simple answers averaging, our model takes all the respondents requests into account, avoiding to fix a pre-determined threshold, but weighting into decreasing order all the scientific products of the evaluated author. Clearly, the first publications will receive more weights, and the relative weight depends on the number of Decision Makers who specified the relative position of this publication into account.

To formalize our model, let us consider a set $\{D_1, D_2,..., D_K\}$ of K Decision Makers or Experts and let n_k the number of publications that the Expert D_k considers as optimal. Moreover, let us consider M scholars to be evaluated, each of them presents n_i, $i = 1,.., M$ publications. Let $\sigma_{i,j}$ be the score of the i−th publication of the j−th scholar, $i = 1,..., n_i$, $j = 1,.., M$. We also suppose that, for practical reasons, an upper limit is considered for the number of publications, say L, so that $n_i \leq L, \forall i = 1,.., M$. Then the (not normalized) OWA weights, defined in the discrete set $S = 1, 2,..., L$, can be obtained as follows:

$$w_i^* = Card\{k : 1 \leq k \leq K, \ n_k \geq i\} \tag{27.2}$$

Finally, the OWA weights are normalized:

$$w_i = \frac{w_i^*}{\sum_k w_k^*} \tag{27.3}$$

27.4   A Simulated Application

In this Section a numerical example is presented. To this purpose, let us suppose the case of 10 Experts, and 40 authors to be evaluated for their best 7 publications, i.e. $K = 10$, $M = 41$, $L = 7$. Figure 27.1 reports the publications of each author ordered in descending order of SNIP factor, i.e. the values $\sigma_{i,j}$. Again, suppose that the values of n_k, $k = 1, ..., 10$, ordered in increasing order, are collected in the vector $\{1, 1, 1, 2, 2, 4, 4, 4, 5, 7\}$. This means that three Experts require only one publication to be considered (thus these Experts are *optimality-prone*), while other two Experts consider the first two publications, and so on, up to the limit case of one Experts which retain to evaluate all the 7 publications (we can say that this Expert is *productivity-prone*). From (27.2) and (27.3) the OWA weights are easily computed and reported in the first row of the table in Fig. 27.1.

The last two columns of Fig. 27.1 report the final scores of all the 40 authors, computed through OWA and (simple) weighting average WA, while Fig. 27.2 reports the author's ranking computed with the two methods, and sorted in descending order

	1	2	3	4	5	6	7	8	9	10		
Weight	w1	w2	w3	w4	w5	w6	w7	w8	w9	w10		
Scale value	0,322581	0,225806	0,16129	0,16129	0,064516	0,032258	0,032258	0	0	0		
	a1	a2	a3	a4	a5	a6	a7	a8	a9	a10	OWA	WA
author01	1,655	1,54	0,609	0,487	0,18	0,1	0,03	0,016	0,006	0	1,074193548	0,4623
author02	4,665	1,099	0,497	0,1	0,086	0,026	0,012	0,011	0,008	0,003	1,856064516	0,6507
author03	1,668	1,148	0,263	0,217	0,128	0,032	0,028	0,017	0,014	0,005	0,884903226	0,352
author04	0,907	0,571	0,121	0,107	0,065	0,012	0,007	0,002	0,001	0,001	0,463096774	0,1794
author05	1,588	1,419	0,66	0,248	0,219	0,01	0,006	0,003	0,003	0	0,993774194	0,4156
author06	2,56	0,843	0,567	0,451	0,102	0,092	0,089	0,061	0,049	0,042	1,192774194	0,4856
author07	0,703	0,68	0,023	0,006	0,003	0,003	0	0	0	0	0,385290323	0,1418
author08	1,784	0,28	0,215	0,083	0,072	0,063	0,017	0,004	0,004	0,003	0,694	0,2525
author09	3,404	1,352	0,338	0,306	0,302	0,259	0,009	0	0	0	1,535354839	0,597
author10	1,964	1,255	0,558	0,455	0,202	0,022	0,008	0,004	0,003	0,003	1,094322581	0,4474
author11	0,574	0,487	0,382	0,177	0,167	0,103	0,012	0	0	0	0,399774194	0,1902
author12	1,594	0,104	0,081	0,074	0	0	0	0	0	0	0,562677419	0,1853
author13	2,467	2,078	1,655	1,613	1,003	0,321	0,151	0,056	0,045	0,029	1,872064516	0,9418
author14	2,421	0,789	0,476	0,094	0,041	0,019	0,001	0	0	0	1,054354839	0,3841
author15	1,394	0,007	0,006	0,002	0	0	0	0	0	0	0,452548387	0,1409
author16	5,779	0,339	0,327	0,089	0,084	0,084	0,062	0,062	0,018	0,001	2,017967742	0,6845
author17	1,024	0,529	0,117	0,079	0,021	0,007	0,001	0	0	0	0,483	0,1778
author18	0,089	0,034	0,012	0,004	0,001	0	0	0	0	0	0,039032258	0,014
author19	0,526	0,162	0,097	0,078	0,054	0,03	0,004	0,003	0,002	0,002	0,239064516	0,0958
author20	4,026	2,52	0,032	0,022	0,005	0,004	0	0	0	0	1,876903226	0,6609
author21	1,234	0,226	0,076	0,029	0,024	0,015	0,002	0,002	0,002	0,001	0,468129032	0,1611
author22	2,225	1,725	1,591	0,904	0,057	0,033	0,015	0,011	0,001	0,001	1,514903226	0,6563
author23	1,665	1,279	0,92	0,529	0,29	0,216	0,151	0,011	0,002	0,002	1,09016129	0,5065
author24	0,06	0,002	0	0	0	0	0	0	0	0	0,019806452	0,0062
author25	0,95	0,715	0,068	0,01	0,001	0,001	0	0	0	0	0,480580645	0,1745
author26	1,582	0,273	0,078	0,075	0,043	0,027	0,015	0,014	0,011	0,007	0,600774194	0,2125
author27	0,187	0,158	0,085	0,047	0,045	0,026	0,014	0,01	0,009	0,003	0,121483871	0,0584
author28	1,089	0,744	0,557	0,049	0,018	0,003	0,003	0	0	0	0,618387097	0,2463
author29	2,158	0,568	0,195	0,131	0,031	0,013	0,002	0,001	0,001	0	0,879451613	0,31
author30	1,872	0,191	0,065	0,012	0,007	0,001	0	0	0	0	0,659903226	0,2148
author31	1,689	1,198	0,85	0,814	0,24	0,161	0,161	0,033	0,011	0,005	1,109612903	0,5162
author32	0,51	0,245	0,237	0,083	0,07	0,058	0,046	0,026	0	0	0,279322581	0,1275
author33	2,025	1,238	0,391	0,034	0,024	0,015	0,009	0,003	0,001	0,001	1,003645161	0,3741
author34	0,065	0,055	0,019	0,007	0,007	0,003	0	0	0	0	0,038129032	0,0156
author35	4,028	1,841	1,244	0,828	0,568	0,523	0,111	0,032	0,028	0	2,106354839	0,9203
author36	1,46	0,82	0,202	0,032	0,015	0,005	0,004	0,002	0	0	0,695129032	0,254
author37	0,605	0,094	0,059	0,053	0,014	0,01	0,005	0,003	0,001	0	0,23583871	0,0844
author38	1,75	0,824	0,013	0	0	0	0	0	0	0	0,752677419	0,2587
author39	4	3,5	2	0	0	0	0	0	0	0	2,403225806	0,95
author40	2	1,5	1,5	1,5	1,5	1,5	1,4	1,3	1,3	1,2	1,658064516	1,47

Fig. 27.1 OWA weights and scholar's data base

of OWA. We can observe that for some authors the rank is reversed. For instance, authors 39 and 40 report a score equal to 0.95 and 1.47 respectively with WA, but 2.40, 1.65 with OWA. The reason is due to author 39 is characterized by low productivity (only 3 papers, see Fig. 27.1) but with high quality (SNIP are 4, 3.5, 2 respectively). Conversely, author 40 is more productive (he presented 10 papers), but with lower quality (the best has SNIP $= 2m$ all the other have SNIP from 1.2 up to 1.5). Namely, WA favors author 40, because for author 30 all the SNIP are null a part the first 3. The converse using OWA, given that the *average* multi-Expert preference, represented by the OWA weights, highlights the excellence. In fact relatively high values are only for the first two best papers: the two first OWA weights, 0.322 and 0.225, sum up to more than 50%. Naturally, different assignment of values n_k will produce different set of OWA weights, and consequently different scores and ranking.

Fig. 27.2 Evaluation

Ranking(WA)		Ranking(OWA)	
Author	WA	Author	OWA
40	1,47000	39	2,40323
39	0,95000	35	2,10635
13	0,94180	16	2,01797
35	0,92030	20	1,87690
16	0,68450	13	1,87206
20	0,66090	2	1,85606
22	0,65630	40	1,65806
2	0,65070	9	1,53535
9	0,59700	22	1,51490
31	0,51620	6	1,19277
23	0,50650	31	1,10961
6	0,48560	10	1,09432
1	0,46230	23	1,09016
10	0,44740	1	1,07419
5	0,41560	14	1,05435
14	0,38410	33	1,00365
33	0,37410	5	0,99377
3	0,35200	3	0,88490
29	0,31000	29	0,87945
38	0,25870	38	0,75268
36	0,25400	36	0,69513
8	0,25250	8	0,69400
28	0,24630	30	0,65990
30	0,21480	28	0,61839
26	0,21250	26	0,60077
11	0,19020	12	0,56268
12	0,18530	17	0,48300
4	0,17940	25	0,48058
17	0,17780	21	0,46813
25	0,17450	4	0,46310
21	0,16110	15	0,45255
7	0,14180	11	0,39977
15	0,14090	7	0,38529
32	0,12750	32	0,27932
19	0,09580	19	0,23906
37	0,08440	37	0,23584
27	0,05840	27	0,12148
34	0,01560	18	0,03903
18	0,01400	34	0,03813
24	0,00620	24	0,01981

27.5 Final Remarks and Conclusion

The model that we have considered in this paper is extremely stylized. It presents the following advantages: it is immediate to be computed, understood and easy to be communicate. For these reasons it could be proposed in situations where the main focus is the selection of the *optimal* number of research products satisfing the trade-off between excellence and productivity. Anywise, more detailed analysis could carried on to include more specific items. For instance, more than one single performance index, as the citation counts could be considered, together with the relative importance of each Expert, depending on their effective role inside the Academy. Again, suitable consensus model together with a refinement of the way by which the information is elicited (web questionnaire, etc.) could be considered in a future work.

References

1. Alonso, S., Cabrerizo, F.J., Herrera-Viedma, E., Herrera, F.: h-Index: a review focused in its variants, computation and standardization for different scientific fields. J. Infom. **3**, 273–289 (2009)
2. Beliakov, G., James, S.: Citation-based journal ranks: the use of fuzzy measure. Fuzzy Sets Syst. **167**, 101–119 (2011)
3. Bouyssou, D., Marchant, T.: Consistent bibliometric rankings of authors and of journals. J. Infom. **4**, 365–378 (2010)
4. Bouyssou, D., Marchant, T.: Bibliometric rankings of journals based on impact factors: an axiomatic approach. J. Infom. **5**, 75–86 (2011)
5. Calvo, T., Mayor, G., Mesiar, R.: Aggregation Operators: New Trends and Applications. Springer (2002)
6. Cardin, M., Corazza, M., Funari, S., Giove, S.: A fuzzy-based scoring rule for author ranking: an alternative of h index. In: Apolloni, B., Bassis, S., Esposito, A., Morabito, C.F. (eds.) Proceedings of the 21st Italian Workshop on Neural Nets Neural Nets WIRN11. Series: Frontiers in Artificial Intelligence and Applications, vol. 234, pp. 36–45. IOS Press (2011)
7. Cardin, M., Giove, S.: A fuzzy method for the assessment of the scientific production. In: Greco, S., Bouchon-Meunier, B., Coletti, G., Fedrizzi, M., Matarazzo, B., Yager, R.R. (eds.) Advances in Computational Intelligence, vol. IV, pp. 37–43. Springer, Heidelberg (2012)
8. Grabisch, M., Marichal, J.L., Mesiar, R., Pap, E.: Aggregation Functions, Encyclopedia of Mathematics and its Applications. Cambridge University Press, Cambridge (2009)
9. Hirsch, J.: An index to quantify an individual's scientific research output. Proc. Natl. Acad. Sci. **102**, 16569–16572 (2005)
10. Llamazares, B.: On generalization of weighet means and OWA operators. In: Proceedings of EUSLAT-LFA 2011 (2011)
11. Marchant, T.: Score-based bibliometric rankings of authors. J. Am. Soc. Inform. Sci. Technol. **60**, 1132–1137 (2009)
12. Waltman, L., van Eck, N.J., Wouters, P.: Counting publications and citations: is more always better? J. Inform. **7**(3), 635–641 (2013)
13. Yager, R.R., Kacprzyk, J., Beliakov, J. (eds.): Recent Developments in the Ordered Weighted Avereging Operators: Theory and Practice. Springer(2011)
14. Yager, R.R.: On ordered weighted averaging aggregation operators in multicriteria decision making. IEEE Trans. Syst. Man Cybern. **18**(1), 183–190 (1988)

Chapter 28
A Cluster Analysis Approach for Rule Base Reduction

Luca Anzilli and Silvio Giove

Abstract In this paper we propose an iterative algorithm for fuzzy rule base simplification based on cluster analysis. The proposed approach uses a dissimilarity measure that allows to assign different importance to values and ambiguities of fuzzy terms in antecedent and consequent parts of fuzzy rules.

Keywords Fuzzy systems · Rule base reduction · Cluster analysis · Ambiguity

28.1 Introduction

The number of rules in a fuzzy system (FIS, Fuzzy Inference System) exponentially increases with the number of the input variables and the number of the linguistic values that these inputs can take (antecedent fuzzy terms) [11, 17]. Several approaches for reducing fuzzy rule base have been proposed using different techniques such as interpolation methods, orthogonal transformation methods, clustering techniques [3, 4, 12, 13, 19, 20]. A typical tool to perform model simplification is merging similar fuzzy sets and rules using similarity measures [2, 5, 6, 9, 10, 16].

In this paper we propose a new clustering procedure for simplifying rule-based fuzzy systems based on a distance (dissimilarity measure) which takes into account the value and the ambiguity of terms. In particular, our proposal allows to assign different importance to values and ambiguities.

The paper is organized as follows. In Sect. 28.2 we illustrate the motivation of our proposal. In Sect. 28.3 we briefly review the basic notions of cluster analysis and prove a result that will be used later. In Sect. 28.4 we illustrate our rule-base reduction method. Finally, in Sect. 28.5 we propose a validity index.

L. Anzilli (✉)
Department of Management, Economics, Mathematics and Statistics,
University of Salento, Lecce, Italy
e-mail: luca.anzilli@unisalento.it

S. Giove
Department of Economics, University Ca' Foscari of Venice, Venice, Italy
e-mail: sgiove@unive.it

© Springer International Publishing AG, part of Springer Nature 2019   307
A. Esposito et al. (eds.), *Neural Advances in Processing Nonlinear Dynamic Signals*, Smart Innovation, Systems and Technologies 102,
https://doi.org/10.1007/978-3-319-95098-3_28

28.2 The Idea

The aim of this paper is to propose a rule-base reduction method based on clustering techniques using a distance (dissimilarity measure) that allows to assign different importance weights to values and ambiguities of antecedent and consequent parts. We recall that the value and the ambiguity of a fuzzy number A with α-cuts $A(\alpha) = [A_L(\alpha), A_R(\alpha)]$, $\alpha \in [0, 1]$, are defined, respectively, as

$$Val_f(A) = \int_0^1 mid(A(\alpha))\, f(\alpha)\, d\alpha\,, \qquad Amb_f(A) = \int_0^1 spr(A(\alpha))\, f(\alpha)\, d\alpha$$

where $mid(A(\alpha)) = (A_L(\alpha) + A_R(\alpha))/2$ and $spr(A(\alpha)) = (A_R(\alpha) - A_L(\alpha))/2$ are the middle point and the spread of $A(\alpha)$ and f is a weighting function.

We start with the distance between two closed intervals $I_1 = [a_1, b_1]$ and $I_2 = [a_2, b_2]$ introduced in [18]

$$d_\theta(I_1, I_2) = \left((mid(I_1) - mid(I_2))^2 + \theta\, (spr(I_1) - spr(I_2))^2\right)^{1/2} \qquad (28.1)$$

where the mid-spread representation of the involved intervals is employed, that is $mid(I) = (a + b)/2$ and $spr(I) = (b - a)/2$ denote the middle point and the spread of the interval $I = [a, b]$, respectively. The parameter $\theta \in [0, 1]$ indicates the relative importance of the spreads against the mids [18]. Note that for $\theta = 1$ we have $d_{\theta=1}^2(I_1, I_2) = \frac{1}{2} d_2^2(I_1, I_2)$, where d_2 is the distance (L_2 distance)

$$d_2(I_1, I_2) = \left((a_1 - a_2)^2 + (b_1 - b_2)^2\right)^{1/2}\,.$$

This means that the distance d_2 gives the same importance to mids and spreads.

The distance d_θ can be extended to the space of all fuzzy number by the distance $dist_\theta(A, B)$ defined by

$$dist_\theta^2(A, B) = \int_0^1 d_\theta^2(A(\alpha), B(\alpha))\, f(\alpha)\, d\alpha \qquad (28.2)$$

where A and B are two arbitrary fuzzy numbers and $A(\alpha)$, $B(\alpha)$ are the α-cuts of A and B, respectively. Taking into account that $midpoints$ are connected with the value of a fuzzy number and $spreads$ with its ambiguity, the parameter $\theta \in [0, 1]$ reflects the relative importance of the ambiguity against the value.

Our proposal is to employ the previous distance between fuzzy numbers as a dissimilarity measure between two fuzzy rules able to give different weight to values and ambiguities of the antecedent and consequent terms. For example (see Sect. 4.3) the distance between j-th fuzzy terms $A_{k,j}$ and $A_{\ell,j}$ of the antecedent part of rules R_k and R_ℓ is

$$dist_\theta^2(A_{k,j}, A_{\ell,j}) = \int_0^1 d_\theta^2(A_{k,j}(\alpha), A_{\ell,j}(\alpha))\, f(\alpha)\, d\alpha$$

that has the following explicit form

$$dist_\theta^2(A_{k,j}, A_{\ell,j}) = \int_0^1 \left(mid(A_{k,j}(\alpha)) - mid(A_{\ell,j}(\alpha))\right)^2 f(\alpha)\, d\alpha$$
$$+ \theta \int_0^1 \left(spr(A_{k,j}(\alpha)) - spr(A_{\ell,j}(\alpha))\right)^2 f(\alpha)\, d\alpha$$

being $A_{k,j}(\alpha)$ the α-cuts of $A_{k,j}$. Moreover, the distance between antecedent parts of two rules can be defined as follows $D_\theta^2(Ant_k, Ant_\ell) = \sum_{j=1}^n dist_\theta^2(A_{k,j}, A_{\ell,j})$, the distance between consequents as $D_\theta^2(Cons_k, Cons_\ell) = dist_\theta^2(B_k, B_\ell)$ and the distance between rules as $D_\theta^2(R_k, R_\ell) = D_\theta^2(Ant_k, Ant_\ell) + D_\theta^2(Cons_k, Cons_\ell)$.

28.3   Cluster Analysis

In this section we present a brief review of the c-means clustering method. Furthermore, we prove a result that will be used later for Rule base reduction.

Clustering groups data objects (instances) into subsets in such a manner that similar objects are grouped together, while different objects belong to different groups. The objects are thereby organized into an efficient representation that characterizes the population being sampled. Since clustering is the grouping of similar instances/objects, some sort of measure that can determine whether two objects are similar or dissimilar is required. There are two main type of measures used to estimate this relation: distance measures and similarity measures. Many clustering methods use distance measures to determine the similarity or dissimilarity between any pair of objects.

Several clustering techniques are developed in literature, such as c-Means algorithm (or K-Means or crisp c-Means), fuzzy c-Means algorithm, entropy-based method. The c-means clustering algorithm belongs to unsupervised classification methods which group a set of input vectors into a previously defined number (c) of classes.

28.3.1   The c-Means Algorithm

We now deal with the c-means algorithm (for a survey on this subject see e.g. [8]). The parameter c is the number of clusters which should be given beforehand. The c-means algorithm starts from a partition of points which may be random or given by an ad hoc rule. The iterative c-means algorithm works as follows. Given the initial partition, the following steps are repeated until convergence, that is, no change of cluster memberships:

1. Select initial location of cluster centers;
2. Generate a (new) partition by assigning each point to its closest cluster center; that is for each center we find the subset of training points (its cluster) that is closer to it than any other center;
3. Calculate new cluster centers as the centroids of the clusters;
4. If the cluster partition is stable, stop; else go to Step 2.

The c-means algorithm can be mathematically formulated as follows. Let $X = \{x_1, \ldots, x_N\}$ be a set of objects and let assume $x_k \in \mathbb{R}^p$. The c-means algorithm classifies objects in X into c disjoint subsets $G_1, \ldots, G_c$, called clusters. In each cluster G_i a center v_i is determined. We denote by $G = \{G_1, \ldots, G_c\}$ the cluster partition and by $V = \{v_1, \ldots, v_c\}$ the set of centers. G is a partition of X, that is such that $\bigcup_{i=1}^c G_i = X$ and $G_i \cap G_j = \emptyset$ for $i \neq j$. The objective function of the minimization problem is

$$J(G, V) = \sum_{i=1}^{c} \sum_{x_k \in G_i} D^2(x_k, v_i) \, .$$

The c-means algorithm is the following:

1. Generate a set of initial centers $\bar{V} = \{\bar{v}_1, \ldots, \bar{v}_c\}$;
2. Allocate all objects x_k to the cluster of the nearest center, that is calculate

$$\bar{G} = \arg \min_{G} J(G, \bar{V}) \tag{28.3}$$

3. Update cluster centers, that is calculate centers (as centroids, or center of gravity) of new clusters $\bar{G}_i$ by solving

$$\bar{V} = \arg \min_{V} J(\bar{G}, V) \tag{28.4}$$

4. If $\bar{G}$ or $\bar{V}$ is convergent, that is the new $\bar{G}$ (or $\bar{V}$) coincides with the last $\bar{G}$ (or $\bar{V}$), stop; else go to Step 2.

28.3.2 Solution of c-Means Algorithm with Euclidean Distance

If we use the Euclidean distance in $\mathbb{R}^p$ defined by $D^2(x, y) = \sum_{j=1}^{p} (x^j - y^j)^2$ for all $x = (x^1, \ldots, x^p)$ and $y = (y^1, \ldots, y^p)$ in $\mathbb{R}^p$, the solution of (28.4) is

$$\bar{v}_i = \frac{1}{|G_i|} \sum_{x_k \in G_i} x_k, \qquad i = 1, \ldots, c$$

that is, for all $i = 1, \ldots, c$, the center $\bar{v}_i = (\bar{v}_i^1, \ldots, \bar{v}_i^p)$ is calculated as

$$\bar{v}_i^j = \frac{1}{|G_i|} \sum_{x_k \in G_i} x_k^j \qquad j = 1, \ldots, p.$$

28.3.3   Solution of c-Means Algorithm with Distance D_γ

We now discuss the solution of c-means algorithm in the case when we use a different distance that we call $D_\gamma(x, y)$. The obtained result will be used later for Rule base reduction. For convenience, we formulate the problem in the space $\mathbb{R}^{2p}$. We consider the following distance in $\mathbb{R}^{2p}$

$$D_\gamma^2(x, y) = \sum_{j=1}^{p} (x^j - y^j)^2 + \gamma \sum_{j=p+1}^{2p} (x^j - y^j)^2$$

for all $x = (x^1, \ldots, x^{2p})$ and $y = (y^1, \ldots, y^{2p})$, where $\gamma > 0$ is a parameter. The c-means algorithm applied to the set $X = \{x_1, \ldots, x_N\}$, with $x_k \in \mathbb{R}^{2p}$, using the distance $D_\gamma(x, y)$ leads to minimize the objective function

$$J_\gamma(G, V) = \sum_{i=1}^{c} \sum_{x_k \in G_i} D_\gamma^2(x_k, v_i). \tag{28.5}$$

Proposition 28.1 *The solution of* (28.4) *(in Step 3 of c-mean algorithm) is*

$$\bar{v}_i = \frac{1}{|G_i|} \sum_{x_k \in G_i} x_k \qquad i = 1, \ldots, c. \tag{28.6}$$

Of course, clusters $G = (G_1, \ldots, G_c)$ will depend on γ.

Proof By computation, we get

$$\frac{\partial J_\gamma}{\partial v_i^j} = \begin{cases} 2 \sum_{x \in G_i} (v_i^j - x_k^j) & j = 1, \ldots, p \\ 2\gamma \sum_{x \in G_i} (v_i^j - x_k^j) & j = p+1, \ldots, 2p. \end{cases}$$

By solving $\dfrac{\partial J_\gamma}{\partial v_i^j} = 0$ and observing that $\gamma > 0$ we obtain (28.6). The assertion follows taking into account the quadratic form of the minimization problem.   □

28.4 A Model for Rules Base Reduction Using Cluster Analysis

We now propose a rule-base reduction method based on clustering techniques.

28.4.1 Fuzzy Systems

The *knowledge* of a FIS can be obtained from available data using some optimization tool as a neural approach, or by direct elicitation from one or a group of Experts. In the latter case, the Experts represent their knowledge by defining a set of inferential rules. The input variables are processed by these rules to generate an appropriate output. In the case of a FIS with n input variables, $x_1, \ldots, x_n$ and a single output y (miso fuzzy system, [11]) every rule has the form

$$R_i : \quad \text{IF } x_1 \text{ is } A_{i,1} \text{ and } \ldots \text{ and } x_n \text{ is } A_{i,n} \text{ THEN } y \text{ is } B_i \qquad i = 1, \ldots, N$$

where $A_{i,j}$ is a fuzzy sets of universe space X_j and B_i is a fuzzy set of universe space Y, and N is the number of rules. The fuzzy set $A_{i,j}$ is the *linguistic label* associated with j-th antecedent in the i-th rule and B_i is the linguistic label associated with the consequent in the i-th rule. We recall that a linguistic label can be easily represented by a fuzzy set [7]. The rule i, R_i, can be represented by the ordered couple $R_i = \left(\bigcap_{j=1}^{n} A_{i,j}(x_j), B_i \right)$, being $A_{i,j}(x_j)$ the j-th component of the antecedent and B_i the consequent, $i = 1, \ldots, N$, and $\bigcap$ is the conjunction operator.

28.4.2 Distance Between Fuzzy Numbers

We consider the distance between two fuzzy numbers defined in (28.2) as

$$dist_\theta^2(A, B) = \int_0^1 d_\theta^2(A_\alpha, B_\alpha) f(\alpha) \, d\alpha$$

being d_θ the distance between intervals defined in (28.1).

We denote $T = T(m, s)$ a symmetric triangular fuzzy number with α-cuts $T_\alpha = [m - s(1 - \alpha), m + s(1 - \alpha)]$, with $\alpha \in [0, 1]$. We observe that if $T_1 = T_1(m_1, s_1)$

and $T_2 = T_2(m_2, s_2)$ are symmetric triangular fuzzy numbers and the weighting function f is defined by $f(\alpha) = 1$ for all α, the distance between T_1 and T_2 is given by

$$dist_\theta^2(T_1, T_2) = (m_1 - m_2)^2 + \frac{\theta}{3}(s_1 - s_2)^2. \qquad (28.7)$$

28.4.3   Distance Between Rules

Let us consider a fuzzy system with n input variables (= number of antecedents of each rule) and N rules

$$R_k = \left(\bigcap_{j=1}^n A_{k,j}(x_j), B_k \right), \qquad k = 1, \ldots, N$$

being $A_{k,j}(x_j)$ the j-th component of the antecedent and B_k the consequent. We define for $\theta \in [0, 1]$ the distance between antecedents as

$$D_\theta^2(Ant_k, Ant_\ell) = \sum_{j=1}^n dist_\theta^2(A_{k,j}, A_{\ell,j})$$

and the distance between consequents as $D_\theta^2(Cons_k, Cons_\ell) = dist_\theta^2(B_k, B_\ell)$. Furthermore, we define the distance between rules (dissimilarity measure) by

$$D_\theta^2(R_k, R_\ell) = D_\theta^2(Ant_k, Ant_\ell) + D_\theta^2(Cons_k, Cons_\ell). \qquad (28.8)$$

We assume that the antecedent and consequent terms of each rule are symmetric triangular fuzzy numbers, that is

$$A_{k,j} = T(m_{k,j}, s_{k,j}) \qquad B_k = T(m_k^B, s_k^B).$$

In this way, we can identify each rule R_k as a vector of $\mathbb{R}^{2n+2}$, that is

$$R_k = (m_{k,1}, \ldots, m_{k,n}, m_k^B, s_{k,1}, \ldots, s_{k,n}, s_k^B).$$

Furthermore, we have

$$D_\theta^2(Ant_k, Ant_\ell) = \sum_{j=1}^n (m_{k,j} - m_{\ell,j})^2 + \frac{\theta}{3} \sum_{j=1}^n (s_{k,j} - s_{\ell,j})^2$$

and

$$D_\theta^2(Cons_k, Cons_\ell) = (m_k^B - m_\ell^B)^2 + \frac{\theta}{3}(s_k^B - s_\ell^B)^2.$$

Thus, the distance between two fuzzy rules R_k and R_ℓ is

$$D_\theta^2(R_k, R_\ell) = \sum_{j=1}^n (m_{k,j} - m_{\ell,j})^2 + (m_k^B - m_\ell^B)^2 + \frac{\theta}{3} \left(\sum_{j=1}^n (s_{k,j} - s_{\ell,j})^2 + (s_k^B - s_\ell^B)^2 \right).$$

28.4.4 Clustering Rules Using Distance

Let $\mathscr{R} = \{R_1, \ldots, R_N\}$ be the set of Rules. The problem is to find a partition $G = \{G_1, \ldots, G_c\}$ of $\mathscr{R}$ and centers $V = \{R_1^v, \ldots, R_c^v\}$ in order to minimize the objective function

$$J_\theta(G, V) = \sum_{i=1}^c \sum_{R_k \in G_i} D_\theta^2(R_k, R_i^v) \qquad (28.9)$$

where $D_\theta(R_k, R_i^v)$ is defined in (28.8). The algorithm we propose is the following:

1. Generate a set of initial centers $\bar{V} = \{\bar{R}_1^v, \ldots, \bar{R}_c^v\}$;
2. Allocate all Rules R_k to the cluster of the nearest center, that is calculate

$$\bar{G} = \arg \min_G J_\theta(G, \bar{V}); \qquad (28.10)$$

3. Update cluster centers, that is calculate centers (as centroids, or center of gravity) of new clusters $\bar{G}_i$ by solving

$$\bar{V} = \arg \min_V J_\theta(\bar{G}, V); \qquad (28.11)$$

4. If $\bar{G}$ or $\bar{V}$ is convergent, that is the new $\bar{G}$ (or $\bar{V}$) coincides with the last $\bar{G}$ (or $\bar{V}$), stop; else go to Step 2.

28.4.5 Solution

Using the result proved in Proposition 28.1 with $p = n + 1$ (and thus $2p = 2n + 2$) and $\gamma = \theta/3$, we can deduce that the solution of Step 3, $\bar{V} = \{\bar{R}_1^v, \ldots, \bar{R}_c^v\}$ with

$$\bar{R}_i^v = \left(\bigcap_{j=1}^n \bar{A}_{i,j}^v(x_j), \bar{B}_i^v \right), \qquad i = 1, \ldots, c$$

is given by

$$\bar{A}_{i,j}^v = T(\bar{m}_{i,j}, \bar{s}_{i,j}) \qquad \bar{B}_i^v = T(\bar{m}_i^B, \bar{s}_i^B)$$

where

$$\bar{m}_{i,j} = \frac{1}{|G_i|} \sum_{R_k \in G_i} m_{k,j}, \qquad \bar{s}_{i,j} = \frac{1}{|G_i|} \sum_{R_k \in G_i} s_{k,j}, \qquad (28.12)$$

and

$$\bar{m}_i^B = \frac{1}{|G_i|} \sum_{R_k \in G_i} m_k^B, \qquad \bar{s}_i^B = \frac{1}{|G_i|} \sum_{R_k \in G_i} s_k^B. \qquad (28.13)$$

28.4.6   Value and Ambiguity

We observe that solutions (28.12) and (28.13) can be rewritten (using arithmetic of triangular fuzzy numbers), respectively, as

$$\bar{A}_{i,j}^v = \frac{1}{|G_i|} \sum_{R_k \in G_i} A_{k,j}, \qquad \bar{B}_i^v = \frac{1}{|G_i|} \sum_{R_k \in G_i} B_k.$$

Using the properties of value Val_f and ambiguity Amb_f, it follows that

$$Val_f(\bar{A}_{i,j}^v) = \frac{1}{|G_i|} \sum_{R_k \in G_i} Val_f(A_{k,j}^v), \qquad Val_f(\bar{B}_i^v) = \frac{1}{|G_i|} \sum_{R_k \in G_i} Val_f(B_k^v)$$

and

$$Amb_f(\bar{A}_{i,j}^v) = \frac{1}{|G_i|} \sum_{R_k \in G_i} Amb_f(A_{k,j}^v), \qquad Amb_f(\bar{B}_i^v) = \frac{1}{|G_i|} \sum_{R_k \in G_i} Amb_f(B_k^v).$$

We observe that clusters $G_1, \ldots, G_c$ are dependent on parameter θ and thus they reflect the relative importance assigned to ambiguities against the values.

28.5   Validity Index

Quality of centroid-based clustering is usually evaluated by internal validity indexes (for a survey see [1]). There are many purposes to employ cluster validity measures; one of the most important applications is to estimate the number of clusters. A reliable validity index for the cluster algorithm must consider both cohesion (compactness) within each cluster and separation between clusters. If only a measure of compactness is considered, the best partition is obtained when each data point is considered as a separate (singleton) cluster. On the other hand, if only a separation measure is considered, the trivial solution corresponding to one cluster is obtained. In this section we first give definitions for the separation and cohesion measures and then we propose a validity index.

The separation measure, based on the distance from the cluster centroids to the global centroid, is given by (note that $\sum_{i=1}^{c} |G_i| = N$)

$$S_\theta = \frac{1}{N \cdot c} \sum_{i=1}^{c} D_\theta(R_i^v, R_{tot})|G_i|$$

where R_{tot} is the centroid of all the clusters. The cohesion (compactness) measure, based on the distance from the points in a cluster to its centroid, is given by

$$C_\theta = \frac{1}{c} \sum_{i=1}^{c} \left(\frac{1}{|G_i|} \sum_{R_k \in G_i}^{N} D_\theta(R_k, R_i^v) \right).$$

We propose the following validity index

$$V_\theta = h(S_\theta, C_\theta)$$

where the function h is increasing with respect to S_θ, decreasing with respect to C_θ and bounded in $[0, 1]$. The higher the value of V_θ, the more suitable the number of clusters. Thus we have to maximize V_θ. Examples of V_θ are

$$V_\theta^1 = \frac{S_\theta}{S_\theta + C_\theta}$$

and (see the score function introduced in [14, 15])

$$V_\theta^2 = 1 - \frac{1}{e^{e^{S_\theta - C_\theta}}}.$$

In order to apply our methodology, we have generated a FIS with 1000 rules. Each rule is described by 5 fuzzy terms in the antecedent part and by 1 fuzzy term in the consequent part. Each fuzzy term is modelled as a symmetric triangular fuzzy number. Figure 28.1 shows the evolution of validity index V_θ^1 with respect to the number of clusters for different values of parameter θ.

In Fig. 28.1 we have plotted the curve of the validity index as a function of number of cluster c, from $c = 2$ to $c = 20$, for different (fixed) values of parameter θ, namely $\theta = 0$ (lower curve), $\theta = 0.1$, $\theta = 0.5$ and $\theta = 1$ (upper curve). From the analysis of the figure we may observe that the global maximum for $\theta = 0$ is achieved at $c = 5$. Furthermore the global maximum for $\theta = 0.1$ and for $\theta = 0.5$ is achieved at $c = 6$. Finally, the global maximum for $\theta = 1$ is achieved at $c = 7$.

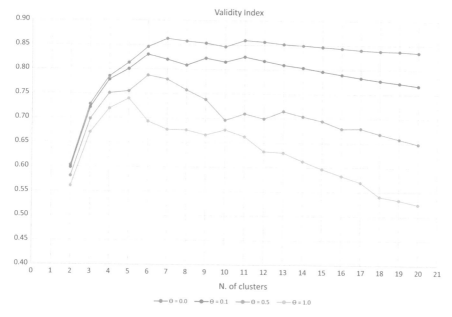

Fig. 28.1 Validity index V_θ^1 for different values of parameter θ

28.6 Conclusion

In this paper we have proposed a methodology for a fuzzy system rule-base reduction based on clustering techniques. As a future development, we intend to investigate the effect of parameter θ on both solutions of cluster analysis and optimal number of clusters.

References

1. Arbelaitz, O., Gurrutxaga, I., Muguerza, J., Pérez, J.M., Perona, I.: An extensive comparative study of cluster validity indices. Pattern Recognit. **46**(1), 243–256 (2013)
2. Babuška, R., Setnes, M., Kaymak, U., van Nauta Lemke, H.R.: Rule base simplification with similarity measures. In: Proceedings of the Fifth IEEE International Conference on Fuzzy Systems, 1996, vol. 3, pp. 1642–1647. IEEE (1996)
3. Baranyi, P., Kóczy, L.T., Gedeon, T.T.D.: A generalized concept for fuzzy rule interpolation. IEEE Trans. Fuzzy Syst. **12**(6), 820–837 (2004)
4. Bellaaj, H., Ketata, R., Chtourou, M.: A new method for fuzzy rule base reduction. J. Intell. Fuzzy Syst. Appl. Eng. Technol. **25**(3), 605–613 (2013)
5. Chao, C.T., Chen, Y.J., Teng, C.C.: Simplification of fuzzy-neural systems using similarity analysis. IEEE Trans. Syst. Man Cybern. B Cybern. **26**(2), 344–354 (1996)
6. Chen, M.Y., Linkens, D.A.: Rule-base self-generation and simplification for data-driven fuzzy models. Fuzzy Sets Syst. **142**, 243–265 (2004)

7. Dubois, D., Prade, H.: Fuzzy Sets and Systems: Theory and Applications, vol. 144. Academic Press (1980)
8. Jain, A.K.: Data clustering: 50 years beyond k-means. Pattern Recogn. Lett. **31**(8), 651–666 (2010)
9. Jin, Y.: Fuzzy modeling of high-dimensional systems: complexity reduction and interpretability improvement. IEEE Trans. Fuzzy Syst. **8**(2), 212–221 (2000)
10. Jin, Y., Von Seelen, W., Sendhoff, B.: On generating FC^3 fuzzy rule systems from data using evolution strategies. IEEE Trans. Syst. Man Cybern. Part B Cybern. **29**(6), 829–845 (1999)
11. Lazzerini, B., Marcelloni, F.: Reducing computation overhead in miso fuzzy systems. Fuzzy Sets Syst. **113**(3), 485–496 (2000)
12. Nefti, S., Oussalah, M., Kaymak, U.: A new fuzzy set merging technique using inclusion-based fuzzy clustering. IEEE Trans. Fuzzy Syst. **16**(1), 145–161 (2008)
13. Riid, A., Rüstern, E.: Adaptability, interpretability and rule weights in fuzzy rule-based systems. Inf. Sci. **257**, 301–312 (2014)
14. Saitta, S., Kripakaran, P., Raphael, B., Smith, I.F.: Improving system identification using clustering. J. Comput. Civ. Eng. **22**(5), 292–302 (2008)
15. Saitta, S., Raphael, B., Smith, I.: A bounded index for cluster validity. In: Machine Learning and Data Mining in Pattern Recognition, pp. 174–187 (2007)
16. Setnes, M., Babuška, R., Kaymak, U., van Nauta Lemke, H.R.: Similarity measures in fuzzy rule base simplification. IEEE Trans. Syst. Man Cybern. Part B Cybern. **28**(3), 376–386 (1998)
17. Simon, D.: Design and rule base reduction of a fuzzy filter for the estimation of motor currents. Int. J. Approx. Reason. **25**(2), 145–167 (2000)
18. Trutschnig, W., González-Rodríguez, G., Colubi, A., Gil, M.Á.: A new family of metrics for compact, convex (fuzzy) sets based on a generalized concept of mid and spread. Inf. Sci. **179**(23), 3964–3972 (2009)
19. Tsekouras, G.E.: Fuzzy rule base simplification using multidimensional scaling and constrained optimization. Fuzzy Sets Syst. (2016, to appear)
20. Wang, H., Kwong, S., Jin, Y., Wei, W., Man, K.F.: Multi-objective hierarchical genetic algorithm for interpretable fuzzy rule-based knowledge extraction. Fuzzy Sets Syst. **149**(1), 149–186 (2005)

Printed in the United States
By Bookmasters